DISSOCIATIVE RECOMBINATION:

Theory, Experiment and Applications

DISSOCIATIVE RECOMBINATION:

Theory, Experiment and Applications

Editors

J. Brian A. Mitchell

The University of Western Ontario, Canada

Steven L. Guberman

Institute for Scientific Research, USA

World Scientific

Singapore • New Jersey • London • Hong Kong

Published by

World Scientific Publishing Co. Pte. Ltd.
P O Box 128, Farrer Road, Singapore 9128

USA office: World Scientific Publishing Co , Inc.
687 Hartwell Street, Teaneck, NJ 07666, USA

UK office: World Scientific Publishing Co. Pte. Ltd.
73 Lynton Mead, Totteridge, London N20 8DH, England

DISSOCIATIVE RECOMBINATION: THEORY, EXPERIMENT AND APPLICATIONS

ISBN: 9971-50-618-1

Printed in Singapore by Utopia Press.

FOREWORD

The International Conference on Dissociative Recombination: Theory, Experiment and Applications was held at Chateau Lake Louise, Alberta, Canada from May 28–31, 1988. This symposium brought together many leading investigators in the fields of molecular ion recombination research, interstellar chemistry, aeronomy, plasma physics, and atomic and molecular theory and experiment. Participants presented talks reviewing their work and these were followed by lively discussion sessions. The papers in this volume are based on these talks with the exception of contributions from H. Nakamura and H. Michels, who were unable to attend. A list of attendees is given at the end of this volume.

We would like to express our gratitude to the United States Air Force Office of Scientific Research and to the National Aeronautics and Space Administration for financial support which contributed greatly to the success of the meeting. Thanks are due to the Centre for Interdisciplinary Studies in Chemical Physics at the University of Western Ontario and to the Institute for Scientific Research for providing administrative assistance. Special mention must also be given to C. Bottcher, A. Giusti-Suzor, T. Morgan, J. Peterson, D. Smith and W. Stwalley for chairing the sessions.

August, 1988

J. Brian A. Mitchell
Department of Physics
The University of Western Ontario
London, Canada N6A 5B9

Steven L. Guberman
Institute for Scientific Research
33 Bedford St.
Lexington, MA 02173
U.S.A.

CONTENTS

INTRODUCTION TO DISSOCIATIVE RECOMBINATION

Steven L. Guberman
Institute for Scientific Research
33 Bedford St., Suite 19A
Lexington, MA 02173

and

J. Brian A. Mitchell
Department of Physics
The University of Western Ontario
London, Canada N6A 5B9

Dissociative recombination (DR) of molecular ions with electrons has important consequences in many areas of physical science. Even though almost 40 years have elapsed since the first experimental and theoretical investigations of DR, it continues to present great challenges to the investigator.

For the diatomic or polyatomic molecular ion AB^+, DR is described by

$$AB^+ + e^- \rightarrow A + B$$

where e^- is an electron and A and B are atoms or molecules that may be in their ground states or vibrationally and/or electronically excited.

The last decade has seen something of a revolution in the study of DR, since ab initio calculations coupled with resonant scattering theory and multichannel quantum defect studies have produced detailed results illuminating the role of ion vibrational excitation, the quantum yields of the DR products, and the role of Rydberg states. The old assumption that for most molecules the DR rate coefficient is about 10^{-7} cm^3/sec is clearly no longer adequate. Indeed, it is now clear that the simplest ions, e.g. H_2^+ and He_2^+, have considerably slower total

DR rate coefficients from their ground vibrational levels. H_3^+, the most important ion in hydrogen plasmas, has been shown theoretically and experimentally to recombine much more slowly when vibrationally cold than once believed. Deep narrow resonances associated with Rydberg state interactions have been predicted and observed. New decay channels have been found and the effects of vibrational excitation are being clarified as theorists probe the details and experimentalists obtain better control of their experiments.

I. Theory

The early experimental results of Biondi and Brown[1] and Holt et al.[2] on rare gas afterglows coupled with the need to explain the electron concentrations in the earth's ionosphere[3] led Sir David Bates to first propose a mechanism for DR in 1950[4] in a seminal paper that was less than one full page in length. The mechanism has stood the test of time and is still the basic mechanism used to describe DR in the absence of the higher order role due to Rydberg states. The proposed mechanism was several years in the making because it had to overcome two important obstacles. First, due to the mass mismatch, a collision with an electron could not be expected to cause a massive molecule to fall apart. The transfer of electronic energy to the nuclear motion is improbable. However, in Bates's proposed mechanism, the electron energy is transferred entirely to the electrons of the target molecule and a neutral state is formed in which the nuclear motion leads to the dissociation of the molecule. The process is illustrated in Figure 1 where a molecular ion, AB^+, shown by the solid potential curve captures an electron of energy ε into repulsive state AB. Once the interparticle distance between A and B becomes sufficiently large, the electron can no longer be ejected and the process is said to be stabilized. The second obstacle, as pointed out in the paper in this volume by O'Malley, is the non-crossing rule which requires that curves of the same molecular symmetry do not cross. According to this rule, the repulsive state labelled AB in Figure 1 could not reach the ion

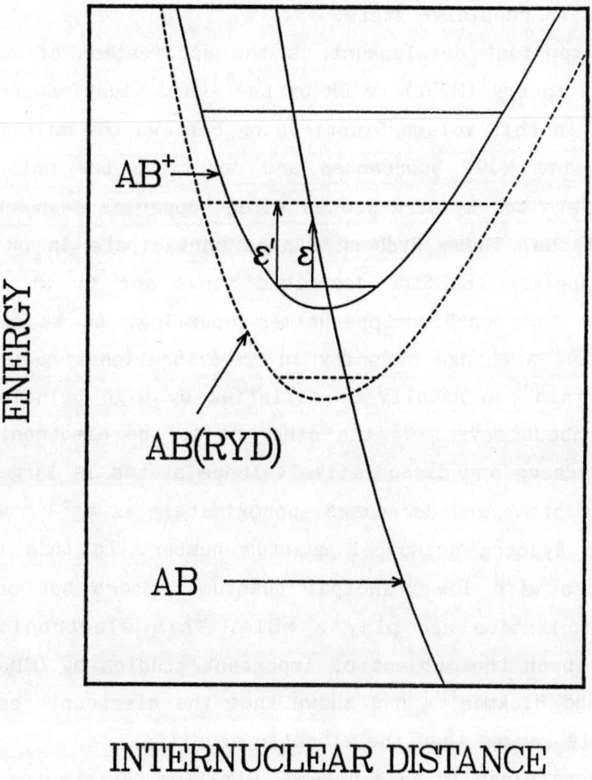

Fig. 1. Dissociative Recombination Mechanisms

because it would have to cross through an infinite number of Rydberg states of the same symmetry as AB and having the ground state of the ion as a series limit. It is now understood however, that the repulsive state, AB, is a diabatic state, i.e., it is not an adiabatic solution of the full Schroedinger equation and therefore does not violate the non-crossing rule.

The first detailed calculations of DR were done by Bardsley[5] who used the configuration mixing (CM) or resonant scattering theory approach. Bardsley also introduced Rydberg states into DR via indirect recombination.[6] In this process, the electron is captured directly into a vibrationally excited Rydberg state by Born-Oppenheimer breakdown (vibronic coupling). In Figure 1, this process is illustrated for an electron of energy ε'. The Rydberg state is then predissociated by the neutral AB repulsive state.

A recent important development is the application of multichannel quantum defect theory (MQDT) to DR by Lee[7] and Giusti-Suzor.[8] In the first article in this volume Giusti-Suzor reviews the main features of both the CM and MQDT approaches and describes the role played by vibrationally excited Rydberg states which appear as resonances in the DR cross sections. These Rydberg states participate in DR by both a non-Born-Oppenheimer coupling described above and by an electronic coupling. For the non-Born-Oppenheimer coupling, the largest matrix elements involve a change of unity in the vibrational quantum number. In practice, this can usually be satisfied by high principal quantum numbers near about seven. On the other hand, the electronic coupling between the Rydberg and dissociative valence states is largest for the lowest Rydberg state and decreases approximately as $n^{*-3/2}$ where n^* is the effective Rydberg principal quantum number. In this mechanism, Rydberg states with low principal quantum numbers but considerable vibrational excitation can play a role. This electronic coupling mechanism has been the subject of important studies by O'Malley[9] and Hickman,[10] and Hickman[10] has shown that the electronic coupling can be considerably larger than the vibronic coupling.

In the second paper in this volume, O'Malley reviews the use of the resonant scattering theory approach to DR with the assumption that the various processes that drive DR are localized in the internuclear distance coordinate. This approximation is especially applicable if the dissociative state does not have a steep slope. O'Malley finds that the electronic coupling mechanism leads to the surprising result that the cross section for the case of a single isolated resonance has a dip at the energy of the Rydberg state vibrational level.

The electronic coupling mechanism between the Rydberg and dissociative states is explored in the paper by Hickman. He shows that the capture is given by a matrix element of a Green function and is second order in the electronic coupling between the electron-ion and dissociative states. These higher order couplings are present in the second and higher terms on the right side of Eq. (5) in the paper by Giusti-Suzor. Hickman also compares the CM and MQDT approaches and points out that for the case of a single isolated resonance, the resonant scattering theory approach in the Independent Resonance Approximation[10] and the MQDT approach lead to the same expression for the DR cross section. The expression takes the form of a Beutler-Fano profile with a non-zero q. A study of DR in H_2^+ shows that the net effect of the indirect process depends upon whether the direct process is fast or slow.

Large scale ab initio results for DR potential curves for O_2 and N_2, and electron capture widths and DR rate coefficients for O_2 are reported in the paper by Guberman. He finds that for O_2, out of the 11 states that can produce $O(^1S)$, the upper state of the green line, only a single state is important for the lowest 9 vibrational levels of the ion. The DR rate coefficient is very sensitive to vibrational excitation. The maximum rate coefficient for generating $O(^1S)$ is from v=2 and is 78 times greater than the rate coefficient from v=0. For $O(^1D)$, three states are important for generating the upper state of the red line from the lowest three vibrational levels of the ion. The $O(^1D)$ rate coefficients are less sensitive to vibrational excitation than the $O(^1S)$ rate coefficients with the rate coefficient from v=2 being a factor of 2 less than the rate coefficient from v=0. The calculated $O(^1D)$ rate coefficients are in good agreement with rate coefficients deduced from satellite measurements. For N_2, DR from v=0 is dominated by four states and an inspection of the dominant configurations indicates that the $^3\Pi_u$ states may have the highest rate coefficients.

Large scale CASSCF calculations are reported by Kraemer and Hazi on the potential curves which describe the DR of HCO^+. Their approach leads to rather accurate values for electronic excitation energies in CO and HCO and an accurate dissociation energy for HCO^+. Since a favorable recombination rate coefficient would require a similar geometry for the linear ion and the neutral state containing the captured electron, they hold the CO internuclear distance at the separation found in HCO^+ while varying the CH distance. In a linear geometry, none of the neutral states give likely dissociative routes from the lowest vibrational level of the ion. While two of the potentials drop significantly after relaxing the CO separation, a favorable dissociative route is still not found in spite of the fact that the laboratory DR rate coefficients are high. They suggest several possible indirect mechanisms.

Nakamura, Takagi, and Nakashima report results for the DR of H_2^+ (and isotopes) and CH^+. For H_2^+, an interesting isotope effect is reported from the v=3 vibrational level of D_2^+. At about 0.5 eV electron energy, the cross section for D_2^+ is very small compared to the cross sections for H_2^+ and HD^+. The shape of the window resembles the DR Franck-Condon type windows found in model calculations.[11] For vibrational excitation from the v=1 state of HD^+, they show that the energy dependence of the cross section is similar to that for DR indicating that the repulsive states are the intermediate states. Above the threshold energy, the v=1 to 2 vibrational excitation cross section is of comparable magnitude to the DR cross section. For CH^+, they find that the DR rate coefficient is about 1×10^{-7} cm^3/sec at T=120K in agreement with the result of the merged beam experiment.[12]

Associative ionization (AI), is related to DR by detailed balance and many of the same theoretical techniques developed for DR can be used for AI. Urbain describes some recent MQDT calculations on AI in H_2 and several recent experimental developments. For the case where the separated atoms limit lies below the lowest vibrational level of the ion, the AI cross section increases with energy above threshold as new vibrational channels become accessible. Above the ion asymptote, Penning ionization competes and the AI cross section decreases. For the case of H^+ colliding with H^-, the entrance channel asymptote is above the lowest vibrational level of the ion and the cross section is large at zero energy decreasing inversely with the increasing energy. Experiments are described which provide for a measurement of the AI cross section by detection of either the emitted electron energy spectrum or of the vibrationally excited product ion. The generation of H(2s) and H(3s) atoms for AI experiments is described.

Since the neutral excited states of HeH lying beneath the ground state of HeH^+ all have the attractive ion core, there are no excited repulsive states which can lead to DR of the ion ground state. The paper by Michels describes the calculation of dissociative routes which can potentially lead to DR of excited states of HeH^+. The lowest ion excited state is only weakly bound and the calculated series of quartet states do not appear to offer favorable routes for DR. An additional doublet series is difficult to calculate since it can autoionize to the ion ground state. A second quartet Rydberg series which can provide DR routes for three higher excited states of the ion but which can also autoionize to a lower ion state is also discussed.

II. Experiment

The merged beam technique is described in the paper of Mitchell and Youssif. Recent measurements of the total DR cross sections for H_2^+, H_3^+, HeH^+, and N_2^+ and of the branching ratio for H_3^+ recombination are discussed. A new channel leading to H_3^* has been identified in the latter measurement. HeH^+ has been measured to have a nonnegligible recombination cross section, although the state of excitation of the ions used in the experiment has yet to be determined. Deep resonant structures have been observed in the cross sections of H_2^+, HeH^+ and N_2^+.

Adams and Smith have brought the Flowing Afterglow Langmuir Probe (FALP) technique to the forefront of recombination research in recent years. An important advantage of this technique is the fact that ion formation and decay occur in different regions of the flow apparatus and so the ions have time to relax internally before their recombination rate is studied. Measurements of H_3^+ recombination with this technique have shown that the rate coefficient for de-excited ions is less than $10^{-8}cm^3/sec$. This has been confirmed by recent merged beam experiments. (See Mitchell and Yousif, this volume). Subsequent studies have found that measured values for He^+,HeH^+ and H_3^+ rate coefficients are indistinguishable. In a plasma, collisional radiative recombination (CRR) must be considered. In the FALP method where electron densities of $10^{10}cm^{-3}$ are encountered, CRR can proceed with a rate coefficient of about $10^{-9}cm^3/sec$, thus setting a lower limit to the measurable dissociative recombination rate coefficient. CRR may also have contributed to recently measured high values of H_3^+ recombination published by Amano.[13] The FALP technique has been extended to the determination of recombination rate coefficients for a wide variety of protonated species and efforts are underway to introduce UV absorption spectroscopy and laser induced fluorescence techniques to measure branching ratios for DR.

Earlier experiments performed using the microwave afterglow technique have been re-examined by Johnsen. It has been found that (a) the presence of molecular additives in a plasma leads to a lowering of the mean electron temperature, and (b) heat conduction in a plasma is insufficient to produce a uniform electron energy distribution over the whole volume. The consequence of these findings is that some of the earlier values for measured recombination rate coefficients must be treated with suspicion. This is particularly true for cluster ions. A reassessment of previous N_2^+ studies is given in the light of recent research on the vibrational deactivation of diatomic ions. Although it is concluded that previous studies applied to vibrationally excited ions, skepticism is expressed concerning the new merged beam results discussed by Mitchell and Yousif. Mention of new studies concerning electron-ion recombination at elevated gas densities and of recent ion-ion recombination work is also given.

Spectroscopic measurements of the product state branching ratios following the DR of N_2^+, O_2^+, CO_2^+, and H_2O^+ in a flowing afterglow, as a function of the initial vibrational excitation of the parent ions, have been performed by Rowe and Queffelec. The H_2O^+ results support the conclusions of Herbst's statistical analysis of branching ratios rather than Bates' theory based on bond breaking.

Translational spectroscopy in which measurement of the kinetic energies of dissociation products allows a determination of their excitation state is a powerful tool which could be used for DR branching ratio measurements. The paper by van der Zande, Koot, Los, and Peterson provides an example of this elegant technique. A study of the energy spectra of the products of O_2^+ + Cs dissociative electron capture is presented. The high resolution inherent in the technique allows the Rydberg states into which the electron is captured to be identified.

A proposed experiment to demonstrate laser stimulated radiative recombination of electrons and protons is described by Morgan and Mitchell. The experiment will employ the Merged Beam Apparatus at the University of Western Ontario and will employ field ionization detection of the highly excited recombination products. The technique also has applications to the measurements of DR branching ratios and dielectronic recombination cross sections.

Recent theoretical and experimental studies by Lundeen on H_2 Rydberg states with high angular momenta have shown that the energies of these states may be accurately calculated using models employing long range interactions only. This is in contrast to the case for low l states where short range interactions dominate and MQDT theories are necessary for energy calculations. Furthermore, long range interaction theories predict short autoionization lifetimes (10^{-9}sec) for the high l states. Short range interaction theories, on the other hand, predict lifetimes of the order of tens of seconds. This implies that long range interactions play a crucial role in the dynamics of high l Rydberg states.

The paper by Graham addresses the subject of electron and ion temperatures in plasma discharges. A knowledge of these is important in order to determine the conditions under which DR acts as a major ion loss mechanism. Generally it is overshadowed by ion diffusion to the walls. DR, however, does play a major role in the production of excited atoms and so a knowledge of product branching ratios is important for the interpretation of plasma spectra which provide diagnostic information concerning the state of the discharge. DR also competes with dissociative attachment for low energy ions and so must be included in models of negative ion sources.

It is generally accepted that the primary formation mechanism for H^- ions in a pure hydrogen discharge is the dissociative attachment of electrons to vibrationally excited H_2 molecules. The source of this vibrational excitation is only partially understood. Hiskes and Karo have examined the role of wall collisions involving H_2^+ and H_3^+ in producing vibrationally excited hydrogen molecules. A description of the processes involved in the neutralization of these molecules at surfaces is given and the authors find that collisions of vibrationally excited H_2^+ and H_3^+ produce copious numbers of $H_2(v'')$ molecules. Indeed, this mechanism appears to be the dominant source of $H_2(v'')$ at low plasma densities.

Bahns and Stwalley provide a review of antimatter cluster ion formation principles. The impetus for this work is to produce antihydrogen in a condensed form for easy transportation and storage. The concept of the pseudowall, a cold, trapped, high density plasma within which many collision processes can occur leading to cluster ion buildup, is discussed. Experimental manifestations of the pseudowall are proposed.

Spectroscopy is an important tool in plasma diagnoses since it is generally non-perturbative. Interpretation, however, requires a good knowledge of the atomic processes leading to photon emission. Shoemaker, Ganguly, and Garscadden describe how a study of the Stark broadening of spectral lines from highly excited Rydberg states within the plasma can provide information on local electric fields. A description of the application of this technique is given.

III. Applications

DR can yield translationally hot products which may have energies that exceed those needed for planetary escape. The role of DR in the escape of N isotopes from Mars is discussed in the paper by Fox. Among several processes that contribute to planetary escape, DR of N_2^+ is the major process at high and average solar activities and is the most important process contributing to the isotope enhancement ratio. The isotope enhancement role of DR disappears for v>0. DR of O_2^+ is an important process in planetary atmospheres, where it generates $O(^1S)$ and $O(^1D)$, the upper states of the green and red lines, respectively. Fox discusses a detailed model of these emissions as a function of altitude in the nightside ionosphere of Venus. The model results for the green line emission are consistent with the measurements made by the Venera space probe.

Yee, Abreu, and Colwell discuss the modelling of the $O(^1S)$ and $O(^1D)$ emissions which result from DR of O_2^+ in the Earth's ionosphere. For the vibrational distribution of the ion they use five models developed by Fox. For the specific rate coefficients yielding ^1D and ^1S atoms, they use the values calculated by Guberman. In order to calculate yields for a distribution of vibrational levels, one must have a specific rate coefficient and a total rate coefficient for each level. Since total rate coefficients have not been measured or calculated for $v>0$, they use the laboratory $v=0$ coefficient for higher v. The adequacy of this assumption must await the results of future calculations or experiments. Their theoretical model shows that the ^1D (^1S) quantum yield decreases slightly (increases strongly) with vibrational excitation. A calculation of the quantum yields using the measured volume emission rates from the Atmosphere Explorer E satellite indicates that at high vibrational excitation, the calculated ^1S quantum yields are higher than the authors' theoretical model results but the ^1D results are in good agreement with the theoretical model.

The products of the DR of polyatomic molecules are important components of the chemistry of dense interstellar clouds. Herbst discusses three methods which have been used to determine these yields and comparative examples are given for H_3O^+, $HCNH^+$, and $C_4H_2{}^+$. The abundances of neutral complex molecules and of small molecules such as OH are sensitive to the model that is used. It is known that the ratio of singly deuterated species to their hydrogenic analogues is often higher than the deuterium to hydrogen ratio in the dense clouds. It is possible that the product branching ratios for DR reactions may account for this, but a definitive statement must await the independent determination of the DR branching ratios.

The role of DR in diffuse interstellar clouds is discussed by Black and van Dishoeck who point out that the concentrations of neutral species can be intimately affected by the products of DR reactions. They stress the need to determine the quantum yields of products resulting from DR of CH_3^+, $C_2H_2^+$, $C_3H_3^+$, and H_3O^+. In diffuse clouds, where the fractional ionization is high, the DR rate coefficient for H_3^+

significantly affects the column density of H_3^+, but in dense molecular clouds the column density is less sensitive to the DR rate coefficient. For HeH^+, the predicted model concentration is greater than the recently observed upper limit, indicating that an important sink is missing from the models.

Several approaches for determining the small ionization fraction in dense interstellar clouds based on deuterium fractionation are discussed by Turner. Reliable values of the DR rate coefficients of H_3^+ and CH_3^+ are needed for these models. The chemistry of cyanopolyynes is discussed and several possibilities are presented to account for the difference between the observed and model values for the ion to neutral concentration ratios. A model of the DR of $H_3C_2O^+$ is presented. Turner argues that the model supports the prediction of Bates for the likely quantum yields of the DR products.

IV. References

1. M. A. Biondi and S. C. Brown, Phys. Rev. 75, 1700 (1949); M. A. Biondi and S. C. Brown, Phys. Rev. 76, 1697 (1949).

2. R. B. Holt, J. M. Richardson, B. Howland and B. T. McClure, Phys. Rev. 77, 239 (1950).

3. D. R. Bates and H. S. W. Massey, Proc. Roy. Soc. A187, 261 (1947).

4. D. R. Bates, Phys. Rev. 78, 492 (1950).

5. J. N. Bardsley, J. Phys. B 1, 349 (1968).

6. J. N. Bardsley, J. Phys. B 1, 365 (1968).

7. C. M. Lee, Phys. Rev. A16, 109 (1977).

8. A. Giusti, J. Phys. B 13, 3867 (1980).

9. T. F. O'Malley, J. Phys. B 14, 1229 (1981).

10. A. P. Hickman, J. Phys. B 20, 2091 (1987).

11. S. L. Guberman, Can. J. Phys. 64, 1621 (1986).

12. J. B. A. Mitchell and J. W. McGowan, Astrophys. J. 222, L77 (1978).

13. T. Amano, Ap. J. 329, L121, (1988).

MULTICHANNEL DYNAMICS OF DISSOCIATIVE RECOMBINATION AND RELATED PROCESSES

Annick GIUSTI-SUZOR
Laboratoire de Photophysique Moléculaire, Bât. 213
Université Paris-Sud, 91405 ORSAY, France

ABSTRACT

Dissociative recombination is presented as a reactive scattering process combining both electron-ion and atom-atom collisions. The aim of a complete dynamical study is to obtain a *scattering matrix* involving the complete set of channels (ionization and dissociation, open and closed). The cross-sections for the various processes (e.g. excitation of the target ion or dissociation) including resonance effects may thus be calculated. A brief review of the alternative theoretical approaches is presented with emphasis on the new developments in this field.

I. DISSOCIATIVE RECOMBINATION : A MULTICHANNEL REACTIVE COLLISION

In this review of some aspects of the theoretical treatment of dissociative recombination (DR), the multichannel character of this process will be emphasized. First, it is a *reactive* collision since the entrance and outgoing channels involve different collision partners (electron-molecular ion and atom-atom respectively). Hence, different collision radial coordinates will be used for the different types of channels (Fig. 1 b) : the distance r between the incoming electron and the ion center for the ionization channels, the internuclear distance R for the dissociation channels. Then, for both types of fragmentation, several channels are generally involved corresponding to different symmetries or different limits. For the electronic collision each channel is defined by the ion electronic state in a given rovibrational level (v^+N^+) and by the energy ε and angular quantum numbers $(\ell\lambda)$ of the external electron. For $\varepsilon > 0$ (electron in the continuum) the channel is open, for $\varepsilon = -\frac{Ry}{n^{*2}} < 0$ the channel is closed and the electron is in a Rydberg orbital with effective principal number n^*. The corresponding wave functions are

$$\psi_{v^+N^+}(E_{tot}) = \left[\Phi^+_{core}(q,R)\chi_{v^+N^+}(R)\right]\varphi_{\ell\lambda}(\varepsilon,r) \qquad (1a)$$

where Φ^+_{core} represents the electronic wave function of the molecular ion (with q the set of electronic coordinates), $\chi_{v^+N^+}$ its rovibrational wave function and $\varphi_{\ell\lambda}$ the radial wave function of the external electron, assumed to be energy normalized. For simplicity we omit here its angular part, and no mention will be made of the rotational couplings except in the last section. The recombination process involves a specific initial level (v_o^+,N_o^+) of the ion but autoionization back to the electronic continuum can put it in any other accessible level, with energy $E_{v^+N^+}$ less than the total energy $E_{tot} = E_{v_o^+N_o^+} + \varepsilon$.

Similarly, several repulsive states of the neutral molecule may play a role in the dissociation process. Their wave functions have the general form

$$\Phi_i(Q,R) \times F_{d_i}(E,R) \qquad (1b)$$

where $Q = \{q,r\}$ represents the set of electronic coordinates, Φ_i is the electronic wave function of one of the repulsive states and F_{d_i} the corresponding continuum wave function, also energy normalized. The energy E of the atomic fragments is related to the energy ε of the incoming electron by

$$E + E_{d_i} = \varepsilon + E_{v^+N^+} = E_{tot} \qquad (2)$$

where E_{d_i} is the dissociation limit for channel (i).

The basic mechanism for efficient dissociative recombination is schematized on Figure 1a. Although the global energy balance (Eq. 2) consists in a transfer of kinetic energy from the incoming electron to the nuclear motion, the mechanism is primarily electronic and proceeds via the excitation of a core electron into an antibonding orbital. The efficiency of the process is thus mainly governed by the matrix elements

$$V_{v^+N^+d_i}(E_{tot}) = \int \chi_{v^+N^+}(R)V_i^{e\ell}(\varepsilon,R)F_{d_i}(E,R)dR$$

where $V_i^{e\ell}(\varepsilon,R)$ is the electrostatic interaction between the monoexcited (Rydberg or continuum) configuration and the diexcited (valence) one. Note that this description rests on a diabatic representation of the molecular states which do not diagonalize the full electronic Hamiltonian.

More details on the mechanism and efficiency of the DR process may be found in several review papers, e.g. in the NATO ASI series [1,2]. Here I will briefly describe and compare the alternative theoretical approaches, with some remarks on the influence of Rydberg states on the DR cross-section.

II. DYNAMICS OF DISSOCIATIVE RECOMBINATION : ALTERNATIVE TREATMENTS

Two steps may be distinguished in the theoretical study of DR. First it requires a good knowledge of the molecular states and couplings which play a role in the process. They can be obtained from electron - ion *scattering calculations*, in which the coupling with the dissociative state is seen as a resonance in the asymptotic phase of the electron wave function[3,4,5], or from Rydberg and valence *bound-state calculations*, with an extrapolation of the Rydberg states properties across the ionization threshold, in order to get the capture width (see the paper by S. Guberman in this volume, and Ref. 6).

The second step is the treatment of the *dynamics* of DR, i.e. the calculation of the cross-section itself. The aim is to obtain a scattering matrix S involving the complete set of channels. From its matrix elements, partial and total cross-sections may be calculated, not only for DR but also for processes involving the same channels and the same molecular interactions, e.g. vibrational excitation of the target ion or associative ionization, the inverse process of DR (details on the experiments and theory for associative ionization are given by X. Urbain in this volume. Here also, two kinds of treatments have been proposed, the configuration mixing (CM) approach and the multichannel quantum defect theory (MQDT), extended to include dissociative channels besides ionization ones. The main contributions to these two methods are summarized in Table I. Both methods start from "uncoupled" channel wave functions, for ionization as well as dissociation. The continuum parts in Eqs (1a) and (1b) are written at large distances (r or R) as

$$\varphi_{\ell\lambda}(\varepsilon,r) = f_{\ell}(\varepsilon,r)\cos\pi\mu_{\ell\lambda}(R) - g_{\ell}(\varepsilon,r)\sin\pi\mu_{\ell\lambda}(R) \qquad (3a)$$

$$F_{d_i}(E,R) = J_o(E,R)\cos\delta\ N_o(E,R)\sin\delta \qquad (3b)$$

where f_ℓ and g_ℓ are regular and irregular Coulomb functions, J_o and N_o are regular and irregular Bessel functions.* The phase-shifts $\pi\mu_{\ell\lambda}$ and δ represent the effect of all short-range interactions except the Rydberg-valence mixing. (Note that for $\varepsilon < 0, \mu_{\ell\lambda}$ is the quantum defect of the Rydberg electron, and may be extrapolated into the phase-shift above threshold).

One then has to take into account the coupling between both types of zero-order channel wave functions. A full CI expansion for the total wave function would be

$$\Psi(E_{tot}) = \sum_{\ell,\lambda,v^+,N^+} \int a_{\varepsilon}^{\ell v^+ N^+} \varphi_{\ell\lambda}(\varepsilon,r)\chi_{v^+N^+}\Phi^+_{core}\,d\varepsilon + \sum_i \int A_E^i F_{d_i}(E,R)\Phi_i\,dE \qquad (4)$$

The unknown coefficients a_ε and A_E could in principle be determined by projecting the Schrödinger equations onto the zero-order basis states, with electronic and nuclear integration. This general approach is impracticable and the CM and MQDT methods use two different ways for handling this equation.

---

* For simplicity we omit the angular part for the nuclear wavefunction also, and include only the J = 0 terms in this schematic equation.

i) in the CM treatment, the electronic continuum is first eliminated by electronic integration and the expansion in the second term is replaced by unknown functions

$$\tilde{F}_{d_i}(E,R) = \int A_E^i F_{d_i}(E,R)dE,$$

solutions of integro-differential equations in the variable R. In the dissociative wave functions thus obtained the coupling to the ionization continuum is exactly taken into account. Nevertheless, the reaction on the ionization channel is not explicitly sorted out. Another problem of this approach is that the terms corresponding to closed ionization channels ($\varepsilon < 0$) in the first expansion of Eq. (4) are replaced from the beginning by a *discrete* summation over the bound Rydberg levels (n,v^+,N^+). This leads to a large number of unknown coefficients a_n^{v+N+} to be determined by solving a linear algebraic system of high dimensionality, on a narrow energy grid due to the rapid energy variations of the resonance structure. The corresponding equations can be found in Ref. (13) and (14), where Hickman describes the "isolated resonance approximation" used to simplify the calculations.

ii) in the MQDT approach, both types of fragmentation channels (ionization/dissociation) are treated symmetrically, leading to a generalized S-matrix. Moreover, open and closed ionization channels (ε positive or negative in Eq. 4) are distinguished only in the last step of the calculations when the asymptotic behavior of the wave functions are defined, and only this part has to be performed on a narrow energy grid. The basic idea of this type of approach is to represent the effects of the short-range interactions (the electronic coupling in the present case) by a "smooth" reaction matrix K calculated at finite distance, i.e. at the boundary of the reaction zone. At larger distances the energy expansions in Eq. (4) are replaced by "on the energy shell" expressions, where irregular parts are adjoined to each channel wave function with reaction matrix elements as coefficients.

Obtaining the reaction matrix K is obviously the difficult part of the calculation. In most of the previous applications a first-order approximation has been performed, which retains only the first term in the Born expansion

$$K(E) = V(E) + \mathcal{P}\int \frac{V(E)V(E')}{E - E'}\, dE' + \dots \qquad (5)$$

Since a unitary S matrix is then built from this first order K-matrix, this type of calculation may be qualified as a "unitarized Born" or "Born II" method. For small couplings, as for example in the NO molecule [18], this approximation has been proved to be accurate [19], but when the interactions are relatively strong higher order terms in the expansion (5) lead to indirect couplings which may have noticeable effects, especially on the resonance shape as we discuss below.

III. THE RESONANCES IN DR CROSS-SECTIONS

First of all, it may be useful to define what is meant by "resonances" in the present context. Indeed, DR (as well as dissociative attachment, the same process with a neutral target) are often called "resonant" processes because they proceed through the formation of a *bound* molecular state (with respect to the electronic motion) embedded in the ionization continuum. Neverthess this "resonance" does not induce a sharp structure in the cross-section because the nuclear motion is unbound, which results in a short life-time for the molecular state. The resonances we comment on now are due to the temporary capture of the incoming electron in a Rydberg orbital (see Fig. 1a), in an excited vibrational level ($v > v_o^+$). Since the continuum and Rydberg configurations differ by a single orbital, the direct coupling for electron capture is purely (ro)vibrational (non-adiabatic couplings) and usually weak.

In the vicinity of these discrete states the system has two ways towards dissociation, the "direct" one through electronic capture into the dissociative state, and the indirect one involving electronic predissociation of the discrete Rydberg levels (Fig. 2). Thus interferences may occur which result in a non-Lorentzian resonance shape. The resonances appear as pure dips in the cross-sections (similar to "window" resonances in photoabsorption) in special cases where (i) there is no vibrational coupling between ionization channels, thus preventing any vibrational mechanism for electron capture, and (ii) indirect electronic couplings arising from high orders of the K-matrix are neglected. Electron capture is thus suppressed at the resonance energy, for which the total wave function is almost entirely concentrated on the discrete Rydberg level.

As soon as vibrational or indirect electronic interactions are strong enough, an asymmetric resonance profile (Fano profile index $q \neq 0$) is obtained, as shown in Fig. 1a of Ref. [13] (effect of vibrational coupling) or in Fig. 2 of Ref. [14] (combination of both effects). Hickman has shown that the modification of the resonance shape due to indirect electronic couplings may lead to a noticeable increase of the averaged cross-section [14].

IV. PRESENT AND FUTURE DEVELOPMENTS

The need for accurate and detailed DR data for modeling interstellar molecular clouds or planetary atmospheres motivates new theoretical developments. Another motivation is the vigorous development of associative ionization experiments, either in thermal beams (see the contribution by X. Urbain [26] in this volume, and references therein) or in optical traps, between ultracold atoms [28]. I list here the main aspects which are being developed, or will be in the near future.

1. MQDT Calculations With An "Exact" Electronic K-Matrix

The main drawback of the previous DR calculations using the MQDT approach of Ref. [11] is the first-order approximation for the K-matrix in Eq. (5). Recently an "exact" *electronic* reaction matrix $K_{e\ell}(R)$ has been determined in the Born-Oppenheimer frame work for the case of the $^1\Sigma_g^+$ states of H_2 [29]. Starting from accurate ab initio calculations, Ross and Jungen have interpreted the double-minima adiabatic curves as resulting from scattering at fixed R of the s, p, d electron partial waves by the ground ($1s\sigma_g$) and excited ($2p\sigma_u$) ion states. The fitted matrix $K_{e\ell}(R)$ is then evaluated in a complete vibrational basis, leading to matrix elements K_{v^+d} and $K_{v^+v'^+}$, thus allowing for indirect electronic couplings between ionization channels. A first calculation is presented in Ref. [26] for associative ionization. Applications to DR and to other molecules than H_2 are planned.

2. Angular Aspects

In most of the previous DR calculations, coupling of angular momenta and rotational interactions have been overlooked. These aspects are being considered carefully in several present works, e.g. by Takagi et al [30] for DR, Dahler et al [31] and Urbain et al [32] for associative ionization.

3. Branching Ratios For Atomic Products of DR

For atmospheric ions, e.g. N_2^+, O_2^+, NO^+, it is very important to predict the relative populations of the atomic states which may be formed in the DR process, when several dissociative curves play a role. It is thus necessary to take into account the interactions between the dissociative states, which means solving a set of coupled equations as for atomic collision studies. Some details are given in Section 2 of Ref. [26] which reports such a calculation for the case of associative ionization. Developments along this line, leading to a complete treatment both of the electron-ion and the atom-atom collision parts of DR and related processes, are necessary and promising.

REFERENCES

[1] Mitchell, J.B.A., in "Atomic processes in electron-ion and ion-ion collisions", NATO ASI series 145, 185 (Plenum NY, 1986).

[2] Giusti-Suzor A., ibid., 223.

[3] Hazi, A.U., Derkits, C., Bardsley, J.N., Phys. Rev. A 27, 1751 (1983).

[4] Takagi, H. and Nakamura, H., J. Chem. Phys. 84, 2431 (1986).

[5] Tennyson, J. and Noble, C.J., J. Phys. B 18, 155 (1985).

[6] Guberman, S.L., in "Physics of Ion-Ion and Electron Ion Collisions" NATO ASI Series 83,167(Plenum NY, 1983).

[7] Bardsley, J.N., J. Phys. B 1, 349 and 365 (1968).

[8] Bottcher, C., Proc. Roy. Soc. London A 340, 301 (1974).

[9] O'Malley, T., J. Phys. B 14, 1229 (1981).

[10] Lee, C.M., Phys. Rev. A 16, 109 (1977).

[11] Giusti, A., J. Phys. B 13, 3867 (1980).

[12] Bottcher, C., J. Phys. B 9, 2889 (1976).

[13] Giusti, A., Bardsley, J.N. and Derkits, C., Phys. Rev. A 28, 682 (1983).

[14] Hickman, A.P., J. Phys. B 20, 2091 (1987).

[15] Nakashima, K., Takagi, H. and Nakamura, H., J. Chem. Phys. 86, 726 (1987).

[16] Guberman, S., this volume and references therein.

[17] Hickman, A.P., Phys. Rev. Lett. 59, 1553 (1987).

[18] Giusti-Suzor, A., and Jungen, C., J. Chem. Phys. 80, 986 (1984).

[19] Raoult, M., J. Chem. Phys. 87, 4736 (1987).

[20] Nakamura, H., J. Phys. Soc. Jpn 31, 574 (1971).

[21] Cohen, J.S., Phys. Rev. A 13, 99 (1976).

[22] Hickman, A.P. and Morgner, H., J. Phys. B 9, 1765 (1976).

[23] Urbain, X., Giusti-Suzor, A., Fussen, D. and Kuback, C., J. Phys. B 19, L273 (1986).

[24] Urbain, X. and Giusti-Suzor, A., proceedings of ICPEAC XV, Brighton book of abstracts 714 (1987).

[25] Takagi H. and Nakamura, H., J. Chem. Phys. 88, L4552 (1988).

[26] Urbain, X., this volume.

[27] Henriet, A., Giusti-Suzor, A., Masnou-Seeuws, F., book of abstracts of contributed papers MOLEC VII (Assisi, 1988).

[28] Gould, P.L., Lett, P.D., Julienne, P.S., Phillips, W.D., Thornstein H. and Weiner, J., Phys. Rev. Let. 60, 788 (1988).

[29] Ross, S. and Jungen, C., Phys. Rev. Lett. 59, 1297 (1987).

[30] Takagi, H. and Nakamura, H., Proceedings of ICPEAC XV, Brighton, book of abstracts 349 (1987).

[31] Jones, D.M. and Dahler, J.S., Phys. Rev. A 37, 2916 (1988).

[32] Urbain, X., Giusti-Suzor, A. and Ross, S., in preparation.

TABLE I : Theoretical treatments of the dynamics of dissociative recombination

CONFIGURATION MIXING	MULTICHANNEL QUANTUM DEFECT THEORY
THEORIES	
Bardsley 1968 [7]	Lee 1977 [10]
Bottcher 1974 [8]	Giusti 1980 [11]
O'Malley 1981 [9]	
APPLICATIONS TO H_2^++e	
Bottcher 1976 [12]	
Giusti, Bardsley, Derkits 1983 [13]	
Hickman 1987 [14]	Nakashima et al 1987 [15]
APPLICATIONS TO OTHER PROCESSES	
1) Direct DR of atmospheric ions (O_2^+,N_2^+)	
	Guberman 1988 [16]
2) Competition between ionization and dissociation	
H_2 : Hickman 1987 [17]	NO : Giusti and Jungen 1984 [18]
	Raoult 1987 [19]
3) Associative ionization	
Nakamura 1971 [20]	H^++H^- : Urbain et al 1986 [23]
He+He : Cohen 1976 [21]	H+H^*(n=2) : Urbain and Giusti 1987 [24]
H+He : Hickman and Morgner 1976 [22]	Takagi et al 1987 [25]
	Urbain 1988 [26]
	Na^*+Na^* : Henriet et al 1988 [27]

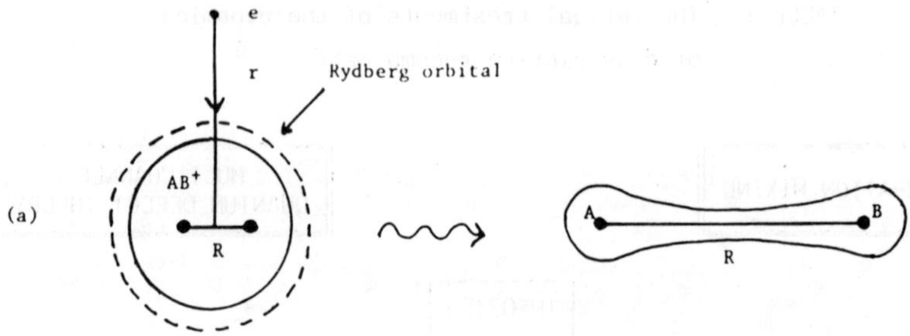

Figure 1 : Basic mechanism for dissociative recombination

(a) deformation of the ion molecular cloud due to the external electron

(b) system of radial coordinates (r, R) for the reactive collision.

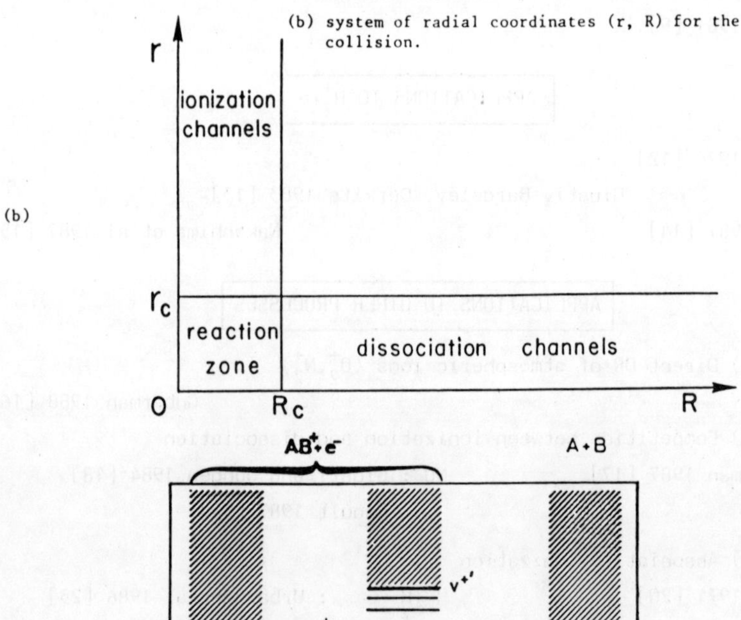

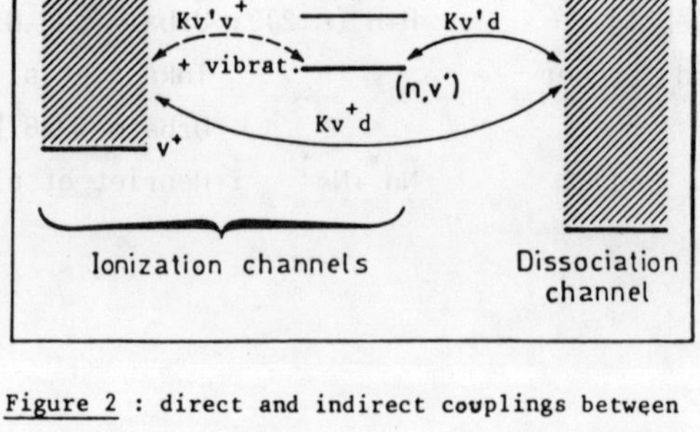

Figure 2 : direct and indirect couplings between ionization channels (open and closed) and a dissociation channel.

DIRECT DISSOCIATIVE RECOMBINATION - ANALYTIC THEORY

Thomas F. O'Malley

2276 Lucretia Ave. #4
San Jose, CA 95122 USA

ABSTRACT

The theory of zeroth-order Direct Dissociative Recombination (DR), i. e. without the Ry States, is briefly reviewed; the distinction between direct and indirect DR is defined and discussed; and finally an analysis of full direct DR (including one Ry state) is presented, with a view toward better understanding of the structure in the cross section.

1. INTRODUCTION

I wish I could play back Sir David Bates' ICPEAC lecture[1] of last year in which he told how he came to understand the mechanism of Dissociative Recombination while trying to model the ionosphere. He first observed, with Massey, that,if they started with the then current belief that physical electronic states of molecules are adiabatic states, the logical conclusion is that the DR rate must be impossibly slow and so they dismissed it as a possibility for several years. His reflection on this situation is memorable. Commenting on the bias toward adiabatic states he said "An authoritative but incorrect formulation is an insidious obstacle to progress".

After rejecting this incorrect (or at least unphysical & inappropriate) formulation, especially as embodied in the non-crossing rule, he then sketched the dissociating potential curve for O_2 in the previously forbidden form, which we have all since become familiar with, one which crosses the O_2^+ curve near its minimum, like the schematic curves shown in Fig. 1.

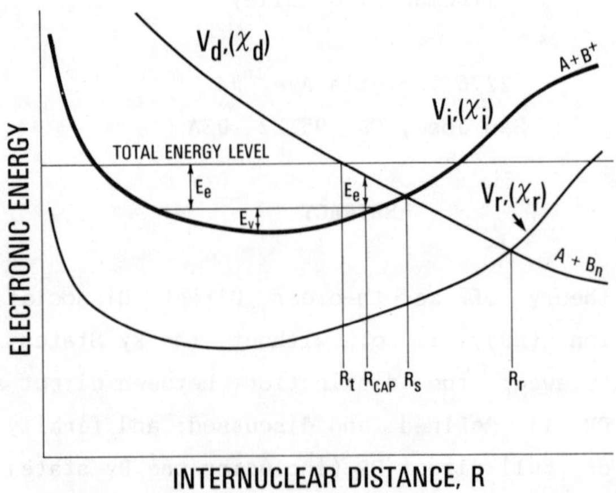

Figure 1. Schematic potential curves, V, (with wave functions χ) for DR. V_i, V_d and V_r are curves for the ion, dissociating, and Ry states. R_{cap} and R_t are the points of Franck-Condon capture into χ_d and its turning point. R_s (for stabilization) and R_r are where it crosses V_i and V_r.

Given the physically appropriate potential curves, he then worked out the essentials of the theory[2], which has stood up to this day, needing only refinements in derivation and in details.

2. ZEROTH-ORDER DR THEORY

Following Bates' work the main help to understanding the physics of dissociative recombination and attachment has come from the explicit recognition that the non-crossing rule did not apply because the dissociating electronic state is not an eigenstate but rather a *resonance* (now also called a quasistationary or diabatic[3] state). The Bardsley-Herzenberg-Mandl[4] model recognized this and started by describing motion in the dissociating state as a Kapur-Peierls resonance arising from a source term. They produced the general cross section formula that is now widely used for DA and DR (see Equation 3 below). The next step[5] was to apply resonance concepts directly to the full Schroedinger equation and derive everything from first principles. The most direct method starts with a projection operator[5]

$$Q = |\phi_d(el)\rangle\langle\phi_d(el)|, \quad P = 1 - Q \tag{1}$$

onto a function which defines electronic motion in the quasistationary dissociating state. Its energy gives the potential curve with an electronic non-local width of magnitude $\Gamma_{el} = 2\pi|\langle\phi_d|QHP|\phi_i\rangle|^2$ and a level shift that can be defined, using the arbitrariness of ϕ_d, to vanish[6]. This operator is then used to expand the wave function into orthogonal subspaces

$$\psi = Q\psi + P\psi. \tag{2}$$

Substituting this in the Schroedinger equation and solving the resulting coupled equations formally, one can extract the cross section in a general way and write it as

$$\sigma_{DR,0} = (\pi/k_i)^2|T_{DR,0}|^2 \simeq \sigma_{el}\ FC\ e^{-\rho} \tag{3}$$

where σ_{el} results from the electron integrals, FC is the Franck Condon nuclear overlap function, and

$$T_{DR,0} = \langle\psi_{i,p}|PHQ|\phi_d\chi_{d,0}\rangle. \tag{4}$$

$\chi_{d,0}$ the nuclear function for the dissociating state and solution of the coupled equations, and ρ, the imaginary part of the phase shift in the same state, are defined as

$$|\chi_{d,0}\rangle = G_Q QHP|\psi_i\rangle \xrightarrow[R>R_S]{} C\ e^{-\rho/2}|\chi_{d,out}\rangle, \tag{5}$$

and

$$\rho/2 = \text{Im}\,\delta_d \simeq \sum_{v=0}^{V}|\langle\phi_d|V_E|\chi_v\rangle|^2 \xrightarrow[(\text{large } V)]{} \int\Gamma_e dt/\hbar. \tag{6}$$

Here δ_d is the exact phase shift for nuclear motion in the state AB_d.

It is complex to represent loss of amplitude to autoionization (AI) after eliminating the open vibrational channels. $\mathfrak{J}_d$ here is a very approximate pure elastic wave function for the dissociating state.

3. DIRECT AND INDIRECT DISSOCIATIVE RECOMBINATION

The foregoing understanding of both disssociative attachment and recombination was already in place by the late 60's. Except for computational realization (far from trivial), it included everything except a complete treatment of the complex non-local AI mechanism which is still being worked out[7], and the Rydberg (Ry) states. Inclusion of the Ry states has been done by two complementary approaches "direct" and "indirect" DR. Their names were fixed in answer to the question "How do we get into the *dissociating* state?" My present purpose is just to define and briefly discuss these two mechanisms. It is helpful further if we subdivide each into a zeroth-order and a full mechanism. Indirect DR was first discussed by Berry and by Bardsley[8].

<u>Zeroth-Order</u> <u>Indirect</u> <u>DR</u>

$$AB^+ + e \xrightarrow{v} AB_{n,v} \xrightarrow{e} AB_d \longrightarrow A^* + B \tag{7}$$

describes the electron as being captured initially into the Ry state n rather than the dissociating state, and with a compensating increase in vibrational energy (Δ v = 1). The mechanism is that of vibronic (v) coupling (matrix element of the nuclear kinetic energy) because there is no electronic matrix element directly connecting the 2 states. This vibronic coupling, since it violates the Franck-Condon principle of nuclear momentum conservation, has a very small amplitude for Δ v = 1 and is considered negligible for larger Δ v's (propensity rule). Energy conservation then limits the allowable n to high values (7 or more for H_2^+). Recombination finally results from electronic coupling to the dissocating state at the crossing point R_r of Fig. 1.

The observable result of zeroth-order indirect DR would be a very narrow peak at the energy of each Ry vibrational level, superimposed on a zero background.

<u>Zeroth-Order Direct DR</u>

$$AB^+ + e \overset{\ell}{\rightleftarrows} AB_d \longrightarrow A^* + B \qquad (8)$$

is exactly the process described in Sec. 2 which is the same for attachment and recombination except that the latter has an E^{-1} energy dependence. It gives the correct background cross section but with no structure, before the Ry state is added.

<u>Full Indirect DR</u>

$$AB^+ + e \left\{ \begin{array}{l} \overset{\ell}{\rightleftarrows} AB_d \longrightarrow A^* + B \\ \qquad\quad \uparrow \ell \\ \overset{v}{\longrightarrow} AB_{n,\, v=v_i+1 \text{ only}} \end{array} \right. \qquad (9)$$

embodied in many calculations, is basically a superposition of the 2 zeroth-order mechanisms above. It moves much closer to the real world by starting with (8) which gives the correct background cross section and adding to it a narrow structure produced by going through a high Ry state, but again limited severely by $\Delta v = 1$. The structure is apparently more complex[9], with peaks as well as dips ($q \neq 0$) but with the same width.

A significant development in calculating this mechanism was Giusti's[10] introduction of dissociating states into MQDT, making it possible for this formalism to describe first indirect DR and ultimately the full process. For a long time full indirect DR was considered the only mechanism to produce structure - until the full direct mechanism was pointed out.

<u>Full Direct DR</u>

$$AB^+ + e \left\{ \begin{array}{l} \overset{\ell}{\rightleftarrows} AB_d \longrightarrow A^* + B \\ \qquad\quad \downarrow\uparrow \ell \\ \qquad\quad AB_{n,\, v \text{ (all)}} \end{array} \right. \qquad (10)$$

is, I believe, the natural extension of DA theory to include the Ry states. It neglects the weak vibronic transition into the Ry state and instead has a 2 way electronic transition between the dissociating and Ry states at R_r. Its structure should be broadest for the lowest Ry states (n=3 for H_2^+), following an n^{-3} law. This structure will be seen in Sec. 4 to occur in the simple form of pure dips or windows, regardless of the strength of autoionization. I wonder if this apparent contrast with the full indirect structure could possibly be due to the fact that the Ry state here *interrupts* dissociation, whereas the indirect second order mechanism *feeds* it.

Of course either (9) or (10) can clearly be extended to include everything, but it is valuable to understand them separately.

4. ANALYSIS OF FULL DIRECT DR

In 1981, I put forth the process of full direct DR[11] and analyzed its dependence on energy and other parameters. In the present section I will briefly summarize this analysis and correct an error in the cross section formula. The hope is to make these broadest structures more understandable and possibly relate them to the physical parameters such as autoionization (AI).

I start with two main approximations, which may be relaxed, the neglect of rotational transitions and of overlapping resonances. Representing the *zeroth-order process* of direct DR, after treating the electronic resonance properly, involved mainly solving the coupled equations for χ_d, the nuclear function. To account now for the presence of a Ry state, it is sufficient to continue looking at χ_d and to ask how it is coupled to the quasi-bound Ry state $\chi_{n,v}$.

<u>Localization</u> <u>of</u> <u>interactions</u>. The most efficient procedure for analyzing χ_d this way relies heavily on the convenient fact that the nuclear wave functions, being semi-classical, interact with other states in a very *localized* way. In particular the capture interaction occurs about R_{cap}, AI occurs between R_t and R_s, and finally interaction with the Ry state occurs only near R_r.

With this localization in mind, we may start given the wave

function $\chi_{d,0}$ of Sec 2 as generated by the incident continuum state, and with AI *already* fully acounted for by the imaginary part, ρ, of δ_d, and all determined in the region $R<R_s$. Then we simply couple it analytically to the Ry state $\chi_{n,v}$ in the neighborhood of $R_r > R_s$. The coupling of a continuum to a quasi-bound state is a straight forward problem of resonance scattering theory[6], for which we need, in addition to the resonant state function $\chi_{n,v}$, 2 functions, χ_p, the "potential scattering" function, and the nuclear Green's function G_d. These functions, consistent with the results of Sec. 2, and needed only for $R > R_s$, are

$$\chi_p = Q_r \chi_{d,out} \tag{11}$$

where Q_r simply orthogonalizes it to $\chi_{n,v}$ by shifting its phase, and

$$\begin{aligned} G_d(R,R') &= \pi \chi_{out}(R_>)\, e^{i\delta_p} \chi_p^*(R_<) \\ &\simeq N^2\pi \exp\left[i(KR_> + \delta_p + \pi/2)\right] \exp\left[-i(KR_< + \delta_p^*)\right]. \end{aligned} \tag{12}$$

δ_p is the full complex phase shift in AB_d with imaginary part $\rho/2$. The tricky thing here (and this was responsible for the error in my previous treatment[11]) is that the complex phase shift, as well as the boundary conditions, rule out the real Green's function which was used in Ref. 11. Then with the help of χ_p and G_d, χ_d can be given exactly as

$$\begin{aligned} |\chi_d\rangle &= \left[1 + (1+G_d V_n)|\chi_{n,v}\rangle (E-E_{n,v}-i\Gamma/2)^{-1} \langle \chi_{n,v}|V_n\right] |\chi_p\rangle \\ &\xrightarrow[R\to\infty]{} Q_r |\chi_{out}\rangle \left[1 + i\Gamma/2\,(E-E_{n,v}-i\Gamma/2)^{-1}\right] \end{aligned} \tag{13}$$

where

$$\begin{aligned} \Gamma/2 &= \mathrm{Im}\langle \chi_{n,v}|V_n G_d V_n|\chi_{n,v}\rangle \\ &\simeq \pi N^2 \left|\langle e^{i(KR+\delta_p)}|V_n|\chi_{n,v}\rangle\right|^2 e^{-\rho} \end{aligned} \tag{14}$$

Note the narrowing of Γ by the autoionization survival factor $e^{-\rho}$.

Then, given the asymptotic form (13) of the dissociating wave function, as fully coupled to a single Ry state, the ratio of the full to the zeroth order direct DR cross section can immediately be found. This follows by noting that χ_d differs from $\chi_{d,0}$ only by the appearance of the bracketed quantity at the end of (13) and so its absolute square gives the ratio as

$$\sigma_{DR} / \sigma_{DR,0} = \frac{(E-E_{n,v})^2}{(E-E_{n,v})^2+\Gamma^2/4} = \cos^2(\eta_{res}). \qquad (15)$$

This, together with Eq. 14 for Γ, is the desired analytic result for *full direct DR* (i.e. coupled to a single Ry state or resonance). Surprisingly it is a pure dip (q = 0), so that the constructive interference I reported previously[11] for weak AI was a mistake. The apparent irrelevance of autoionization to the shape of (15) is surprising. However it appears now that Γ itself (14) is always narrowed by the survival factor $e^{-\rho}$, which shrinks with increasing E or v_i.

Integral Equation Method For Complex Non-Local Potential V_d

W. Huo presented results calculated by the above method at this Conference. Most striking among the results were the presence of resonance windows about each autoionizing (AI) threshold, even without including Ry states, and a smoothing or simplifying of the resonance structure from the high Ry states. Both apparently are related to the magnitude Γ_{el} of the imaginary part of the non-local V_d. The surprising AI structure can be reproduced non-rigorously by applying the above analysis with $\chi_{n,v}$ replaced by the threshold AI channel function χ_{vf}. But the needed closed channel electron function is not defined. If however one uses the tour de force of substituting for ϕ_{el} a packet or pseudo state of width Γ_{el} to reflect the uncertainty in the electronic energy V_d of AB_d, the analysis leading to (15) goes through formally as before, predicting a window resonance at each AI threshold, as calculated. The physical relation of the structure to Γ_{el} suggests that this pseudo state could tend to wipe out the narrower structures above $E_{el} = -\Gamma_{el}/2$ (or $n^* > 5$ for H_2^+).

5. RELATION TO CALCULATIONS AND EXPERIMENT

The present formula (15) for the full direct DR process appears to be consistent with a formula for Fano's q found by Giusti-Suzor et al[9] which owes its origin to vibronic coupling ($W_{v,v'}$) only. I would therefore expect calculations to give pure dips whenever $\Delta v > 1$ (only direct mechanism possible), but with the possibiility of peaks if $\Delta v = 1$. In looking at some calculated H_2^+ results, those of Giusti-Suzor et al[9] who did 2 different calculations (CI & MQDT), and of Nakashima et al[12] and Hickman[13], I noticed that all 4 calculations found pure dips when $\Delta v>1$ in very good agreement with the preceding analysis. But when $\Delta v=1$ only 2 of the 4 found the allowable peaks to occur. This point is discussed in this volume by Hickman who argues that the peaks are more correct. Therefore, where calculations do not identify the mechanism causing structure, it seems reasonable to identify the mechanism by the absence or presence of peaks.

The MQDT calculation of Ref 9 for a n=4 state illustrates the larger relative width of the low n resonances, even though reduced somewhat by the $e^{-\rho}$ of Eq. 14. And the possible impact of the integral equation method[7] on both the narrow and broad structures could be an interesting future development.

Comparisons with experiment can only be rough at the present time, but both the original results of Auerbach et al[14], which inspired much theoretical analysis, and the recent, more refined results of Mitchell and his collaborators[15] are generally consistent with the dips of Eq. 15, even going to zero for isolated resonances. We hope for even better resolution in both vibrational states and energy in the future.

REFERENCES

1) Bates, D. R, in "Electronic & Atomic Collisions", Gilbody, Newell, Read & Smith eds. (North Holland, 1987, p. 3)

2) Bates, D. R., Phys. Rev. **78**, 492 (1950).

3) Lichten, W., Phys. Rev. **131**, 229 (1963).
O'Malley, T.F., in Adv. At. Mol. Phys., **7**, 223 (1971).

4) Bardsley, J.N., Herzenberg, A., & Mandl, F. in "Atomic Collision Processes", M.R.C. McDowell,ed., (North Holland, p. 415, 1964).

5) O'Malley, T.F., Phys. Rev. **150**, 14 (1966).

6) Lippmann, B.A. & O'Malley, T.F., Phys. Rev. **A2**, 2115 (1970).

7) See also works of W. Domke
& co-workers, e.g. J. Phys. Chem., **88**, 4862 (1984).

8) Berry, R.S., J. Chem. Phys. **27**, 1288 (1957).
Bardsley, J.N, J. Phys. B: At. Mol. Phys. **1**, 349 (1968).

9) Giusti-Suzor, A., Bardsley, J.N. and Derkits, C., Phys. Rev. A, **28**, 682 (1983).

10) Giusti, A., J. Phys. B **13**, 3867 (1980).

11) O'Malley, T.F., J. Phys. B **14**, 1229 (1981).

12) Nakashima, K., Takagi, H. and Nakamura, H., J. Chem. Phys. **86** 726 (1987).

13) Hickman, A. P., J. Phys. B: At. Mol. Phys. **20**, 2091 (1987).

14) Auerbach, D., Cacak, R., Caudano, R., Gaily, T.D., Keyser, J., McGowan, J.W, Mitchel, J.B.A. and Wilk, S.F.J., J. Phys. B: At. Mol. Phys. **10**, 3797 (1977).

15) Hus, H., Yousif, F., Noren, C., Sen, A., and Mitchell, J.B.A., Phys. Rev. Lett., 14 March, 1988.

Resonances in Dissociative Recombination of Electrons with H_2^+

A. P. Hickman
Molecular Physics Department
SRI International
Menlo Park, CA 94025

ABSTRACT

Dissociative recombination of electrons with diatomic molecular ions is reviewed, with emphasis on the qualitative description of the "direct" and "indirect" mechanisms involved. Calculations for H_2^+ are presented and compared with the recent experimental results of Hus et al. The calculations confirm the importance of the "indirect" mechanism, which accounts for the resonant structure observed in the cross section.

1. INTRODUCTION

Dissociative recombination (DR) involves the capture of an electron by a molecular ion, the formation of a compound state, and the subsequent dissociation into fragments. This paper discusses the mechanisms that come into play as the process unfolds and presents specific calculations for $e + H_2^+$. This system is an important prototype because it involves the simplest molecular ion, because experimental data[1,2] are available, and because the relevant excited state potential curves of H_2 have been calculated.[3,4]

Section 2 discusses the so-called "direct" and "indirect" mechanisms. Section 3 summarizes the theory. A single resonance in the DR cross section may be described by a Beutler-Fano profile,[5] but the general approach treats overlapping resonances. Section 4 presents calculations for H_2^+. Section 5 contains concluding remarks.

2. MECHANISMS FOR DISSOCIATIVE RECOMBINATION

2.1 Direct Recombination

The physical picture of direct DR, treated by Bardsley[6] in 1968, may be understood using the potential curves sketched in Fig. 1a and the following symbolic representation:

$$e + AB^{+} \rightleftarrows AB^{**} \rightarrow A + B^{*} \quad (1)$$

After capture into the dissociating state AB^{**}, the nuclei follow the dissociating potential curve V^*. The excess electronic energy of the resonant state is converted to kinetic energy of the dissociating products. These products must separate to the point where V^* crosses V^+ for the recombination to be complete. Competing with the stabilization of the resonant state by dissociation is electronic autoionization [indicated by the leftward pointing arrow in Eq. (1)]. For H_2, classical mechanics predicts that a time on the order of 10^{-15} sec is required for dissociation. Typical calculated values[4] of the autoionization lifetime are also on the order of 10^{-15} sec. The conclusion is that dissociation provides a stabilization mechanism fast enough to compete effectively with autoionization.

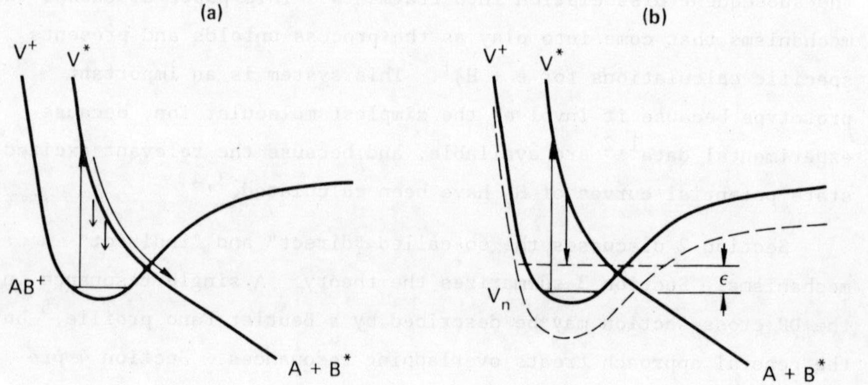

Figure 1. Illustration of direct (a) and indirect (b) recombination.

The details of the theoretical analysis of direct DR are available in the literature.[6-10] The result is that the cross section is approximately proportional to the square of the following integral:

$$\xi_{v_0} = \left(\frac{2M\pi}{k}\right)^{1/2} \int \chi_{v_0}(R)\, V_{el}(R)\, F(R)\, dR \qquad (2)$$

χ_{v_0} is the wave function of the initial vibrational level of the molecular ion, V_{el} is the matrix element for electron capture (or autoionization), and F(R) is the regular solution to the Schroedinger equation for the final state potential $V^*(R)$ at energy $E = \hbar^2k^2/2M$. This integral is essentially a Franck-Condon factor between the initial and final states, mediated by the electronic coupling V_{el}.

Direct DR for thermal energy electrons is expected to have a large cross section when the Franck-Condon factor is large. This situation usually occurs when the dissociating potential V^* crosses the ionic potential V^+ near the minimum of the latter. Conversely, when the dissociating curve does not cross V^+ near the minimum, one expects the thermal energy cross section for direct DR to be small. However, even when direct DR is small, other mechanisms may be significant and may lead to a large total cross section. A qualitative discussion of this possibility is presented in the next subsection.

2.2 Indirect Recombination

At certain well-defined energies, the incident electron may be captured into compound resonant states, here denoted $AB^{\#}(n,v)$. These states provide an alternative pathway to the final state $A + B^*$. The situation represented by Eq.(1) must therefore be generalized:

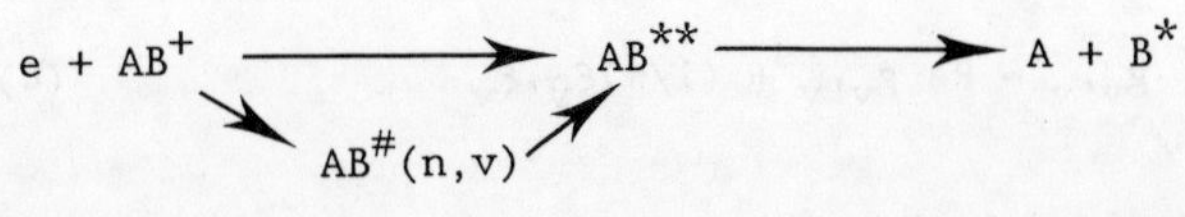

(3)

The additional states are labelled by the indices n and v. Their physical form is an ion core AB^+ in vibrational level v, and a Rydberg electron in an orbital whose principle quantum number is n. (We omit discussion of the electron's angular momentum).

One way to consider these resonant states is the following. Beginning with the ionic potential $V^+(R)$, define the series of Rydberg potential energy curves $V^n(R) = V^+(R) - 0.5\,[n-\mu(R)]^{-2}$. The quantum defect $\mu(R)$ defines the binding energy of the Rydberg electron. Each potential $V^n(R)$ of the molecule AB has a set of vibrational energy levels. One of the $V^n(R)$ is shown in Fig. 1b, and one of its vibrational energy levels is drawn. For an incident electron of the proper energy, the total energy of the system will be close to that of this particular $AB^{\#}(n,v)$. One expects the cross section for recombination to have additional structure near this energy.

3. THEORY

The theory of DR may be formulated[6-8] using the matrix elements

$$g_{v'v} = \frac{2M}{k} \int \chi_{v'}(R')\, V_{el}(R')\, G(R',R)\, V_{el}(R)\, \chi_v(R)\, dR'\, dR$$

where $G(R',R)$, defined explicitly in ref. 8, is a Green function for the Schroedinger equation with potential V^*. The matrix element $g_{v'v}$ is related to the amplitude for capture into the resonant state $AB^{\#}(n,v')$ from the initial vibrational state v of AB^+. The mechanism for this capture is a second order (in V_{el}) electronic coupling via the resonant state AB^{**}. The momentum transfer from the light electron to the heavy nuclei via potential scattering is neglected. This approximation is well-fulfilled in practice for $H_2{}^+$.

For the present discussion, it is sufficient to note that G corresponds to outgoing-wave boundary conditions (i.e., dissociation) and is therefore a complex-valued function. We may write

$$g_{v'v} = \mathrm{Re}\; g_{v'v} + (i/\pi)\xi_{v'}\xi_v \qquad (4)$$

where the ξ_v are defined in Eq. (2).

Giusti[9] first treated DR including both direct and indirect mechanisms, using "first order" Multichannel Quantum Defect Theory (MQDT). The terms Re $g_{v'v}$ were neglected. Later calculations for

$e + H_2^+$ by Giusti et al.[7] and Nakashima et al.[10] also neglected Re $g_{v'v}$. Hickman[8] included Re $g_{v'v}$ in an approach based on the Independent Resonance Approximation (IRA) and discovered that the previously neglected terms were quite important for $e + H_2^+$. However, the relation between the two theories (MQDT and IRA) was obscure. Our recent work, summarized here, clarifies this relation and unifies the two theories.

MQDT is formulated in terms of a reactance matrix coupling several different arrangement channels. These channels have the asymptotic forms $A + B^*$ (labelled d), and $AB^+(v) + e$, for v=0,1,... The specific form of the matrix that we use is

$$
\begin{array}{c|c:ccccc}
 & d & v=0 & v=1 & v=2 & v=3 & \\
\hline
d & 0 & \xi_0 & \xi_1 & \xi_2 & \xi_3 & \cdots \\
\hdashline
v=0 & \xi_0 & & & & & \\
v=1 & \xi_1 & & \pi \,\mathrm{Re}\, g_{v'v} & & & \\
v=2 & \xi_2 & & & & & \\
 & \vdots & & & & &
\end{array}
\qquad (5)
$$

This matrix may be "contracted" into an S matrix using the general MQDT prescription.[11] S may have a smaller dimension than the reactance matrix, because some of the arrangement channels are closed.

We have found a limiting case in which the MQDT contraction can be explicitly compared with the IRA. Consider the channels d, v=0 (the initial channel), and v=1 (a closed channel). The effect of a single resonance on the recombination may then be described in terms of a background process (direct DR) that is modified in a restricted energy range by the resonance. Identical result are obtained with the IRA and MQDT:

$$\sigma(\epsilon) = \sigma^{\mathrm{direct}}(\epsilon)\ (x + q)^2/(x^2 + 1) \qquad (6)$$

where $x = (E-E_{res})/(\Gamma/2)$, and $\Gamma = 2\xi_v^2/\pi n^3$ is the width of the resonance. We find that q is given by $-(\mathrm{Re}\ g_{01})/(\mathrm{Im}\ g_{01})$. This result demonstrates the equivalence of MQDT and the IRA for single resonances, provided that one includes the effect of the full Green function as shown in Eq. (5). The "first-order" limit of MQDT discussed by Giusti[9] corresponds to neglecting the block of terms in

Eq. (5) whose values are π Re $g_{v'v}$. For the calculations reported here, we use the full MQDT approach based on the reactance matrix, Eq. (5).

4. RESULTS FOR e + H_2^+

Fig. 2 shows the intermediate states $AB^{\#}(n,v)$ introduced in 2.2. The vibrational levels of the ion AB^+ are drawn, as well as an electronic Rydberg series converging to each ionization limit. Clearly, for many incident electron energies, contributions from more than one intermediate state must be included.

Two limiting cases may be identified: small Δv and large n (the left hand side of Fig. 2), and large Δv and small n (the right hand side of Fig. 2). The discussion in 2.2 and Fig. 1b correspond to resonances of the first type. One may wonder whether resonances of the second type, involving up to seven quanta of vibrational energy transfer, are important. Following the arguments of O'Malley,[12] one

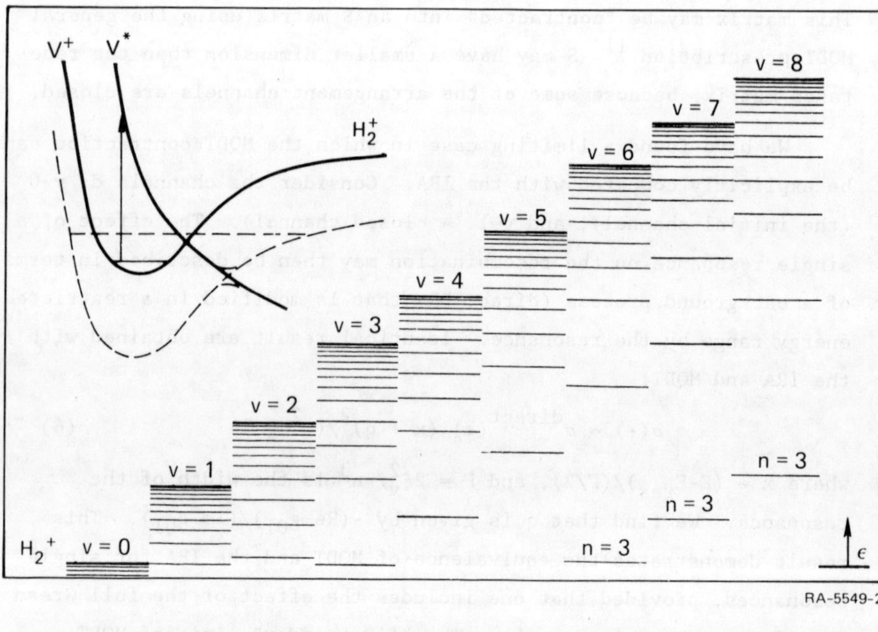

Figure 2. Rydberg energy levels of H_2. Inset: Curve crossing mechanism for high v, low n resonances.

sees that such resonances may be interpreted in terms of a curve crossing as shown in the upper left of Fig. 2. The electron is captured into the dissociating state, but during dissociation the system undergoes a curve crossing to the potential curve $V^n(R)$, where n=3 (the smallest value of n possible for the dσ waves considered here).

The effect of the overlapping resonances may be probed by model calculations that include only selected resonances. Two series of such calculations are presented in Fig. 3. First, we consider the case $v_0=0$. (v_0 denotes the initial vibrational level of H_2^+). In this case the Franck-Condon factor is low and one expects the direct mechanism to be small. Panel (a) shows a dotted line corresponding to the direct process; the transition probability is indeed small, about 10%. This panel also shows the regular structure arising from the Rydberg series converging to v=1 (cf. Fig. 2). The structure defined by Eq. (6), which could be obtained by the IRA, is essentially repeated, properly scaled, by each member of the Rydberg series. Panel (b) shows the result if one considers a single resonance corresponding to v=7 and n=3. Panel (c) shows the combined results. The overall pattern is more complicated, but still reminiscent of the simpler results obtained in (a) and (b). Note that the deep minimum seen in (b) near 0.20 eV persists.

Panels (d)-(f) show a parallel calculation for the case $v_0=1$. In this case the Franck-Condon factor is large, and the direct mechanism (dotted line) is much larger, about 70%. The series of resonances for v=1 to v=2 excitation [panel (d)] are "window" resonances, that is, purely destructive interference. The single resonance for v=1 to v=8 excitation is also of this type. Again, the composite case [panel (f)] exhibits features that are similar to those of each contributing resonance.

Our results suggest that the indirect mechanism tends to inhibit recombination when the direct mechanism is large. When the direct mechanism is small, however, the indirect mechanism may enhance or inhibit the cross section. Having a smaller cross section when the indirect mechanism is included is consistent with the physical picture

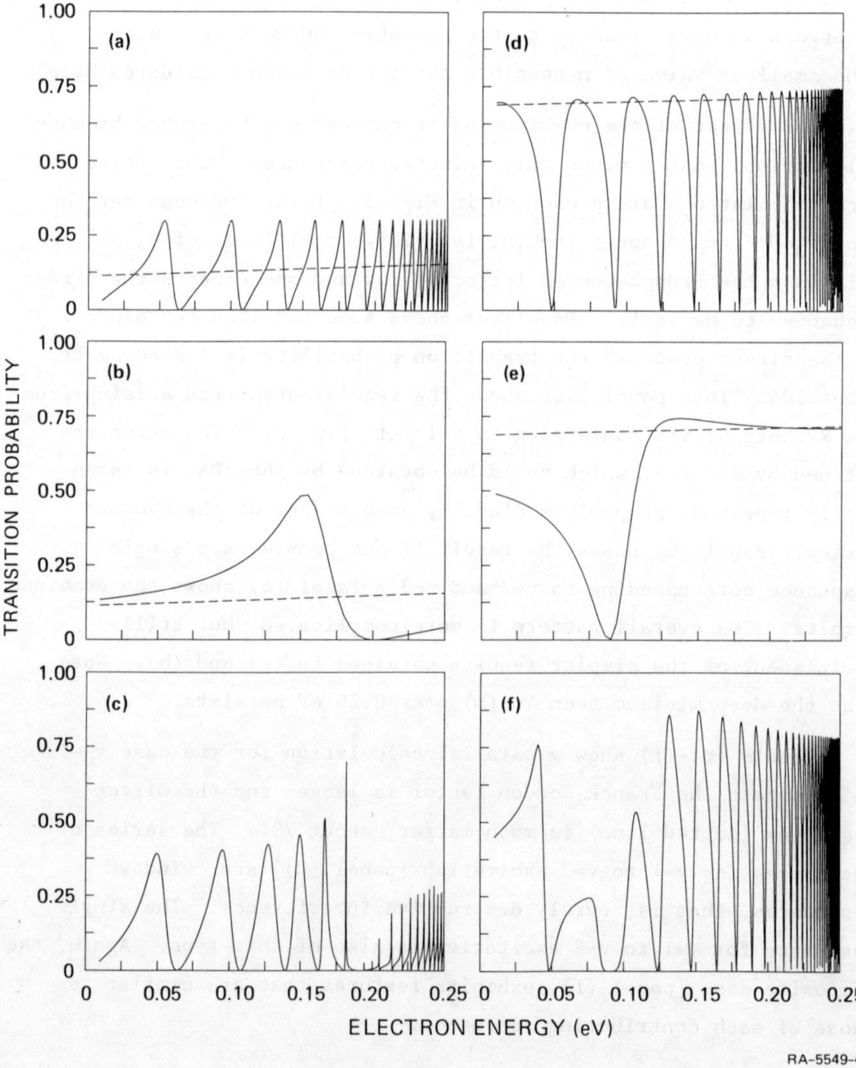

Figure 3. Effect of selected resonant states for $e + H_2^+$.

developed by O'Malley based on classical trajectories. The curve crossings in the outgoing channel make dissociation more difficult.

The exact energies of the resonances corresponding to large Δv are sensitive to the molecular potential curves. We find that different values of $\mu(R)$ from the literature lead to different values of the resonance energies. For $\Delta v \sim 7\text{-}8$, the discrepancies are large, on the order of tenths of eV. (Another contribution to the resonance energy comes from Re g_{vv}). Since the effect of these large v resonances is often to reduce substantially the cross section over a particular energy range, it is essential to calculate accurately the location of each resonance.

These uncertainties preclude a definitive calculation even though the MQDT formalism provides the computational framework. Nevertheless, Fig. 4 presents calculations including the vibrational levels v=0-9 for H_2^+. Separate calculations for v_0=0 and v_0=1 were performed; the experimental values of Hus et al., shown for comparison in both panels, correspond to a mixture of these two initial states. Some testing of the effects of using different functions $\mu(R)$ from the literature was carried out. Our results tend to favor the $\mu(R)$ of Nakashima et al.,[10] which is numerically much smaller than the other functions. Those results are shown.

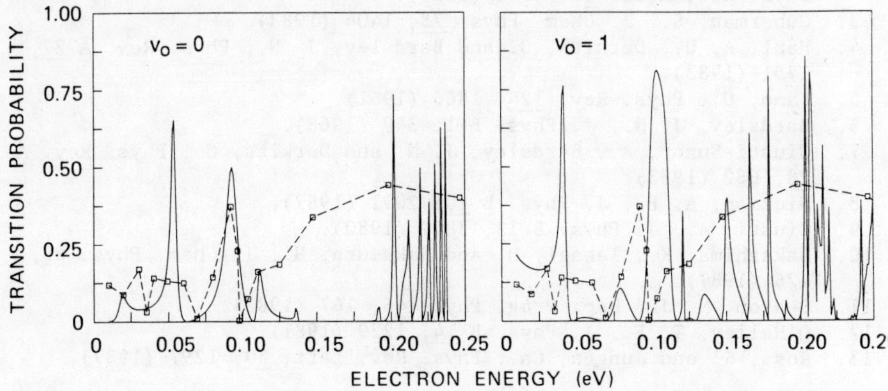

Figure 4. Comparison with experiment (Hus et al.[2]) for e + H_2^+.

5. CONCLUDING REMARKS

There has been considerable progress in the understanding of DR of molecular ions in the past few years. The recent experimental studies of Hus et al. have an electron energy resolution of about 0.02 eV, and reveal a complex structure of the cross section. Theoretical calculations confirm that the structure arises from the pattern of overlapping Rydberg series converging to the various vibrational levels of the target molecular ion. These Rydberg states collectively contribute to the "indirect" mechanism of recombination, which is of prime importance for e + H_2^+. This indirect mechanism should be assessed in future calculations on other systems, particularly in cases where Franck-Condon arguments suggest that the direct mechanism will be small.

ACKNOWLEDGMENT

This work was supported by AFOSR contract No. F49620-88-K-0005.

REFERENCES

1. Auerbach, D., Cacak, R., Caudano, R., Gaily, T. D., Keyser, C. J., McGowan, J. W., Mitchell, J.B.A. and Wilk, S.F.J., J. Phys. B 10, 3797 (1977).
2. Hus, H., Yousif, F., Noren, C., Sen, A. and Mitchell, J.B.A., Phys. Rev. Lett. 60, 1006 (1988).
3. Guberman, S., J. Chem. Phys. 78, 1404 (1983).
4. Hazi, A. U., Derkits, C. and Bardsley, J. N., Phys. Rev. A 27, 1751 (1983).
5. Fano, U., Phys. Rev. 124, 1866 (1961).
6. Bardsley, J. N., J. Phys. B 1, 349 (1968).
7. Giusti-Suzor, A., Bardsley, J. N. and Derkits, C., Phys. Rev. A 28, 682 (1983).
8. Hickman, A. P., J. Phys. B 20, 2091 (1987).
9. Giusti, A., J. Phys. B 13, 3867 (1980).
10. Nakashima, K., Takagi, H. and Nakamura, H., J. Chem. Phys. 86, 726 (1987).
11. Seaton, M. J., Rep. Prog. Phys. 46, 167 (1983).
12. O'Malley, T. F., J. Phys. B 14, 1229 (1981).
13. Ross, S. and Jungen, Ch., Phys. Rev. Lett. 59, 1297 (1987).

AB INITIO STUDIES OF DISSOCIATIVE RECOMBINATION

Steven L. Guberman
Institute for Scientific Research
33 Bedford St., Suite 19A
Lexington, MA 02173

ABSTRACT

Quantum chemical calculations of the dissociative recombination of O_2^+ and N_2^+ are reported. An approach for calculating autoionization widths from high principal quantum number Rydberg states is summarized and an example is presented for the lowest $^3\Pi_g$ dissociative state of O_2. For O_2^+, the $^1\Sigma_u^+$ state is the sole source of $O(^1S)$ from the lowest 10 vibrational levels of the ion. For $O(^1D)$, the $B^3\Sigma_u^-$, $^1\Sigma_u^+$, and $1^1\Delta_u$ states are the sources of $O(^1D)$ from the lowest three vibrational levels. Rate coefficients for generating $O(^1S)$ and $O(^1D)$ at ionospheric temperatures are reported. For N_2^+, states of $^3\Pi_u$, $^1\Sigma_g^+$ and $^3\Delta_g$ symmetries intersect the ion at the v=0 level. The configurational structure of these states indicates that the $^3\Pi_u$ states are likely to be the most important routes for dissociative recombination from v=0.

1. INTRODUCTION

Interest in dissociative recombination (DR) first arose out of the need to understand the properties of the Earth's ionosphere.[1],[2] Initial attention focused on the DR of O_2^+, N_2^+ and NO^+. While there has been much progress in the measurement of the total DR rate coefficients of these ions, the dependence of the quantum yield of excited and ground state atoms upon the electron temperature and ion vibrational level is only currently being unraveled. Impediments to experimental progress have included the difficulty of generating ions in specific vibrational levels and simultaneously identifying the atomic products.

Nevertheless there has been recent progress toward this goal.[3],[4],[5] An alternative theoretical approach based on spectroscopic identification of the potential curves which provide routes for DR is difficult since these states often do not have dipole allowed transitions from the ground state. Furthermore, electron capture occurs at high energies on the repulsive wall, well above the dissociation asymptote and out of range of the RKR approach. Additional difficulties include the paucity of information on electron capture widths although progress has been made in spectroscopic analysis of Rydberg-valence interactions[6],[7],[8],[9] from which widths can be derived.

Here, I review recent theoretical ab initio results obtained in this laboratory for the DR of O_2^+ and describe some preliminary results on the DR of N_2^+. For O_2^+, potential curves, electron capture widths and rate constants have been obtained for generating ^1S and ^1D atoms, the upper states of the green and red lines respectively. For N_2^+, potential curves for DR of the v=0 ion level have been computed.

2. WAVE FUNCTIONS

The wave functions calculated here are based on orbitals determined in either multiconfiguration or complete active space self consistent field (MCSCF or CASSCF) calculations. Each of the orbitals, ϕ_j, is expanded in a large set of Gaussian basis functions, χ_i,

$$\phi_j = \sum_i c_{j,i}\, \chi_i \tag{1}$$

Some of the χ_i are individual Gaussian primitives[10] while others are fixed combinations (contractions) of Gaussian basis functions determined from atomic calculations.[11] The orbitals are used to construct large scale configuration interaction (CI) wave functions containing many configurations, each of which describes a possible distribution of the electrons in the molecular orbitals for a particular state symmetry. For example, for the $^1\Sigma_g^+$ ground state of N_2, the most important configuration near the equilibrium separation is:

$$(1\sigma_g)^2(1\sigma_u)^2(2\sigma_g)^2(2\sigma_u)^2(3\sigma_g)^2(1\pi_u)^4 \quad (2)$$

The 1σ and 2σ orbitals are mostly N atom 1s and 2s orbitals, respectively. The $3\sigma_g$ orbital is a bonding combination consisting mostly of the N atom 2p orbitals lying along the bond axis. The $1\pi_u$ orbitals are out of plane bond orbitals consisting mostly of the N 2p orbitals that are perpendicular to the bond axis. Each configuration can have one or more spin couplings (referred to below as terms) for a given total spin symmetry. A wave function is constructed consisting of a superposition of many terms, Φ_i, each of which has the same total spin and spatial symmetry,

$$\psi_j = \sum_{i=1}^{N} C_{j,i}\,\Phi_i. \quad (3)$$

In the CASSCF[12] approach a set of active orbitals is defined and N(CASSCF) terms are generated in (3) by taking all possible excitations of the electrons within this set while restricting each term to have the total spin and spatial symmetry of interest. Both the $c_{j,i}$ and the $C_{j,i}$ are determined by variationally optimizing the total energy of ψ_j. The MCSCF approach is identical except that usually a subset of the full active space is used. Since the orbitals are expanded in large basis sets, χ_i, the number of orbitals used in the CASSCF is only a subset of the orbitals that can be formed from the basis set. The CI wave functions are constructed by exciting electrons out of a small set of reference configurations into the remaining (virtual) orbitals. The reference configurations often consist of all the configurations needed to properly dissociate that particular state to the proper separated atoms limit. This approach leads to wave functions that are not biased to favor particular internuclear distances. Some of the wave functions reported here for N_2 include the 2σ orbitals in the active space and use the entire CASSCF wave function as a reference set. Taking at most one or two electrons from each of the reference configurations and promoting them to the virtual and other valence orbitals leads to a singles and doubles CI denoted CISD. The CI wave functions are expanded over the CASSCF orbitals and have the form of (3) with

N(CI)>>N(CASSCF). Some of calculations used here employ large scale wave functions having more than 100,000 terms with the CI coefficients determined by direct CI techniques.[13] Calculations have recently been completed on several states of N_2 and O_2 which provide examples of the sort of accuracy which can be expected with these techniques. For the ground state of N_2, 14 reference configurations (needed to properly dissociate the ground state to ^4S atoms) are used in the CI with a basis set consisting of 6s,3p,2d and 1f type contracted Gaussians leading to a CI wave function having 176,536 terms. The calculated (experimental[14]) spectroscopic constants are 2340.57(2356.56cm^{-1}) for ω_e, 14.27(14.3244cm^{-1}) for $\omega_e x_e$, and 2.0856(2.074347a$_0$) for the equilibrium distance, R_e. Similar calculations on the ground state of O_2 using a basis set of the same size leads to a 228,036 term wave function. The calculated (experimental[15]) results are 1565.47(1580.19) for ω_e, 10.93(11.98cm^{-1}) for $\omega_e x_e$, and 2.2999(2.2819a$_0$) for R_e. For O_2, a series of vertical excitation energies have been calculated. The calculated (experimentally derived) results are 5.8982(5.90±0.05eV) for $^1\Sigma_u^-$, 6.1318(6.11±0.05eV) for $^3\Delta_u$,6.2766(6.19±0.05eV) for $^3\Sigma_u^+$, and 8.66(8.61±0.05eV[16]) for $^3\Sigma_u^-$. All the experimental energies are derived from Ref.15 (except for $^3\Sigma_u^-$ by extending the RKR results with an ab initio repulsive wall.

Recent results have been obtained for N_2 using a [4s,3p,2d,1f] contracted Gaussian basis set and a CISD reference set consisting of the CASSCF configurations with 2σ in the active space. For the $C^3\Pi_u$ state the calculated (experimental)[14] results are 11.01(11.05eV) for T_e, 1986(2047cm^{-1}) for ω_e, 25.05(28.44cm^{-1}) for $\omega_e x_e$, and 2.1932(2.1707a$_0$) for R_e. For the $w^1\Delta_u$ state the results are 8.83(8.84eV) for T_e, 1532(1559cm^{-1}) for ω_e, 11.67(12.01cm^{-1}) for $\omega_e x_e$, and 2.4192(2.3977a$_0$) for R_e. The highly accurate values of the excitation energies and the other spectroscopic constants give us confidence that these repulsive curves will be located quite accurately relative to the ion potential curve.

3. POTENTIAL CURVES FOR O_2

Potential curves describing DR in O_2 and based on a [3s,2p,1d] contracted Gaussian basis set and first order CI wave functions have been described previously.[17] These calculations showed that the $^1\Sigma_u^+$ state is the only state that can generate O(^{1}S) from the low vibrational levels. It crosses the ion between the outer turning points of the v=1 and v=2 vibrational levels and dissociates to the ^{1}D + ^{1}S asymptote. The next accessible state is $5^3\Pi_g$ which crosses the outer turning point of v=10. Among the states dissociating to O(^{1}S) there are three routes which cross the $a^4\Pi_u$ metastable state between the turning points for v=0. These routes with their dissociation limits shown in parentheses are $5^3\Pi_g$(^{3}P + ^{1}S), $5^3\Pi_u$(^{3}P + ^{1}S), and $4^1\Pi_g$(^{1}D + ^{1}S). Because of its spin symmetry, the $4^1\Pi_g$ state will have only a small DR rate coefficient from the metastable state. The remaining two states can lead to hot O atoms, each with 3.4eV kinetic energy. Future experiments designed to measure the ^{1}S quantum yield must be careful to be certain that no $a^4\Pi_u$ O_2^+ is generated which would interfere with the determination of the ^{1}S generated from the ion ground state.

For generating O(^{1}D), in addition to the $^1\Sigma_u^+$ state, the $B^3\Sigma_u^-$ state crosses the ion near the inner turning point of v=0 and leads to ^{3}P + ^{1}D atoms. A $^1\Delta_u$ state leading to two ^{1}D atoms crosses the ion near the outer turning point of v=0. The lowest valence state of $^1\Pi_g$ symmetry crosses the ion near the inner turning point of v=0 and can lead to ^{1}D atoms via a curve crossing with the $2^1\Pi_g$ state. At higher energies, the lowest $^3\Pi_g$ state crosses the ion near the inner turning point of the v=3 ion level and can lead to ^{1}D atoms via an avoided crossing with the $2^3\Pi_g$ state. The next accessible states for ^{1}D production are $2^3\Sigma_g^-$(^{1}D + ^{3}P), $2^3\Pi_g$(^{1}D + ^{3}P), and 2 $^3\Pi_u$(^{1}D + ^{3}P) near the outer turning point of v=3, $3^3\Pi_g$(^{1}D + ^{3}P) and $2^1\Pi_g$(^{1}D + ^{1}D) near the outer turning point of v=4. As has been pointed out earlier[17], several of these routes are expected to have small electron capture widths.

The reader is referred to Ref. 17 for a discussion of additional routes that lead to only ground state atoms.

The important $B^3\Sigma_u^-$, $^1\Delta_u$, and $^1\Sigma_u^+$ states identified above have been the subject of larger scale calculations involving [6s,3p,2d,1f] contracted Gaussian basis sets and CI wave functions that include all single and double excitations to the virtual space.[18],[19] Second order

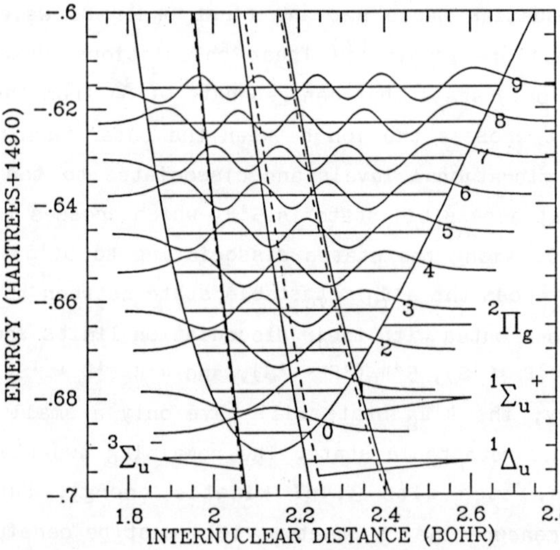

Fig. 1. Dissociative first order (dashed) and second order (solid) potentials for O_2 plotted with the ion potential from Ref. 15.

curves for the Π_g states have not been obtained because of their small widths (see below). Both the earlier curves and the large scale curves are plotted in Figure 1. The ion curve is the RKR[15] curve plotted at the experimental ionization potential above the large scale calculated ground state. The ion curve has been shifted to larger R to compensate for the difference between the calculated and experimental ground state R_e. In order to plot the first order curves on the same figure they have been shifted to larger R to correct for the difference between the first order and second order calculated R_e's. Also, the first order curves are shifted in energy so as to be plotted relative to the first order ground state. Relative to the ion, after the shift in the energy and distance coordinates, the curves are qualitatively similar. The largest difference is for $^1\Delta_u$ for which the second order curve is shifted by about $0.02a_o$ to smaller R near v=0 relative to the first order result.

4. ELECTRON CAPTURE WIDTHS FOR O_2

The DR cross section is approximately directly proportional to the electron capture or autoionization width, Γ, a matrix element of the Hamiltonian operator, H, given by Fermi's golden rule:

$$\Gamma(R) = 2\pi\rho|\langle\Psi_f|H|\Psi_i^c\rangle|^2. \quad (4)$$

Ψ_i^c is the wave function of the properly antisymmetrized product of a "free" electron in a coulomb like orbital and the wave function for the ion. Ψ_f is the electronic wave function for the neutral state which describes the DR products and ρ is the density of states. Rydberg orbitals with high principal quantum numbers, n, have been used to describe the coulomb like orbital. Except for the Rydberg orbital, all the orbitals in (4) are bound and have large amplitude only near the molecule. As a result, it is only important to know the amplitude of the Rydberg orbital near the molecule. The inner part of coulomb orbitals are well known to be very similar to the inner part of high Rydberg orbitals.[20] With the coulomb orbital represented as a high Rydberg orbital all the orbitals in (4) are bound and are represented in terms of Gaussian basis functions.

For $n^*>>\ell$, where ℓ is the angular momentum quantum number, the matrix element in (4) can be closely represented by

$$|\langle\Psi_f|H|\Psi_i^c\rangle|^2 = (1/n^{*3})k\exp(-c/n^*) \quad (5)$$

where k and c are constants determined from the calculation of the matrix elements on the left side of (5) for the highest Rydberg states. These Rydberg states are nearly hydrogenic and the energies can be represented as $E=-(1/2(n^*)^2)$ where $n^*=n-\delta$ and δ is the quantum defect. The density of states can be written as

$$\rho = 1/(E(n^*-1/2) - E(n^*+1/2)) \quad (6)$$

which can be closely represented as

$$\rho = n^{*3}\exp(-1/(2n^{*2})). \quad (7)$$

Substituting (7) and (5) into (4) gives

$$\Gamma = 2\pi k \exp(-(c/n^*)-(.5/n^{*2})). \tag{8}$$

Taking the limit of $n^* \rightarrow \infty$ gives the threshold capture width,

$$\Gamma = 2\pi k. \tag{9}$$

The Rydberg wave functions have been refined further by optimizing them in the field of an optical potential due to the Ψ_f states. Feshbach projection operators[21] have been defined which partition the total space such that $\Psi = P\Psi + Q\Psi$ where $P\Psi$ contains the Rydberg states to be optimized and $Q\Psi$ are the Ψ_f states described above. In the usual Feshbach projection operator formalism, an eigenvalue equation is derived for $P\Psi$ that is difficult to solve because it contains an energy-dependent optical potential. However, using partitioning techniques,[22] $P\Psi$ can easily be determined by solving the usual CI problem. Writing the total wave function as $\Psi = P\Psi + Q\Psi$, we can write the Schroedinger equation in matrix form as

$$\begin{pmatrix} H_{PP} & H_{PQ} \\ H_{QP} & H_{QQ} \end{pmatrix} \begin{pmatrix} P\Psi \\ Q\Psi \end{pmatrix} = E\ T \begin{pmatrix} P\Psi \\ Q\Psi \end{pmatrix} \tag{10}$$

where $H_{PP} = PHP$, $H_{QP} = QHP$, etc.

Multiplying the matrices in (10) leads to:

$$H_{PP}P\Psi + H_{PQ}Q\Psi = EP\Psi \tag{11}$$

and

$$H_{QP}P\Psi + H_{QQ}Q\Psi = EQ\Psi. \tag{12}$$

From Eq.(12) we have,

$$Q\Psi = H_{QP}P\Psi/(E - H_{QQ}). \tag{13}$$

Substituting Eq.(13) into Eq.(11) leads to a matrix optical potential for $P\Psi$,

$$(H_{PP} + H_{PQ}H_{QP}/(E - H_{QQ}))\ P\Psi = EP\Psi. \tag{14}$$

It is difficult to solve (14) directly for $P\Psi$ since E is on both sides of the equation. Nevertheless, $P\Psi$ can be determined by retaining the coefficients of the $P\Psi$ terms resulting from the diagonalization of the H matrix in Eq.(10). Before diagonalizing H we must project out the low energy $Q\Psi$ roots in order to prevent them from mixing into $P\Psi$. Therefore, it is necessary to first solve for the $Q\Psi$ roots by diagonalizing H_{QQ}. The H_{QQ} portion of the H matrix in Eq.(10) is transformed to project out the physically meaningful $Q\Psi$ roots. $P\Psi$ is then determined by diagonalizing the new H matrix with the transformed H_{QQ}. The net effect is to provide additional correlation to the $P\Psi$ space from the nonphysical roots in the Q space. $P\Psi$'s for successive principal quantum numbers are used to represent Ψ_1^c in Eq.(4). The widths obtained by this procedure are then extrapolated to the continuum as described above. The importance of the correlation added to the P space from the higher terms in Q space has been emphasized by Hazi who has described an approach similar to that used here.[23]

An application of this approach to the calculation of the widths of the B and L $^2\Pi$ states of NO has already been discussed.[24] An additional example of this approach is given in Table 1. for the O_2 $^3\Pi_g$ width. This is the only O_2 width for which there are experimentally derived matrix elements for comparison. An analysis of the released kinetic energy line width for the v=1 level of the $C^3\Pi_g$ state of O_2 indicates that the electronic matrix element between the C Rydberg state and the lowest valence $^3\Pi_g$ state is 0.079eV.[8] An additional determination based on the experimental line widths of the lowest 4 vibrational levels of the C state obtains a matrix element of 0.0625(+0.006,-0.004)eV.[9]

All calculations in Table 1. have been done at $R=2.2819a_0$. A [3s,2p,1d] basis set on each center was supplemented with a set a 18 diffuse s Gaussians at the midpoint and the Rydberg orbitals were determined by the Improved Virtual Orbital method.[25] The $^3\Pi_g$ valence state was described with a full valence CI consisting of 46 terms. For the C state, a valence CI on the ion ground state was coupled with the Rydberg orbital and each of the three valence virtual σ_g orbitals leading to a 636 term CI. Using the procedure outlined above leads to the widths shown in Table 1. Using the widths for n=8,9 gives the expression, $\Gamma = 0.00217\exp(1.39/n^* - (0.5/n^{*2}))$. The extrapolated width for $n^* \to \infty$ is 0.00217eV. This width is about two orders of magnitude smaller than the widths for the three important dissociative routes identified above. As a result, the $^3\Pi_g$ state was not included in the DR rate coefficient calculations discussed below.

Table 1. Electron-Ion Capture Widths (Γ) from $^3\Pi_g$ Rydberg States

n	n*	$\rho(eV^{-1})$	Γ(eV)
9	7.8897	17.903	0.00257
8	6.8896	11.892	0.00263
7	5.8902	7.4022	0.00272
6	4.8913	4.2110	0.00285
5	3.8930	2.0972	0.00311
4	2.8959	0.84005	0.00367
3	1.8984	0.21774	0.00522

The matrix element calculated here for n=3 is 0.0617eV. Since the electronic matrix element varies with R, the experimentally derived matrix element is an averaged value. However, since the variation of the matrix element with R over the relevant range of nuclear vibrations is estimated to be less than about 10%, the results show that relatively small wave functions can yield interaction matrix elements that are in good agreement with experiment.

5. DR RATE COEFFICIENTS FOR O_2

The expression for the direct DR cross section derived by Giusti[26] has been used to calculate cross sections and rate coefficients for generation of $O(^1S)$ and $O(^1D)$ along the $^1\Sigma_u^+$, $B^3\Sigma_u^-$, and $^1\Delta_u$ diabatic potential curves described above. For DR leading to $O(^1S)$,[18] the v=2 level of the ion has the highest rate coefficient for the lowest eight vibrational levels and is a factor of 78 greater than the coefficient from v=0 at an electron temperature (T_e) of 300K. A comparison of the calculated rate coefficients to those deduced from atmospheric models shows that the atmospheric models apply to molecular ions which are not in a Boltzmann distribution but which are instead significantly populated in the v>0 levels. In contrast to the situation for $O(^1S)$ generation, the total rate coefficient[19] for producing $O(^1D)$ is greatest from v=0 and is a factor of 1.9 greater than the coefficient from v=2 at T_e=300K. The rate coefficient for v=0 is in good agreement with those deduced from atmospheric models. The calculations show that the dominant route for producing $O(^1D)$ from v=0 is $^1\Delta_u$ and $^3\Sigma_u^-$ is the

Table 2. DR Rate Coefficients for Production of $O(^1S)$ for $600 \leq T_e \leq 1000$K.

Ion Vibrational Level	Rate Coefficient(cm^3/sec)^a
0	0.24(+.12,-.08)[-0.14]
1	5.0(+1.2,-1.1)[-0.36]
2	15.(+2.,-3.)[-0.50]
3	4.8(+3.3,-2.3)[-0.77]
4	4.9(+0.6,-1.2)[-0.32]
5	1.2(+1.9,-0.9)[-0.90]
6	5.1(+0.6,-1.1)[-0.56]
7	1.4(+0.9,-0.9)[+0.06]

a. All rate coefficients have been multiplied by 10^9. The exponent of the temperature dependence is in square brackets. The rate coefficient for v=0 is $2.4(+1.2,-0.8)\times10^{-10}\times((T_e/800)^{-0.14})cm^3/sec$.

dominant route from v=1 and v=2.

Since the electron temperatures in ionospheric models are near 800K, we list in Table 2. the rate coefficients for producing $O(^1S)$ for $600 \leq T_e \leq 1000$K for each of the lowest eight ion vibrational levels.

Table 3. DR Rate Coefficients for Production of $O(^1D)$ for $600 \leq T_e \leq 1000$K.

Ion Vibrational Level	Rate Coefficient(cm^3/sec)^a
0	1.37(+.14,-.16)[-0.51]
1	1.09(+.12,-.14)[-0.55]
2	0.78(+.13,-.16)[-0.41]

a. All rate coefficients have been multiplied by 10^7. The exponent of the temperature dependence is in square brackets. The rate coefficient for v=0 is $1.37(+.14,-.16)\times10^{-7}\times((T_e/800)^{-0.51})cm^3/sec$.

Because of the position of the intersection of the dissociative curve with the v=5 level, the coefficient from v=5 is highly sensitive to an energetic displacement of the dissociative route within its expected uncertainty (0.1eV). Table 3 has the $O(^1D)$ rate coefficients in the same temperature range from each of the lowest three ion vibrational levels.

6. DISSOCIATIVE ROUTES FOR N_2

Preliminary results are reported here which identify the important dissociative routes for the v=0 level of the ground state of N_2^+. These results are based on large scale second order CI calculations using atomic natural orbital basis sets.[27] The details will be reported separately.

The dominant dissociative routes from v=0, $2^1\Sigma_g^+(^2D + {}^2D)$, $C'^3\Pi_u(^4S + {}^2P)$, $3^3\Pi_u(^2D + {}^2D)$, and $G^3\Delta_g(^4S + {}^2D)$, are shown in Fig. 2.

The dominant configuration of the $2^1\Sigma_g^+$ state is

$$....(3\sigma_g)^2(1\pi_{ux})\ (1\pi_{uy})\ (1\pi_{gx})\ (1\pi_{gy})$$

where .... denotes $(1\sigma_g)^2(1\sigma_u)^2(2\sigma_g)^2(2\sigma_u)^2$. This configuration differs by a triple excitation from the dominant configuration of the ion plus a "free" or Rydberg electron,

$$....(3\sigma_g)\ (1\pi_{ux})^2(1\pi_{uy})^2(\mathrm{RYD}\sigma_g).$$

The matrix element of the Hamiltonian between these two configurations is zero and these dominant terms will not contribute to the electron capture width. The magnitude of the width will be determined by matrix elements between secondary terms in both states and may be small. As a result, $2^2\Sigma_g^+$ may make only a small contribution to the total DR rate coefficient from v=0.

The G state also passes through the v=0 level of the ion and dissociates to $^4S + {}^2D$ giving atoms with 1.72eV kinetic energy from v=0. The dominant configurations of this state,

$$....(3\sigma_g)^2(1\pi_{ux})\ (1\pi_{uy})\ (1\pi_{gx})^2,$$
$$....(3\sigma_g)^2(1\pi_{ux})\ (1\pi_{uy})\ (1\pi_{gy})^2,$$
$$....(3\sigma_g)^2(1\pi_{ux})^2(1\pi_{gx})\ (1\pi_{gy}),$$
$$....(3\sigma_g)^2(1\pi_{gx})^2(1\pi_{gx})\ (1\pi_{gy}),$$

differ from the ion plus a continuum electron by a triple excitation. Therefore, as for $2^2\Sigma_g^+$, the dominant configurations will not contribute

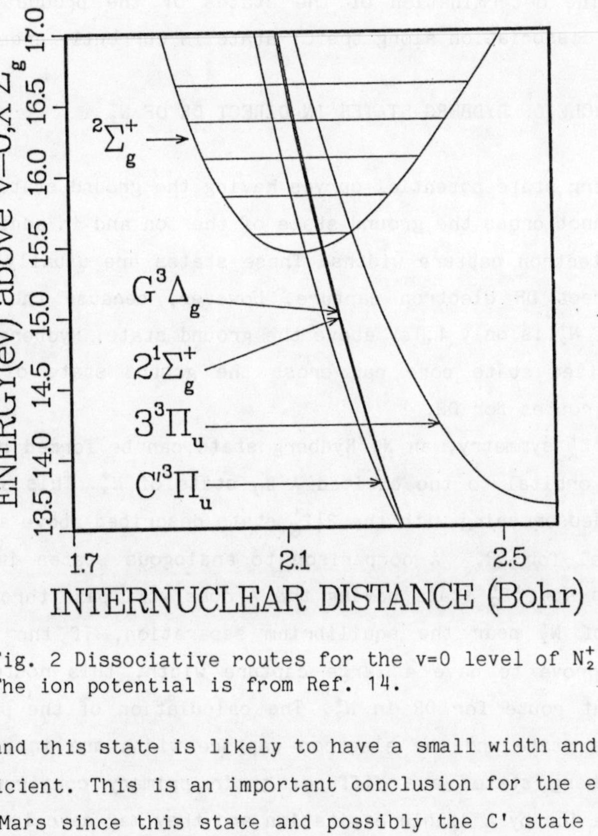

Fig. 2 Dissociative routes for the v=0 level of N_2^+. The ion potential is from Ref. 14.

to the width and this state is likely to have a small width and a small DR rate coefficient. This is an important conclusion for the escape of N atoms from Mars since this state and possibly the C' state (through an avoided crossing with the C state) are the only states that can supply enough kinetic energy for escape from v=0.

The primary configuration of the C' and $3^3\Pi_u$ states,

$$....(3\sigma_g)\ (1\pi_{ux})^2(1\pi_{uy})\ (1\pi_{gx})\ (1\pi_{gy}),$$

differs by a double excitation from the ion ground state plus a continuum electron. The capture width for this state may be larger than that for $2^1\Sigma_g^+$ and $G^3\Delta_g$ and it appears that these $^3\Pi_u$ states may be the dominant states for DR from v=0. The importance of the $^3\Pi_u$ states is in good agreement with the results of Michels.[28] The C' state has an avoided crossing with the lower $C^3\Pi_u$ state which dissociates to ^4S + ^2D atoms. The determination of the states of the product atoms due to initial dissociation along the C' state is currently under study.

7. THE ROLE OF RYDBERG STATES IN DIRECT DR OF N_2^+

Rydberg state potential curves having the ground state of the ion as core cannot cross the ground state of the ion and in addition have very small electron capture widths. These states are usually not important for direct DR electron capture. However, because the first excited state of N_2^+ is only 1.1eV above the ground state, Rydberg states having the excited state core can cross the ground state of the ion and provide routes for DR.

For $^1\Sigma_g^+$ symmetry, an N_2 Rydberg state can be formed by binding a π_u Rydberg orbital to the excited $A^2\Pi_u$ state of N_2^+. This state will have an avoided crossing with the $2^1\Sigma_g^+$ state described above and can provide a channel for DR. A comparison to analogous states in O_2 indicates that the $4p\pi$ $^1\Sigma_g^+$ state having the A core will pass through the ground state of N_2^+ near the equilibrium separation. If the $4p\pi$ $^1\Sigma_g^+$ state should prove to have a large capture width, this route could be an important route for DR in N_2^+. The calculation of the potential curve for this state and the electron capture width are in progress. While the Rydberg states are diffuse, their primary configurations differ from the ion by a double excitation and they may have larger electronic widths than the $^1\Sigma_g^+$ valence state which has primary terms that differ from the ion plus a continuum electron by a triple excitation. It appears that a multistate treatment involving both the valence states and the excited core Rydberg states may be needed to calculate the cross sections for DR of N_2^+.

8. ACKNOWLEDGEMENT

This research was supported by the National Science Foundation under grant ATM-8616776 and by the National Center for Atmospheric Research which is sponsored by the National Science Foundation. Support was also provided by NASA Grant NAGW-1404, by the Air Force Office of Scientific Research under grant AFOSR-84-0109 and by NASA Cooperative Agreement NCC 2-308.

9. REFERENCES

1. J. Kaplan, Phys. Rev. 38, 1048(1931).

2. D. R. Bates and H. S. W. Massey, Proc. Roy. Soc.(Lond.)192, 1(1947).

3. J. L. Queffelec, B. R. Rowe, M. Morlais, J. C. Gomet, and F. Vallee, Planet. Space Sci. 33, 263(1985).

4. D. Kley, G. M. Lawrence, and E. C. Stone, J. Chem. Phys. 66, 4157(1977).

5. E. C. Zipf, Planet. Space Sci. 36, 621(1988).

6. R. Galluser and K. Dressler, J. Chem. Phys. 76, 4311(1982); M. Raoult, J. Chem. Phys. 87, 4736(1987).

7. H. Lefebvre-Brion, Can. J. Phys. 47, 541(1969); D. Stahel, M. Leoni, and K. Dressler, J. Chem. Phys. 79, 2541 (1983).

8. W. J. van der Zande, W. Koot, J. Los, and J. R. Peterson, Predissociation of 3s $^{1,3}\Pi_g$ and $^{3,5}\Pi_u$ Rydberg States of O_2: Mechanisms and Pathways, this volume.

9. R. Friedman, private communication.

10. S. Huzinaga, J. Chem. Phys. 42, 1293(1965).

11. T. H. Dunning, Jr., J. Chem. Phys. 55, 716(1971).

12. P. Siegbahn, A. Heiberg, B. Roos, and B. Levy, Phys. Scr. 21, 323 (1980); P. E. M. Siegbahn, J. Almhof, A. Heiberg, and B.O. Roos, J. Chem. Phys. 74, 2384 (1981).

13. P. E. M. Siegbahn, J. Chem. Phys. 72, 1647(1980).

14. A. Lofthus and P. H. Krupenie, J. Phys. Chem. Ref. Data 6, 113(1977).

15. P. H. Krupenie, J. Phys. Chem. Ref. Data,1, 423(1972).

16. A. C. Allison, S. L. Guberman, and A. Dalgarno, J. Geophys. Res. in press,(1986).

17. S. L. Guberman, Potential Energy Curves for Dissociative Recombination, in Physics of Ion-Ion and Electron-Ion Collisions ed. by F. Brouillard (Plenum, New York, 1983) pp. 167-200.

18. S. L. Guberman, Nature 327, 408 (1987).

19. S. L. Guberman, Planet. Space Sci. 36, 47 (1988).

20. M. J. Seaton, Rep. Prog. Phys. 46, 167(1983).

21. H. Feshbach, Ann. Phys. (N. Y.) 5, 357(1958).

22. P. O. Lowdin, in Perturbation Theory and Its Application in Quantum Mechanics, ed. by C. H. Wilcox (John Wiley and Sons, N. Y., 1966), p.255; P. O. Lowdin, Advan. Chem. Phys. 2, 207 (1959); P. O. Lowdin, J. Math. Phys. 3, 969 (1962).

23. A. U. Hazi, Molecular Resonance Phenomena, in Electron-Atom and Electron-Molecule Collisions, ed. by J. Hinze,(Plenum Press, New York, 1983), p.103.

24. S. L. Guberman, Theoretical Studies of Dissociative Recombination, in Thermophysical Aspects of Re-entry Flows, ed. by J. N. Moss and C. D. Scott, (American Institute of Aeronautics and Astronautics, New York, 1986) pp.225-242.

25. W. J. Hunt and W. A. Goddard III, Chem. Phys. Lett. 6, 414(1969).

26. A. Giusti, J. Phys. B 13, 3867 (1980).

27. J. Almlof and P. R. Taylor, J. Chem. Phys. 86, 4070 (1987).

28. H. H. Michels, Electronic Structure of Excited States of Selected Atmospheric Systems, in The Excited State in Chemical Physics, ed. by J. M. McGowan (Wiley, New York, 1981), Vol 2; H. H. Michels, Abstracts of the Third Int. Conference on Atomic Physics, 173 (1972).

Dissociative Recombination of HCO^+: Complete Active Space (CAS) SCF Electronic Structure Calculations

W. P. Kraemer
Max-Planck-Institute of Astrophysics
D-8046 Garching, West Germany

and

A. U. Hazi
High Temperature Physics Division
Lawrence Livermore National Laboratory
Livermore, California

ABSTRACT

We have carried out ab initio calculations of those electronic states of HCO that may play a role in the dissociative recombination of electrons with HCO^+ ions under interstellar conditions. The Complete Active Space (CAS) SCF method was used to study the potential energy curves of HCO^+ ($X\ ^1\Sigma^+$) and of the relevant valence states of HCO in linear geometries as a function of the H–C and C–O bond distances. The present calculations suggest that the recombination of vibrationally cold HCO^+ ions with low-energy electrons can only proceed via an indirect mechanism, and that three different dissociation channels are potentially available to stabilize the intermediate states formed by electron capture.

INTRODUCTION

Laboratory measurements of the dissociative recombination of electrons with HCO^+ ions were performed using the stationary microwave afterglow technique [1,2] as well as in a flowing afterglow Langmuir probe (FALP) experiment [3]. In a critical review of the early microwave afterglow experiments, Johnsen [4] recently pointed out that the electron temperature dependences determined in some of these measurements are subject to experimental complications and uncertainties in interpretation due to the effects of vibrational excitation of the recombining the ions. Reinvestigation [2] of the electron temperature dependence of the recombination coefficient for HCO^+ gave a power law of the form

$$\alpha(HCO^+)(cm^3/s) = (2.4 \pm 0.4) \times 10^{-7} \, [T_e \,(K)/300]^{-0.69 \pm 0.07}$$

in the range $293K \leq T_e \leq 5500K$. Lower temperatures were excluded from this measurement in order to avoid complications due to possible clustering of HCO^+ ions with CO molecules. Direct information about the magnitude of the recombination coefficient of HCO^+ at the lower temperatures relevant for interstellar molecular clouds was obtained by Smith and Adams in a flowing afterglow/Langmuir probe experiment [3]. Their measurements gave $\alpha = 1.1 \times 10^{-7}$ cm^3/s and 2.9×10^{-7} cm^3/s at 300° K and 95° K, respectively. The value at 300° K appears to be smaller, by about a factor of 2, than the corresponding result of the stationary afterglow experiments. Apart from this discrepancy, the recombination coefficients of HCO^+ determined experimentally are similar in magnitude to those observed for simple diatomic ions, while the temperature dependence is somewhat stronger than the $T_e^{-0.5}$ dependence predicted by a simplified theory.[5]

So far the experiments have not provided any information about the identity of the neutral products resulting from the dissociative recombination reaction of HCO^+ or about the possible reaction pathways. Previously we have performed ab initio electronic structure calculations [6] at the SCF-CI level of theory to investigate the molecular potentials that are relevant for dissociative recombination of electrons with HCO^+ ions at low temperatures. Potential energy curves of the ground state of HCO^+ and of possible HCO resonance states were calculated in linear geometries as a function of CH distance for a fixed CO bond distance of R(CO) = 2.1 a_o, a value which is close to the optimum CO bond length in the HCO^+ ion. However, this study of HCO in the linear geometry provides information about only those resonance states that are accessible in a direct reaction mechanism, i.e., direct electron capture by linear HCO^+ ions. From the results of these calculations [6], we have concluded that there is a sparsity of HCO resonance states available to low-energy electrons and, therefore, dissociative recombination of HCO^+ ions under interstellar conditions should occur with smaller probability than previously assumed.

In view of the apparent contradiction of these theoretical results by the experimental findings, especially the FALP data at 95K, improved ab initio calculations were undertaken to re-examine the possibility of resonant electron capture by HCO^+ ions under the low temperature conditions present in dense interstellar clouds. Complete active space (CAS) SCF calculations [7] were performed employing fairly large atomic basis sets. The results of these CASSCF calculations are summarized in the present contribution and they will be used in a further study of HCO using the multi-reference configuration interaction method.

CALCULATIONS

In order to assess the accuracy of the present CASSCF calculations, various molecular properties of HCO and the HCO^+ ion, as well as of the CO fragment, were calculated using identical computational approaches. Some of the results of these calculations are compared below to available experimental data. Electronic excitation energies (T_e) were determined generally with an error of about 0.1 eV with respect to the corresponding experimental results (given in parenthesis):

$$T_e\ (\mathrm{CO},\ X\ ^1\Sigma^+ - a\ ^3\Pi) = 6.13\ \mathrm{eV}\ (6.04\ \mathrm{eV})$$
$$T_e\ (\mathrm{CO},\ X\ ^1\Sigma^+ - a'\ ^3\Sigma^+) = 7.03\ \mathrm{eV}\ (6.92\ \mathrm{eV})$$
$$T_e\ (\mathrm{HCO},\ X\ ^2A' - A\ ^2\Pi) = 2.11\ \mathrm{eV}\ (2.20\ \mathrm{eV})$$

and a similar discrepancy was obtained for the dissociation energy (D_e) of the HCO^+ ion:

$$D_e\,(HCO^+,\ X\,^1\Sigma^+) = 6.51\ \text{eV}\ (6.45\ \text{eV}).$$

Correcting this D_e value for the effect of the difference between the zeroth vibrational energies of HCO^+ and CO, the proton affinity of CO is obtained as PA(CO) = 141.4 Kcal/mol in very good agreement with the generally accepted laboratory value of 141.9 Kcal/mol. The ionization potentials (IP_e) derived from the calculated CASSCF energies show slightly larger errors in comparison to the corresponding experimental values for the adiabatic IP's (given in parenthesis):

$$IP_e\,(H) \;\;= 13.60\ \text{eV}\ (13.6058\ \text{eV})$$
$$IP_e\,(CO) = 13.81\ \text{eV}\ (14.0139\ \text{eV})$$
$$IP_e\,(HCO) = 7.85\ \text{eV}\ (8.10\ \text{eV, from Ref. 8})$$

To determine all the possible neutral valence states that are accessible through resonant electron capture in the direct reaction mechanism

$$AB^+ + e \rightleftharpoons (AB)^{**} \rightarrow A + B,$$

one can formally consider the formation of the "doubly excited" intermediate state (AB)** as a two-step process which, under low temperature conditions, can be characterized as: (i) excitation of a valence electron of the AB^+ ion into an empty low-lying orbital, and (ii) accommodation of the unbound electron in one of the available low-lying valence orbitals. In the case of the linear, electronic ground state of $HCO^+(X^1\Sigma^+,\ \ldots 4\sigma^2\ 5\sigma^2\ 1\pi^4)$, there exist three different possibilities for the excitation step and each of them can be followed by different ways to accommodate the unbound electron:

$(HCO^+)^*$	HCO	
	$... 4\sigma^2\ 5\sigma^2\ 6\sigma\ 1\pi^4$	HCO $(X\ ^2\Sigma^+)$
$...4\sigma^2\ 5\sigma\ 6\sigma\ 1\pi^4 + e \rightarrow$	$... 4\sigma^2\ 5\sigma\ 6\sigma^2\ 1\pi^4$	HCO $(^2\Sigma^+)_{C.T.}$
	$... 4\sigma^2\ 5\sigma\ 6\sigma\ 1\pi^4\ 2\pi$	HCO $(2\ ^2\Pi)$
$...4\sigma^2\ 5\sigma\ 1\pi^4\ 2\pi + e \rightarrow$	$... 4\sigma^2\ 5\sigma^2\ 1\pi^4\ 2\pi$	HCO $(A\ ^2\Pi)$
	$... 4\sigma^2\ 5\sigma\ 6\sigma\ 1\pi^4\ 2\pi$	HCO $(2\ ^2\Pi)$
$...4\sigma^2\ 5\sigma^2\ 1\pi^3\ 2\pi + e \rightarrow$	$... 4\sigma^2\ 5\sigma^2\ 6\sigma\ 1\pi^3\ 2\pi$	HCO $(2\ ^2\Sigma^+)$

Following the $\sigma\sigma^*$-excitation, the unbound electron can enter three different valence orbitals. In the first case the repulsive HCO $(X\ ^2\Sigma^+)$ state is formed which dissociates into $H(^2S)$ + $CO(X\ ^1\Sigma^+)$. The second $^2\Sigma^+$ state formed in this scheme is a charge transfer state dissociating into $H^-(^1S) + CO^+ (X\ ^2\Sigma^+)$. In order to be able to describe this state properly, a fairly large optimized hydrogen basis set was used in the present calculations. Finally, the third valence state that can be formed after $\sigma\sigma^*$-excitation is the first excited $^2\Pi$ state of HCO. In linear geometries this is a repulsive state with the dissociation limit $H(^2S)$ + CO $(a\ ^3\Pi)$.

In case of the $\sigma\pi^*$-excitation, the above scheme leads to the formation of either the lowest $A\ ^2\Pi$ or the first excited $^2\Pi$ state of HCO, which has been already mentioned above. The $A\ ^2\Pi$ state is known[9] to be Renner-Teller degenerate with the $X\ ^2A'$ ground state in linear geometry and it dissociates diabatically to H (^2P) + CO $(X\ ^1\Sigma^+)$, the dissociation energy being approximately 5.0 eV. This state is well separated from the HCO^+ $(X\ ^1\Sigma^+)$ ground state and therefore will not be considered further in the present context.

The $\pi\pi^*$-excitation leads to the first excited $^2\Sigma^+$ state of HCO which, in linear geometries, is a repulsive state dissociating into $H(^2S)$ + CO $(a'\ ^3\Sigma^+)$.

There is one other possibility of resonance state formation, namely, the Rydberg states belonging to the series converging to the repulsive HCO^+ ($^3\Sigma^+$) state which itself dissociates into $H(^2S)$ + CO^+ ($X\ ^2\Sigma^+$). At the dissociation limit, the $^3\Sigma^+$ state of HCO^+ is almost degenerate with the $X\ ^1\Sigma^+$ ground state. As a result, the crossing between the C-H dissociation potentials of the lowest members of this Rydberg series and of the HCO^+ ground state occur far above the minimum energy region of the HCO^+ ground state. Thus these Rydberg states are not accessible under low energy conditions.

RESULTS

All the valence states of HCO, except the $A\ ^2\Pi$ state, derived in the above scheme are potential candidates for the resonance state formed by direct electron capture, and their C-H dissociation potentials were calculated, at the CASSCF level of theory, in linear geometries with the CO bond distance held fixed at R(CO) = 2.1 a_0. The results of these calculations are shown in Fig. 1. In the CASSCF calculations the orbital space was partitioned into inactive and active orbitals such that the carbon and oxygen 1s orbitals together with the oxygen lone-pair orbital were kept inactive, and the active space contained six σ-orbitals and three π-orbitals in each degenerate direction. The active σ electrons are thus fully valence correlated.

Turning now to the results for the repulsive HCO ($X\ ^2\Sigma^+$) valence state, the present CASSCF calculations show that its C-H dissociation potential is <u>below</u> the ionic potential over the entire range of CH distances (see Fig. 1). Its relative position to the HCO^+ potential can directly be checked at the dissociation limit where the separation is equal to the ionization potential of the hydrogen atom. This is well reproduced by the present calculations. At CH distances in the range of $1.5\ a_0 \leq R(CH) \leq 2.0\ a_0$, the HCO ($X\ ^2\Sigma^+$) lies below HCO^+ ($X\ ^1\Sigma^+$) by approximately 3.5 eV. Changes in the CO bond length within the Franck-Condon region of the zeroth vibrational level of HCO^+ do not change the separation between the two potentials significantly. This is in general agreement with previous SCF-CI calculations by Tanaka and Davidson[9]] and by Bruna et al[10]]. On the other hand, the present CASSCF results contradict the recent ab initio study by Talbi et al.[11]] in which the C-H dissociation potential of the HCO (X $^2\Sigma^+$) state, obtained in a so-called quasi-diabatic approximation runs right through the minimum of the HCO^+ potential in the linear geometry. However, the procedure used to determine this approximation seems to be ambiguous and the large upward shift of the diabatic potential is most probably due to a poor description of the repulsive HCO (X $^2\Sigma^+$) state relative to the HCO^+ ground state.

It should be noted that the present CASSCF calculations were done using a triple-zeta quality valence basis set on each atomic center except for the hydrogen, for which a more extended basis set was chosen in order to properly account for the $^2\Sigma^+$ charge transfer state. The present CASSCF wavefunction for the HCO $^2\Sigma^+$ state in the range 1.5 $a_o \leq$ R (CH) $\leq$ 2.0 a_o contains only very small contributions from Rydberg states (coefficients less than 0.03). Thus, any additional mixing with $^2\Sigma^+$ Rydberg states that would result from repeating the CASSCF calculations with an enlarged atomic basis set containing more diffuse orbitals, would tend to lower the dissociation potential of the X $^2\Sigma^+$ state even further relative to the ground state of HCO^+.

The dissociation potential of the $^2\Sigma^+$ charge transfer state of HCO derived in the above scheme has a $\frac{1}{R}$ dependence at large R(CH) values with a fairly broad minimum around R(CH) = 5.0 a_o, and it rises very steeply at smaller R(CH) values. Since the equilibrium CO bond length in CO^+ (X $^2\Sigma^+$) is close to the R(CO) distance used in the present CASSCF calculations, variation of R(CO) did not lower the dissociation potential at all. Due to the strongly repulsive shape of the potential for R(CH) < 4.0 a_o, the crossing with the ionic potential occurs high above the minimum of the HCO^+ curve (see Fig. 1). Therefore the charge transfer state of HCO can be eliminated as candidate for low energy electron capture by HCO^+.

The charge transfer state, however, cannot be totally ignored because it has a strong configuration mixing with the first excited $^2\Sigma^+$-state of HCO near R(CH) = 5.0a_o. This $^2\Sigma^+$ state is repulsive for linear geometries and dissociates into H(^{2}S) + CO (a' $^3\Sigma^+$). The C-H dissociation potential calculated with fixed R(CO) = 2.1 a_o crosses the HCO^+ potential approximately at R(CH) = 3.5 a_o about 2.7 eV above the minimum energy of the ion (see Fig. 1). A similar unfavourable crossing also occurs for the first excited $^2\Pi$ state of HCO which is also repulsive as a function of R(CH) for linear geometries. For R(CO) = 2.1 a_o the crossing of the 2 $^2\Pi$ state with the HCO^+ potential occurs about 1.6 eV above the ion minimum (see Fig. 1). Thus, the present CASSCF results for linear geometries agree with the findings of the previous SCF-CI calculations [5], which showed that there is no HCO resonance state which is directly accessible from the zeroth vibrational level of HCO^+ in a low-energy electron recombination reaction.

In order to study the possibility of an indirect recombination mechanism, additional CASSCF calculations were performed relaxing the constraint of keeping the CO bond distance fixed at R(CO) = 2.1 a_o. (Further CASSCF calculations will also be done to optimize the bond angle as a function of the CH distance.) We have already pointed out that variations of the CO bond distance did not lower the X $^2\Sigma^+$ and the $^2\Sigma^+_{C.T.}$ states of HCO relative to HCO^+ (X $^1\Sigma^+$). However, a significant lowering of the C-H dissociation potentials relative to the ionic potential was obtained for the first excited $^2\Sigma^+$ and $^2\Pi$ states. As expected, optimization of the CO bond distance for the first excited $^2\Sigma^+$ state gave a bond length close to the equilibrium value in CO (a' $^3\Sigma^+$), i.e., R_e (CO) = 2.557 a_o. Due to this drastic change in the optimum R(CO), the lowering of the C-H dissociation potential is substantial: 2.3 eV at R(CH) = 5.0 a_o and 1.4 eV at R(CH) = 2.0 a_o. Similarly for the first excited $^2\Pi$ state, the optimized CO bond distance turns out to be close to the equilibrium bond length in CO (a $^3\Pi$), R_e (CO) = 2.279 a_o, and the lowering of the dissociation potential varies from 0.5 eV to about

0.3 eV in the range 5.0 $a_o \leq R(CH) \leq 2.0\ a_o$. The changes of these two potential curves due to CO bond optimization are shown in Figs. 2 and 3, respectively. Further lowering of these two excited states can be expected when the bond angles are optimized. Unfortunately, due to the large changes in molecular geometry, these states of HCO are not accessible in direct electron capture by the HCO^+ ion, which is linear, with $R_e(CO) = 2.1\ a_o$, in its equilibrium configuration.

SUMMARY

In conclusion, the present CASSCF calculations suggest that the recombination of vibrationally cold HCO^+ ions with low-energy electrons can only proceed via an indirect reaction mechanism. Three different dissociation channels are in principle available to stabilize the intermediate states formed by electron capture. Dissociation can occur along the repulsive potentials of the X $^2\Sigma^+$ and of the first excited $^2\Sigma^+$ and $^2\Pi$ states of HCO. Different electronic states of CO are produced in the three different dissociation channels and their exothermicities vary from ΔE_e = 7.1 eV for CO (X $^1\Sigma^+$) to ΔE_e = 1.1 eV for CO (a $^3\Pi$) and finally to ΔE_e = 0.2 eV for CO (a' $^3\Sigma^+$). It is clear that laboratory measurements identifying the nature of the dissociation products would be very valuable in elucidating the dominant dissociative recombination mechanism of HCO^+ at low temperatures.

ACKNOWLEDGEMENTS

This work was carried out in part under the auspices of the U. S. Department of Energy by the Lawrence Livermore National Laboratory under contract number W-7405-ENG-48.

REFERENCES

1. M. T. Leu, M. A. Biondi, and R. Johnsen, Phys. Rev. A 8, 420 (1973).
2. B. Ganguli, M. A. Biondi, R. Johnsen, and J. L. Dulaney, Phys. Rev. A 37, 2543 (1988).
3. D. Smith and N. G. Adams, Ap. J. (Letters) 284, L13 (1984); N. G. Adams, D. Smith, and E. Alge, J. Chem. Phys. 81, 1778 (1984)

4. R. Johnsen, Int. J. Mass Spectrosc. Ion Proc. 81, 67 (1987).
5. J. N. Bardsley and M. A. Biondi, Adv. Atom. and Mol. Phys. 6, 1 (1970).
6. W. P. Kraemer and A. U. Hazi, in "Molecular Astrophysics," ed. G. H. F. Diercksen et. al., (D. Reidel Publ. Comp.) 1985.
7. B. O. Roos, P. R. Taylor, and P. E. M. Siegbahn, Chem. Phys. 48, 157 (1980).
8. J. C. Traeger, Int. J. Mass Spectrosc. Ion Proc. 66, 271 (1985).
9. K. Tanaka and E. R. Davidson, J. Chem. Phys. 70, 2904 (1979).
10. P. J.Bruna, S. D. Peyerimhoff and R. J. Buenker, Chem. Phys. 10, 323 (1975); P. J. Bruna, R. J. Buenker, and S. D. Peyerimhoff, J. Mol. Structure 32, 217 (1976).
11. D. Talbi, A. P. Hickman, F. Pauzat, Y. Ellinger, and G. Berthier, preprint (1988).

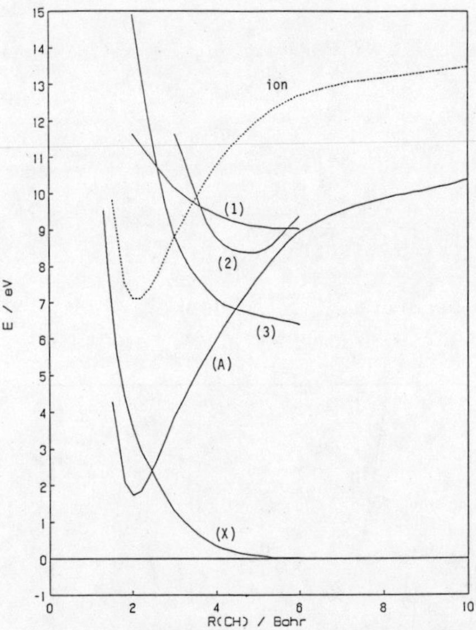

Figure 1: C-H dissociation potentials calculated (CASSCF) at linear geometries and fixed CO bond distance of 2.1 a_o for the HCO^+ ground electronic state (dotted curve) and various valence states of HCO : X $^2\Sigma^+$ (X); A $^2\Pi$ (A); first excited $^2\Sigma^+$ state (1); charge-transfer $^2\Sigma^+$ state (2); first excited $^2\Pi$ state (3).

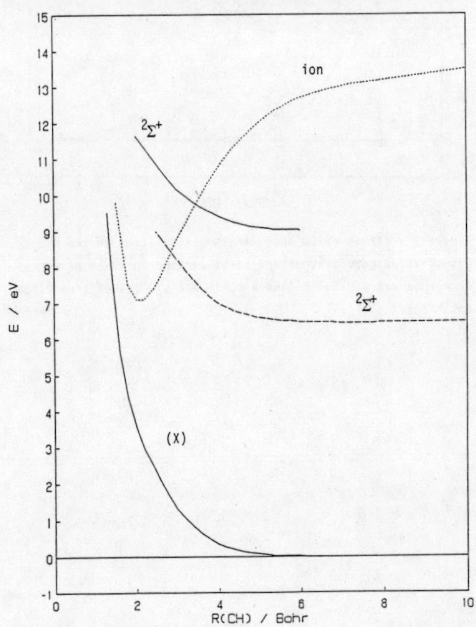

Figure 2: Effect of CO bond length optimization on the C-H dissociation potential of the first excited $^2\Sigma^+$ state of HCO in linear geometry. Solid line R(CO) = 2.1 a_o, dashed line R(CO) optimized.

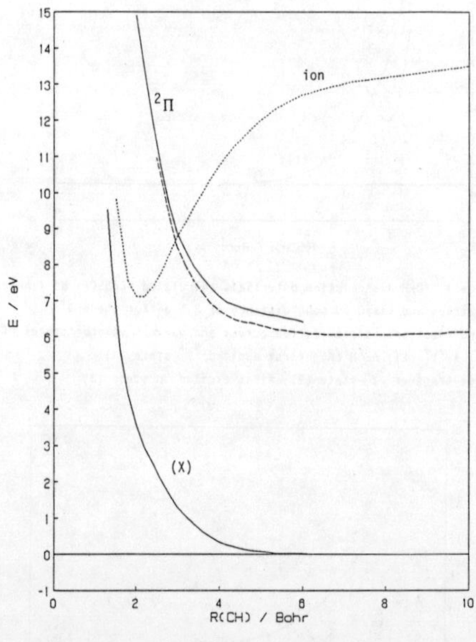

Figure 3. Effect of CO bond length optimization on the C-H dissociation potential of the first excited $^2\Pi$ state of HCO in linear geometry. Solid line R(CO) = 2.1 a_o, dashed line R(CO) optimized.

THEORETICAL STUDY OF DISSOCIATIVE RECOMBINATION OF ELECTRONS WITH H_2^+ and CH^+

Hiroki NAKAMURA
Division of Theoretical Studies, Institute for Molecular Science
Myodaiji, Okazaki 444, Japan

Hideki TAKAGI
Physics Laboratory, School of Medicine, Kitasato University
Sagamihara, Kanagawa 228, Japan

Keiji NAKASHIMA
Department of Molecular Science and Technology
Kyushu University, Hakozaki, Fukuoka 812, Japan

ABSTRACT

Theoretical studies of the title processes are reported. Use was made of the multichannel quantum defect theory (MQDT). The results are in good agreement with the recent merging beam experiment. Mechanisms of the processes are discussed, and the importance of dissociative superexcited state in various dynamic processes is pointed out.

I. WHAT ARE THE BASIC MECHANISMS ?

In order for dissociative recombination to occur effectively there should exist a curve crossing between molecular ion state and dissociative state near the initial vibrational level of molecular ion. In the so called direct process, an incident electron interacts directly with the electrons of molecular ion, and is captured into the dissociative state by exciting one of the core electrons. This interaction is essentially the electron correlation, and is called configuration mixing or electronic coupling $V(R)$ between the ionization continuum and the dissociative state. Since the dissociative state is embedded in the ionization continuum (i.e., an autoionizing superexcited state) at interatomic distance smaller than the crossing point, and can not be an eigenstate of electronic Hamiltonian, the electronic coupling not only survives, but also plays a decisive role. These dissociative superexcited states are expected to exist in various diatomic molecules[1]. The important physical parameter in this direct process is thus the following bound-free Franck-Condon like factor:

$$\xi_v = \pi\langle v|V(R)|F_d\rangle, \qquad (1-1)$$

where $\langle v|$ represents the vibrational state, and F_d is the energy normalized radial wave function in the dissociation channel. In addition to the above mentioned direct process, there exists an indirect process in which vibrationally excited Rydberg state plays a role of intermediate resonance state. An incident electron excites the target molecular ion vibrationally and itself is trapped into one of the Rydberg orbits. The electrostatic interaction responsible for this vibrational excitation can be represented by the R-dependent quantum defect $\mu(R)$. The important physical parameter in this process is $\langle v|$ cos or sin $[\pi\mu(R)]|v'\rangle$. Dissociation in the indirect process occurs again through the electronic coupling between the vibrationally excited Rydberg resonance state and the dissociative superexcited state.

II. MQDT APPROACH TO DISSOCIATIVE RECOMBINATION

The MQDT provides us with a powerful tool to deal with ionization continuum and Rydberg manifold in a unified way. This theory was first developed by Seaton[2] for atomic problems, and extended to molecular problems by Fano[3], and by Jungen and Atabek[4]. Being initiated by Lee[5], Giusti[6] proposed a two-step treatment which can incorporate the dissociation channel and thus can deal with dissociative recombination. Within the framework of this two-step formalism, the rotationally averaged cross section is given by

$$\sigma_{DR}(v_i) = g\frac{\pi}{k^2}\sum_{l,\Lambda}|S^{\Lambda}_{d,v_il}|^2, \qquad (2-1)$$

where g is a statistical factor, k is the incident electron wave number, v_i is the initial vibrational quantum number of molecular ion, l is the angular momentum of the incident electron, Λ is the molecular axis component of the total electronic angular momentum, and S^{Λ}_{d,v_il} represents the scattering matrix element for the dissociative recombination $(v_il) \rightarrow d$.

In order to estimate the scattering matrix we have employed the Seaton's procedure[7] instead of the eigenchannel method used by Giusti et.al.[8] This procedure has the following advantages in addition to its simplicity: (1) The unitarity of S-matrix is analytically guaranteed, and (2) we can easily interpolate the boundary-condition-free reactance matrix $\mathcal{R}$ (eq. (2-4) below) with respect to energy, since this quantity weekly and smoothly depends on energy. The scattering matrix S is given as follows:[6,7,9)]

$$S = (1 - iR)^{-1}(1 + iR), \tag{2-2}$$

where

$$R = \mathcal{R}_{OO} - \mathcal{R}_{OC}(\tan \pi\nu + \mathcal{R}_{CC})^{-1}\mathcal{R}_{CO}, \tag{2-3}$$

$$\mathcal{R} = \mathcal{S}\mathcal{C}^{-1}, \tag{2-4}$$

$$\mathcal{C}_{vl,\alpha}[\mathcal{S}_{vl,d}] = \sum_{v'l'} U_{v'l',\alpha}\langle v|\cos[\sin](\pi\mu_{l'\Lambda}(R) + \eta_\alpha)X_{ll'}(R)|v'\rangle, \tag{2-5a}$$

and

$$\mathcal{C}_{d,\alpha}[\mathcal{S}_{d,\alpha}] = U_{d,\alpha}\cos[\sin]\eta_\alpha. \tag{2-5b}$$

The suffix $O(C)$ means open (closed) channel, $\tan \pi\nu$ is a diagonal matrix with ν_j representing a parameter to measure energy from the j-th closed level [energy $= -(2\nu_j^2)^{-1}$ in a.u.], and the label α designates the eigenchannel in the reaction zone. The quantities $\{ U_{j\alpha} \}$ and $-\frac{1}{\pi}\tan \eta_\alpha$ are, respectively, the unitary diagonalization matrix and eigenvalue of the K-matrix (reaction matrix) associated with the electronic coupling $V(R)$, namely,

$$\sum_{j'} K_{jj'}U_{j'\alpha} = -\frac{1}{\pi}\tan \eta_\alpha U_{j\alpha}. \tag{2-6}$$

In the first order perturbation theory, we have

$$K_{jj'} = \begin{cases} \langle v|V(R)|F_d\rangle = \xi_v/\pi & for\ j = v\ and\ j' = d, \\ 0 & otherwise, \end{cases} \tag{2-7}$$

and

$$\eta_\alpha = \pm \tan^{-1}\xi\ , 0, ..., 0, \tag{2-8}$$

where

$$\xi^2 = \sum_v \xi_v^2. \tag{2-9}$$

The quantity $X_{ll'}(R)$ is a unitary matrix to diagonalize the partial wave coupling matrix, whose eigenvalues are equal to $-\frac{1}{\pi}\tan \pi\mu_{l'\Lambda}(R)$.

In order to have a deeper insight into the mechanisms, it is instructive to derive the following simplified analytical expressions for S-matrix. If we neglect all the closed channels and assume only one partial wave, then we obtain from eqs. (2-2) - (2-5)[9)]

$$S_{d,v_i} = \frac{-2i}{1+\xi^2}\sum_v \xi_v M_{vv_i} \tag{2-10}$$

and

$$S_{v_f,v_i} = \sum_v M_{v_f v} M_{vv_i} - \frac{2}{1+\xi^2}\sum_{v,v'} M_{v_f v'}\xi_{v'}\xi_v M_{vv_i}, \tag{2-11}$$

where S_{v_f,v_i} is the scattering matrix element for the vibrational transition $v_i \to v_f$ and

$$M_{vv'} = \langle v|\exp(i\pi\mu(R))|v'\rangle. \tag{2-12}$$

III. DISSOCIATIVE RECOMBINATION OF H_2^+

At low electron energies (lower than or around 1.0 ev) the lowest doubly excited $^1\Sigma_g\ (2p\sigma_u)^2$ state of H_2 is the only candidate for the dissociative state (see Fig.1). Quite accurate information on the electronic state, electronic coupling, and quantum defect is available for the $^1\Sigma_g$ symmetry. We have taken these quantities from our previous calculations of the elastic scattering of electrons from H_2^+.[10)] The $^1\Sigma_g\ (2p\sigma_u)^2$ superexcited state is known to couple almost exclusively to the $d\sigma_g$-partial wave ($l = 2, \Lambda = 0$) of electron. Since only the singlet scattering contributes to the dissociative recombination, the statistical factor g in eq. (2-1) is equal to 1/4 and the rotationally averaged cross sections are given by

$$\sigma_{DR}(v_i) = \frac{\pi}{4k^2}|S_{d,v_i}|^2 \quad for\ D.R., \tag{3-1}$$

and

$$\sigma_{VIB}(v_i \to v_f) = \frac{\pi}{4k^2}|S_{v_f,v_i}|^2 \quad for\ vibrational\ transition. \tag{3-2}$$

Since the quantum defect required in the MQDT treatment should be the "diabatic" one which corresponds to the Rydberg states diabatically crossing the dissociative state, we have subtracted the resonance contribution from the "adiabatic" quantum defect obtained in our scattering calculations. Our quantum defect thus obtained is quite different from that of Giusti et.al.[8]; and this difference mainly causes the discrepancy in the final results of cross section.

Along the theoretical framework described in the previous section we carried out the numerical calculations of the dissociative recombination and the vibrational transitions for H_2^+, HD^+ and D_2^+ at incident electron energies $E \simeq 0.02 \sim 1.0\ eV$. The number of vibrational states included in the calculation is 10 for H_2 and HD, and 13 for D_2, most of which are closed channels. Numerical results for the dissociative recombination of $H_2^+(v_i = 0,\ 1)$ are shown in Figs. 2-3. These results show very rich resonance structure due to the closed channel vibrationally excited Rydberg states. Positions of some of these Rydberg states (n, v) are indicated by single arrows. Double arrow shows the opening of a new vibrationally inelastic channel. The dashed lines are the envelopes of the curves at lower energies.

A comparison with the other calculation by Giusti et.al.[8] and the experimental result of Auerbach et.al.[12] is shown in Fig. 4. The theoretical results shown here are the ones vibrationally averaged by assuming the relative population as $1 : 2 : 2$ for $v_i = 0,\ 1,\ 2$, and also convoluted by assuming the triangular apparatus function of half-width 0.04 eV. The present result seems to agree better with the experiment than that of Giusti et.al. in respect to the energy dependence at low energies. Especially the position of the first dip at $E \simeq 0.1\ eV$ agrees well with that of the experiment. Very recently, a new experiment was carried out by the same group.[13] Although a detailed mutual comparison is still premature, the agreement with our theoretical result seems to be further improved. Not only the sharp dip at $E \simeq 0.1\ eV$, but also the absolute value of the cross section seems to be in better accordance with each other. The discrepancy between our result and that of Giusti et.al. is attributed to the difference in quantum defect. We think that their values are a little too large.

We have also carried out the calculations for HD^+ and D_2^+, and found some interesting isotope effects.[9] One example is shown in Fig. 5. These are again the convoluted results. This interesting difference in energy dependence can be attributed to the difference in energy dependence of the parameter ξ_v. Since $|M_{vv}| \gg |M_{vv'}|$ $(v \neq v')$ holds quite well in general, the scattering matrix element for dissociative recombination given by eq. (2-10) can further be simplified to

$$S_{d,v_i} \simeq \frac{-2i}{1+\xi^2}\, \xi_{v_i}\, M_{v_i v_i}.$$

Thus the energy dependence of ξ_{v_i} is directly reflected in the cross section. The sharp dip at $E \simeq 0.5\ eV$ in the case of D_2^+ ($v_i = 3$) is actually due to the zero of $\xi_{v_i=3}$ (D_2^+) there. On the other hand, $|\xi_{v_i=3}$ (HD^+)$|$ increases strongly with E in the present energy region, and cancels the decreasing factor $1/k^2$ in eq.(3-1), leading to the flat overall energy dependence. It would be very interesting, if this kind of isotope effect could be experimentally revealed.

It should also be noted that the resonances seen in Figs. 2-3, which are the manifestions of the interference between the direct and indirect processes, appear almost exclusively as dips. This can also be explained by the fact that the off-diagonal elements $M_{vv'}$ are much smaller than ξ_v's, since the "Fano's resonance profile index q" corresponds to $M_{v_i v_f} \ / \ \xi_{v_i}\xi_{v_f}$ in our case.[6]

In order to have a very rough estimate of the rate constants, we have employed the straight-line envelopes shown in Figs. 2-3.[9] If we approximate the energy dependence of the cross sections simply by $\sigma_{DR} = AE^{-\beta}$, then the rate constant α can be calculated from

$$\alpha = A\left(\frac{8}{\pi m}\right)^{\frac{1}{2}} \Gamma(2-\beta)\ (\kappa T)^{\frac{1}{2}-\beta}, \qquad (3-3)$$

where m is the electron mass, κ is the Boltsmann constant and T is the tempara-ture. The results are summarized in Table 1.

IV. DISSOCIATIVE RECOMBINATION OF CH^+

This process has been attracting a considerable interest, since this is an important process in the interstellar cloud. There are two sets of estimate of the recombination rate, namely $\sim 10^{-10} cm^3/s$ and $\sim 10^{-7} cm^3/s$. This order of magnitude difference comes from whether a curve crossing between dissociative state and molecular ion state is assumed or not. Having carried out a quite elaborate calculation for this system, recently Takagi et.al.[14] concluded that there exists a crossing near the left turning point of the first excited vibrational state of CH^+, and estimated the rate constant to be $\sim 1.12 \times 10^{-7} cm^3/s$ at $T = 120K$, which is in fairly good agreement with the merging beam experiment ($\sim 3.1 \pm 0.3 \times 10^{-7} cm^3/s$)[15]. Their calculations are briefly outlined here. The relevant electronic states were calculated by adopting the self-consistent-field and the configuration mixing method.

A particular attention was paid to the diffuse character of the two-electron excited $^2\Pi$ dissociative state. Not only the potential energy curves, but also the information on the quantum defect and electronic coupling were deduced from this calculation. The partial waves contributing to the process are $p\pi$ and $d\pi$, and the statistical factor g in eq. (2-1) is unity in this case. The diabatic potential curves obtained are shown in Fig. 6. The calaculated result of the cross section for $v_i = 0$ is shown in Fig. 7. Recently, calibration of the merging beam experiment was repeated[16], and the correct values of rate constant and cross section turned out to be one-half of the previous ones.[15] Thus the agreement of the theoretical result and the experiment becomes much better.

V. MORE REMARKS ON THE DYNAMICS OF SUPEREXCITED STATES.

A comparative study on the role of $V(R)$ and $\mu(R)$ in dissociative recombination was made by doing some model calculations. Although no figure can be presented here because of the limit of space, the following facts were confirmed: (1) The magnitude of cross section is essentially determined by $V(R)$. (2) Interestingly, there appears a rich resonance structure even in the case of $\mu(R) = 0$. This is because the intermediate resonance states (vibrationally excited Rydberg states) can be formed by the two-step mechanism via dissociative state through the electronic coupling. (3) However, the detailed resonance structure is, of course, determined by $\mu(R)$.

The $d\sigma_g$-wave partial cross sections for vibrational transitions of H_2^+, HD^+ and D_2^+ by electron impact were also obtained as a byproduct from our MQDT calculations.[9] Some of the results are shown in Fig. 8 for the case of HD^+ ($v_i = 1$). It is clearly seen from this figure that the energy dependence is very similar for dissociative recombination and vibrational transition. This indicates that the two-step mechanism via dissociative state, $v_i \to d \to v_f$, is the main pathway in the vibrational transitions. This is because the factors $|M_{vv'}|$ ($v \neq v'$) are much smaller and less important than $|\xi_v|$'s. The second term of eq. (2-11) with $v = v_i$ and $v' = v_f$ dominates the first term. The similarity in energy dependence thus originates in the common factor ξ_{v_i} appearing in eqs. (2-10) and (2-11).

The fact that $|\xi_v|$ is much larger than $|M_{vv'}|$ is expected to be quite general, and implies the physical significance of dissociative superexcited state in various dynamic processes. For instance, resonant photoionization via Rydberg state or autoionization of vibrationally excited Rydberg state occurs dominantly via dissociative superexcited state; and the propensity rule becomes very different from that of the vibrational autoionization. The importance of dissociative superexcited state was actually found in the photoionization of NO[17,18] and was discussed generally in relation to photoelectron spectra.[19]

In any case the various dynamic processes in the energy region above the lowest ionization threshold are governed by the two basic physical quantities $V(R)$ and $\mu(R)$, and can be treated in a unified way by the MQDT. Essential physics occuring in the various processes should be the same, and the main differences among them lie only in the boundary conditions.[20]

VI. CONCLUSION

We have reported the results of calculation for the dissociative recombination of H_2^+ and CH^+. They are in good agreement with the recent merging beam experiment. Although the rotational motions were completely disregarded in the present treatment, we do not expect that the general features found in these calculations would be affected much by this disregard. The detailed analysis of the dissociation products and the resonances would be interesting subjects to be investigated in future. It is also desirable that the isotope effects such as those found in our calculations could be investigated experimentally.

Table 1. Dissociative recombination rate $\alpha = \alpha_0\,[100/T(k)]^\gamma\ (cm^3/s)$

v_i	H_2^+		HD^+		D_2^+	
	α_0	γ	α_0	γ	α	γ
0	2.31×10^{-8}	0.29	2.26×10^{-8}	0.44	4.11×10^{-9}	0.12
1	1.81×10^{-7}	0.50	1.54×10^{-7}	0.47	1.19×10^{-7}	0.49
2	1.34×10^{-8}	0.66	1.58×10^{-7}	0.60	1.54×10^{-7}	0.52
3	1.98×10^{-8}	0.32	6.46×10^{-10}	-0.16	6.34×10^{-8}	0.83
4	3.26×10^{-8}	0.77	5.55×10^{-8}	0.58	4.02×10^{-8}	0.43

References

1). Mitchels, H.H., Adv. in Chem. Phys. **XLV**, 225 (1981).

2). Seaton, M.J., Proc. Phys. Soc. London **88**, 801, 805 (1966).

3). Fano, U., Phys. Rev. **A2**, 353 (1970).

4). Jungen, Ch. and Atabek, O., J. Chem. Phys. **66**, 5584 (1977).

5). Lee, C.M., Phys. Rev. **A16**, 109 (1977).

6). Giusti, A., J. Phys. **B13**, 3867 (1980).

7). Seaton, M.J., Rep. Prog. Phys. **46**, 167 (1983).

8). Giusti-Suzor, A., Bardsley, J.N., and Derkits, C., Phys. Rev. **A28**, 682 (1983).

9). Nakashima, K., Takagi, H., and Nakamura, H., J. Chem. Phys. **86**, 726 (1987).

10). Takagi, H. and Nakamura, H., Phys. Rev. **A27**, 691 (1983).

11). Takagi, H. and Nakamura, H., J. Chem. Phys. **84**, 2431 (1986).

12). Auerbach, D., Cacak, R., Caudano, R., Gaily, T.D., Keyser, C.J., McGowan, J.W., Mitchell, J.B.A., and Wilk, S.F.J., J. Phys. **B10**, 3797 (1977).

13). Hus, H., Yousif, F., Noren, C., Sen, A., and Mitchell, J.B.A., Phys. Rev. Lett. **60**, 1006 (1988).

14). Takagi, H., Kosugi, N., and LeDourneuf, M., J. Phys. B (submitted).

15). Mitchell, J.B.A. and McGowan, J.W., Astrophys. J. **222**, L77 (1978).

16). Mitchell, J.B.A., private communication.

17). Giusti-suzor, A. and Jungen, Ch., J. Chem. Phys. **80**, 986 (1984).

18). Achiba, Y., Sato, K., and Kimura, K., J. Chem. Phys. **82**, 3959 (1985).

19). Nakamura, H., Chem. Phys. **10**, 271 (1975).

20). Nakamura, H., J. Phys. Chem. **88**, 4812 (1984).

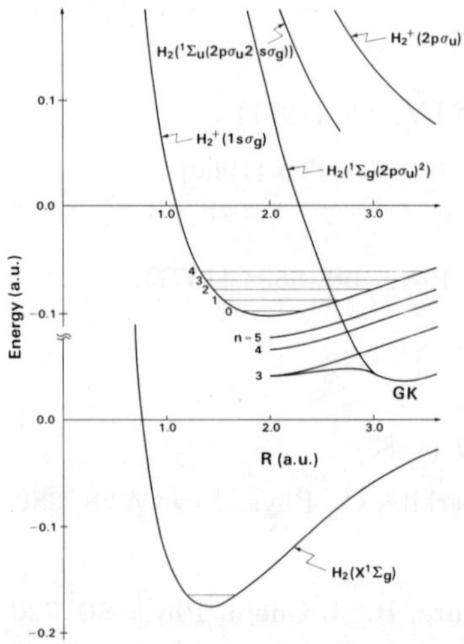

Fig.1 Potential energy curves of H_2 near ionization threshold.

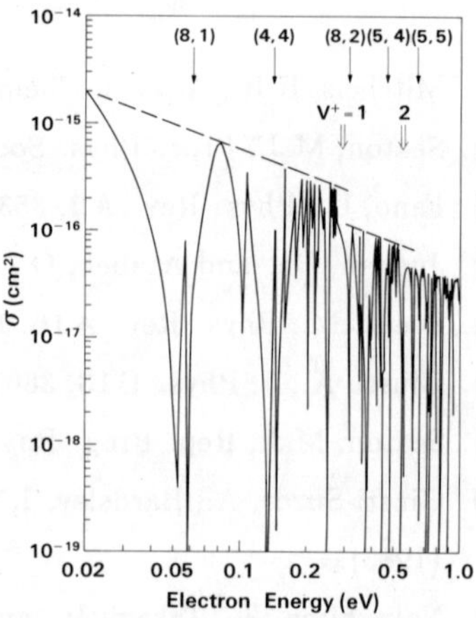

Fig.2 Dissociative recombination cross section for H_2^+ $(v_i = 0)$.

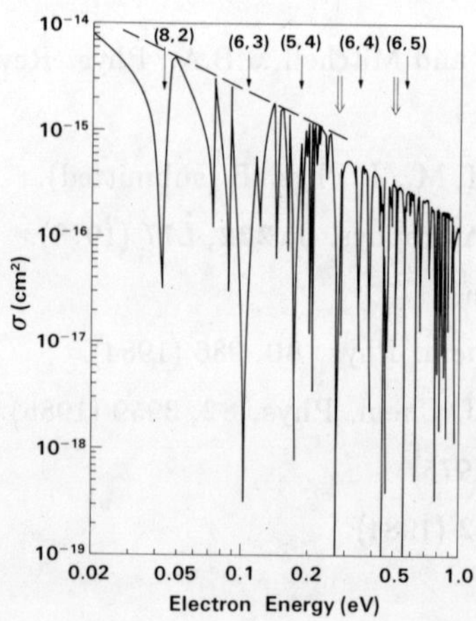

Fig.3 Same as Fig.2 for $H_2^+(v_i = 1)$

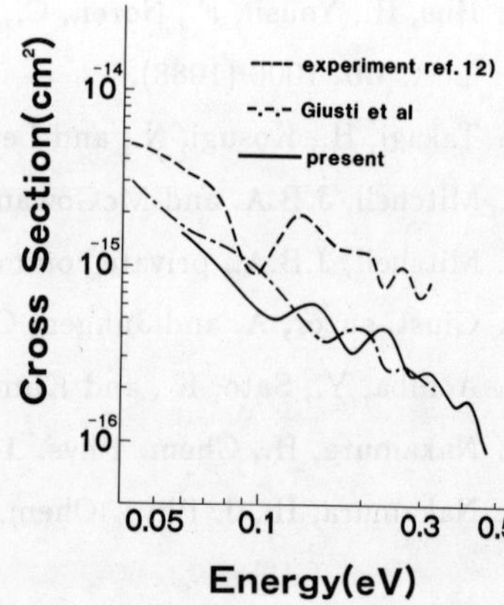

Fig.4 Averaged cross sections for H_2^+.

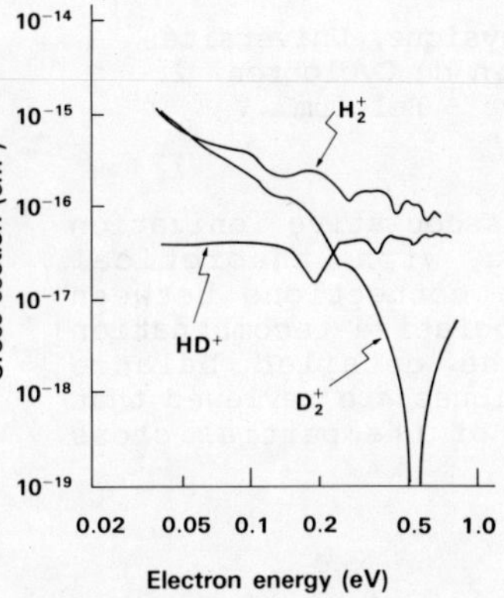

Fig.5 Convoluted D.R.cross section for H_2^+, HD^+ and $D_2^+(v_i = 3)$

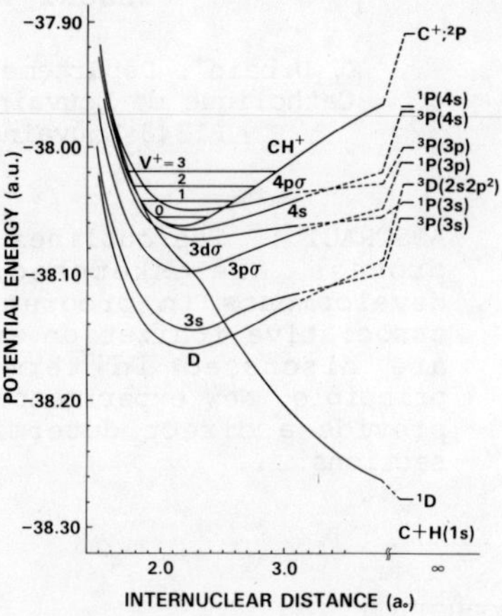

Fig.6 Diabatic potential curves of CH

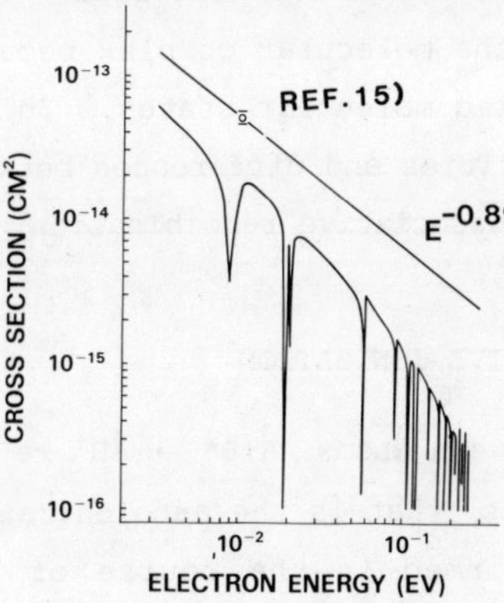

Fig.7 Dissociative recombination cross section of $CH^+(v_i = 0)$

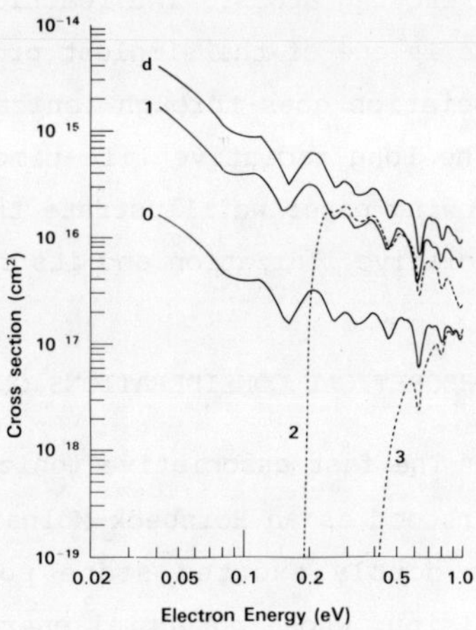

Fig.8 Convoluted cross sections for D.R. (d) and vibrational transitions $(v_f = 0 \sim 3)$ in the case of $HD^+(v_i = 1)$

COMMENTS ON ASSOCIATIVE IONIZATION : $A+B^* \rightarrow AB^+ + e^-$ THEORY AND EXPERIMENT

X. Urbain[+], Département de Physique, Université Catholique de Louvain - Chemin du Cyclotron, 2 B1348 Louvain-la-Neuve - Belgium

ABSTRACT : The outlines of the associative ionization process are sketched, along with theoretical developments in progress. The connections between associative ionization and dissociative recombination are discussed in terms of the detailed balance principle. New experimental techniques are reviewed that provide a direct determination of the partial cross sections.

INTRODUCTION

In the field of atomic and molecular physics, the reactive collisions are given more and more attention, as they constitute a crucial test case of the formalisms developed for stable molecules and non reacting atoms. The reaction of association between two colliding atoms is one of the simplest processes leading to chemical bond. This association goes through ionization of the molecular complex because of the long radiative life-time of excited molecular states. In the following paper we illustrate the similarities and differences between associative ionization and its reverse, dissociative recombination.

1. THEORETICAL CONSIDERATIONS ON ASSOCIATIVE IONIZATION

The fast associative ionization between atoms $A+B^* \rightarrow AB^+ + e^-$ is understood as an Hornbeck-Molnar[1)] process, that is the autoionization of a doubly excited state possibly formed in the course of the collision, whose potential energy curve crosses the Rydberg manifolds converging to the molecular ion limit.

[+] *aspirant du Fonds National Belge de la Recherche Scientifique*

A general expression of the total cross section can be given as :

$$\sigma_d = \frac{\pi}{K^2} \sum_{J;v} (2J+1) \left| S_{Jv\leftarrow d} \right|^2 \qquad (1)$$

where **J** stands for the total angular momentum, and the sum over **v** runs up to a maximum value depending on the centrifugal distortion of the potential curve of the ion.

Besides the fact that the final state is one of the vibrational levels of the ion, the vibrational motion is essential to the description of the dynamics of the process. Indeed, the discrete vibrational levels of the Rydberg states hinder the associative ionization in a resonant way, while Franck-Condon factors directly modulate the autoionization probability.

Furthermore, the global rotation of the collisional system moves all the levels up, changing Franck-Condon factors and resonances positions, and distributing the final products over a whole rotational vibrational spectrum. The upper limit in **J** is reached when the ground vibrational level of the rotating ion lies above the collisional energy. Because of statistical weights, i.e. **(2J+1)**, the major contribution comes from the highest **J** value.

The dependance on collisional energy may strongly differ from case to case. If the separated atom limit lies below the molecular ground state, as it occurs in the reaction $H(1s)+H(2s) \rightarrow H_2^+ + e^-$, then the cross section is monotonically increasing with energy, due to the accumulation of partial waves and the growing number of vibrational levels open to autoionization of the molecule (see fig.1). Above the dissociation limit of the molecular ion, Penning ionization competes, and no new rovibrational level contributes anymore, so that the cross section falls down, according to the simultaneous decrease of the Franck-Condon factors. The energy behaviour is expected to be similar in the case where no threshold exists, except at low energy. In the Na+Na ultra cold laser trap studies, cross sections as large as 10^{-13} cm^2 have been reported[2].

In the special case of associative ionization between H^+ and H^- ions, the atomic limit is far above the ground state of the molecular ion. It follows that many partial waves contribute even at zero energy, and up to 9 vibrational levels are populated at threshold. The total cross section then simply decreases as inverse power of the energy, as expected for a coulombic entrance channel (see fig.2).

Inclusion of the Rydberg states in the calculation results in an average absorption of flux through a dense superposition of resonances, one series being associated to each vibrational level and each orbital momentum of the outer electron. When the threshold corresponding to an higher excitation of the collisional partners is reached, the dissociation into this new channel can occur and then compete with the autoionization process. This dissociation proceeds via transitions between the doubly excited state and the Rydberg states correlated to this separated atom limit.

The detailed balance principle allows us to connect the cross sections for associative ionization and dissociative recombination. Stating the balance principle as

$$K_i^2 \sigma_{j\to i} = K_j^2 \sigma_{i\to j} \tag{2}$$

or in terms of the S matrix,

$$\left|S_{j\to i}\right| = \left|S_{i\to j}\right| \tag{3}$$

we see that the **S** matrix is the same for both processes, as soon as the states involved are fully detailed, i.e. defined in terms of all appropriate quantum numbers. Thus the balance no longer holds for total cross sections, for which the summation and weighting of partial waves, as well as the contribution of all open channels (either dissociative or ionizing) involve different **S** matrix elements for different processes.

At this point, some approximations are usually introduced. The total cross section for dissociative recombination can still be obtained by totally neglecting the intricate problem of dynamical correlation of the molecular states to the atomic ones (down to fine structure). At the same level of approximation, theoreticians often neglect the detailed autoionization process by just including a decay width in the radial equations solved in the **R** coordinate (complex potential method)[3]. Both approximations are best justified when the final states lie far below the collisional energy, so that the dynamics in that arrangement do not influence the rearrangement probability itself (from electron-ion to atom-atom).

2. THEORETICAL DEVELOPMENTS IN PROGRESS

In order to give some interpretation of the past and coming experimental results of the group of Louvain-la-Neuve, we began to study the associative ionization by using the same computer codes that had been developed for dissociative recombination by Giusti-Suzor and coworkers[4]. Indeed, only the asymptotic conditions change from one process to the reverse, the **S** matrix being the same as already said.

Our construction of the **S** matrix closely follows the treatment of dissociative recombination proposed by Giusti-Suzor[5] in 1980, based on a multichannel quantum defect approach of the closed channels resonant contribution. Commuting to associative ionization first requires the partial wave expansion of the nuclear motion, but also the treatment of long range interactions, that determine the flux coming onto the autoionizing states also responsible for the inverse process. A first attempt[6] was made for the case of $H^+ + H^- \rightarrow H_2^+ + e^-$, where we assumed, supported by the results of charge exchange studies[7], that the system behaves either diabatically or adiabatically around the pseudo-crossings taking place at large internuclear distance. We selected a single entrance channel, coupled to first order with the lowest doubly excited state of H_2. To go one step further and treat the atomic collision with the same accuracy as could be done for electron-ion scattering (see below), one has to construct the **K** matrix of the atom-atom collision and the corresponding distorted radial wave functions. This is done by solving coupled equations that formally include all interactions except those with the ionization channels. The vibrational wave functions are obtained by propagating inwards the equations after the asymptotic **K** matrix has been constructed by fitting the solutions of the outward propagation to the reference potential functions, i.e. the solutions of the Schrödinger equation for the separate channels. The new radial functions are used in place of the unperturbed ones to calculate the vibronic matrix elements

$$\int F(R)\ K_{el}(R)\ \chi_v(R)\ dR \qquad (4)$$

with F(R) being the dissociative vibrational wave function, $\chi_v(R)$ the vibrational function of the ion in the **v** level, and $K_{el}(R)$ the matrix element of the electronic **K** matrix at fixed **R**.

This improvement is required in order to study the reaction $H(1s)+H(2s) \rightarrow H_2^+ + e^-$ [8] where the incoming channel now consists of two asymptotically degenerate potentials in the $^1\Sigma_g$ symmetry, one pertaining to the $\mathbf{1s\sigma_g\ ns\sigma_g}$ Rydberg series, the second being adiabatically correlated to the $(\mathbf{2p\sigma_u})^2$ autoionizing state (see fig.3). The difference between the two paths leads to different asymptotic phase shifts, bringing about interferences, or Stückelberg oscillations, in the partial wave cross section. However, this oscillatory behaviour drastically changes with the **J** value, so that the average coming from the sum over many partial waves washes out even totally destructive interferences. The results of preliminary calculations are shown in fig.4, exhibiting a dense resonant pattern superimposed on a smoothly modulated increase of the cross section, corresponding to the accumulation of partial waves and open vibrational states. The wide structure around 1.5 eV comes from interferences between the two Rydberg series included in the calculation, namely $\mathbf{1s\sigma_g\ ns\sigma_g}$ and $\mathbf{1s\sigma_g\ nd\sigma_g}$.

The aforementioned electronic **K** matrix corresponds to electron-molecular ion scattering at fixed **R**, and fully describes all the electronic processes in the Born-Oppenheimer approximation. S. Ross and Ch. Jungen[9] have obtained this matrix for hydrogen by considering the adiabatic energies as the result of the scattering of the outer electron by the $\mathbf{1s\sigma_g}$ and $\mathbf{2p\sigma_u}$ core orbitals in the **s,p** and **d** waves, and then making use of an original M.Q.D.T. approach. Projecting these **K** matrix elements onto a complete vibrational basis set relaxes the locality and couples nuclear to electronic motion. The formalism of A. Giusti-Suzor has been developed for such a **K** matrix which can be restricted to its first order approximation, **V**(R), in the case of weak coupling. Although molecular hydrogen has been the test case for this M.Q.D.T. approach, we admit that the **V** approximation was very coarse, in view of the fairly large interactions that perturb the lowest part of the spectrum. However, the autoionization probability tends to saturate over a wide range of coupling strength, so that the cross sections could be weakly sensitive to higher order corrections.

3. MEASUREMENT OF TOTAL AND PARTIAL CROSS SECTIONS FOR ASSOCIATIVE IONIZATION : EXPERIMENTAL TECHNIQUES

When theoreticians ask the experimentalists for cross sections, several criteria are to be fulfilled by the experimental results. The total cross section should be measured in absolute magnitude, over a wide range of collisional energy, and at least with well specified partners. These requirements are matched in different ways depending on the technique (fast, thermal or trapped atoms) and their possible extensions (merged beams, ejected electron spectra, translational spectroscopy, photodissociation,...).

Contrary to the not yet completely settled question of the vibrational cooling of the molecular ions, atomic species can be produced with quantum states perfectly determined down to fine structure. With the use of thermal beams, one can easily excite any state with high efficiency. More surprisingly, collisions between ground state atoms are harder to study, because no velocity selection by Doppler shifted excitation would then be possible. The merged beams are produced mostly in the ground state, through neutralization of ions on a target gas. Excited states can only be reached if a metastable species is formed at first, and pumped with intense laser light afterwards. That has been realized for hydrogen with a cesium vapour oven as neutralization cell, the H(2s) atoms being excited to **3s**,**3p** and **3d** levels with a dye laser[10].

The velocity selection is obtained easily with merged beams, due to the addition of the velocities that gives a much lower relative energy in the center-of-mass frame than in the laboratory frame. Collisional energies as low as a few meV are reached currently, and the lower the collision energy is, the narrower the velocity distribution will be. Thermal beams produce only rate coefficients averaged over velocity distributions far from Maxwellian but Doppler selection of a velocity class in the beam allows the measurement of true cross sections[11].

The determination of absolute cross sections in beam studies requires the measurement of the form factor and of the intensity of the parent beams. We just point out the relatively hard task of measuring currents of neutral atoms instead of ions, keeping in mind the tricky problem of superimposing a 50 eV electron beam on a 500keV ion beam to measure dissociative recombination.

The experimental efforts are now concentrated on the measurement of partial and differential cross sections, in order to address the question of the final state distribution over the rovibrational spectrum of the ion. This can be achieved in two ways, the energy balance giving complementary amounts of energy to the products.

The energy spectrum of the ejected electron is measured currently by Hotop's group in Kaiserslautern and is sufficiently resolved to assign its structures to specific rotational and vibrational levels[12] (see fig.5). Such spectra provide at least a direct measurement of the shape of the potential well of the molecular ion. In addition, the relative amplitudes of the peaks corresponding each to a single **v**, J final state are related to the partial cross sections through some transmission function of the spectrometer. This is probably the reason why the only comparison between theory and experiment has been conducted in the case of He* +H [13], the initial state lying 6.3 eV above the dissociation threshold of HeH^+. In such a case, the less energetic electrons are produced by Penning ionization, i.e. formation of HeH^+ in its vibrational continuum, He +H^+, while the faster electrons are the signature of the stabilization of HeH^+, or associative ionization (see fig.6).

The same information about partial cross sections could be extracted from the distribution of internal energy in the molecular ion. J. Weiner and coworkers[14] designed an experiment where an efficient photodissociation of the ions previously extracted from the interaction region was combined with a position sensitive detection of the photofragments. This method provides a spectrum that can be related to the partial cross sections with the help of some model only.

4. EXPERIMENTAL DEVELOPMENTS IN PROGRESS

Our first calculations on associative ionization in hydrogen predict small cross sections in the case of reactions between neutrals. Therefore the measurements ask for high intensity beams of H(1s), H(2s) and H(3s) to provide accurate results on the $H(1s)+H(2s) \rightarrow H_2^+ + e^-$ and $H(1s)+H(3s) \rightarrow H_2^+ + e^-$ reactions. The choice of these reactions is motivated by the fact that they cover, with the $H^+ + H^-$ results[15], all the possible cases of associative ionization, i.e. endothermic, exothermic and exothermic with coulombic entrance channel. Furthermore, these three reactions are the first members of the alkali-alkali series of reactions that are extensively studied with thermal and supersonic beams by several groups : Na(3p)+Na(3p), K(4p)+K(4p), ... The comparison of the preliminary results of the potassium experiment[16] with our calculations for H_2 indicates a similar shape of the cross section as a function of energy all through the alkali-alkali series, provided that the energy balance is the same.

A general description of the set-up can be found elsewhere[17]. The production of metastable atoms in the **2s** state is achieved in the quasi-resonant electron capture by fast protons (several keV) on Cesium[18]. The metastable fraction tends to 25% at low energy, while the remaining fraction radiates very quickly to the ground state.

The **2s-3s** optical transition is forbidden, unless some perturbation is applied to the system[19]. An axial electric field (200 V/cm) surrounds the laser-atom interaction, mixing all the n=2, n=3 states into Stark states. When the wavelength corresponding to the Stark state correlated to **3s** in the zero field region is selected, a fairly large yield of excitation to **3s** can be achieved (up to 80%). The relative depopulation of the **2s** level is measured by induced fluorescence in front of a channeltron, where a strong axial electric field quenches the **2s** by mixing with the short-lived **2p** state.

A second neutralization cell is added to the Cesium oven to produce H(1s) atoms with a different velocity from the H(2s,3s). A bias voltage acts on the velocity of the protons inside the cell. The produced neutrals keep that different velocity when coming out of the cell, while the outgoing protons recover their initial velocity.

CONCLUSION

The new techniques of measurement of partial cross sections for associative ionization and for dissociative recombination, clearly open a new field of investigation of these reactive processes. The close connection between one reaction and its reverse should convince the theoreticians to combine the formalisms developed for both processes which are not always explicitly related.

REFERENCES

1) Hornbeck, J.A., and Molnar, J.P., Phys. Rev. 84, 615 (1951).

2) Weiner, J., *Associative ionization* $A+B \rightarrow AB^{+}+e^{-}$. *A case study in collision dynamics* , in the proceedings of the *Workshop on reactive collisions of laser excited Na atoms* , Utrecht, 11-13 january 1988, p.7.

3) Miller, W.H., J. Chem. Phys. 52, 3563 (1970).

4) Giusti-Suzor, A., Bardsley, J.N. and Derkits, C., Phys.Rev. A28, 682 (1983).

5) Giusti, A., J. Phys. B13, 3867 (1980).

6) Urbain, X., Giusti-Suzor, A., Fussen, D. and Kubach, Ch., J. Phys. B19, L273 (1986).

7) Fussen, D. and Kubach, Ch., J. Phys. B19, L31 (1986)

8) Urbain, X. and Giusti-Suzor, A., proceedings of the *XVth Conference on Physics of Electronic and Atomic Collisions* , Brighton, abstracts of contributed papers, 714 (1987).

9) Ross, S. and Jungen, Ch., Phys. Rev. Lett. 59, 1297 (1987).

10) Lorent, V., Claeys, W., Cornet, A. and Fussen, D., J. Phys. B18, 3667 (1985).

11) Wang, M.-X., de Vries, M.S., Keller, J. and Weiner, J., Phys. Rev A32, 681 (1985).

12) Lorenzen, J., Hotop, H., Ruf, M.-W. and Morgner, H., Z. Phys. A297, 19 (1980).

13) Waibel, H., Bußert, W., Ruf, M.-W. and Hotop, H., proceedings of the *XIVth Conference on Physics of Electronic and Atomic Collisions* , Palo Alto, abstracts of contributed papers, 376 (1985).

14) Keller, J., Bonnano, R., Wang, M.-X., de Vries, M.S. and Weiner, J., Phys. Rev. A33, 1612 (1986).

15) Poulaert, G., Brouillard, F., Claeys, W., Mc Gowan, J.W. and Van Wassenhove, G., J. Phys. B11, L671 (1978).

16) Brencher, L., Nawracala, B. and Pauly, H., *Energy and polarization dependance of ionization processes in K(4P)+K(4P) collisions* , in the proceedings of the *Workshop on reactive collisions of laser excited Na atoms* , Utrecht, 11-13 january 1988, p.187.

17) Fussen, D., Claeys, W., Cornet, A., Jureta, J. and Defrance, P., J. Phys. B15, L715 (1982).

18) Brouillard, F., Claeys, W. and Van Wassenhove, G., J. Phys. B10, 687 (1977).

19) Fussen, D., Claeys, W., Cornet, A. and Lorent, V., J. Phys. B18, 3673 (1985).

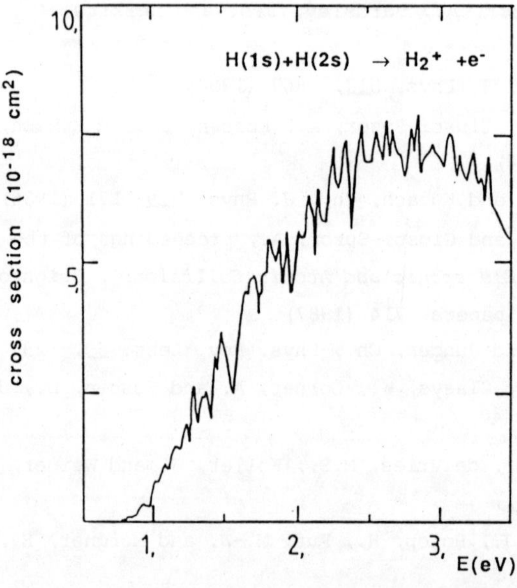

Fig.1 : Total cross section for the reaction $H(1s)+H(2s) \rightarrow H_2^+ + e^-$. The threshold energy is 0.75 eV. The dissociation limit of H_2^+ is reached at 3.4 eV.

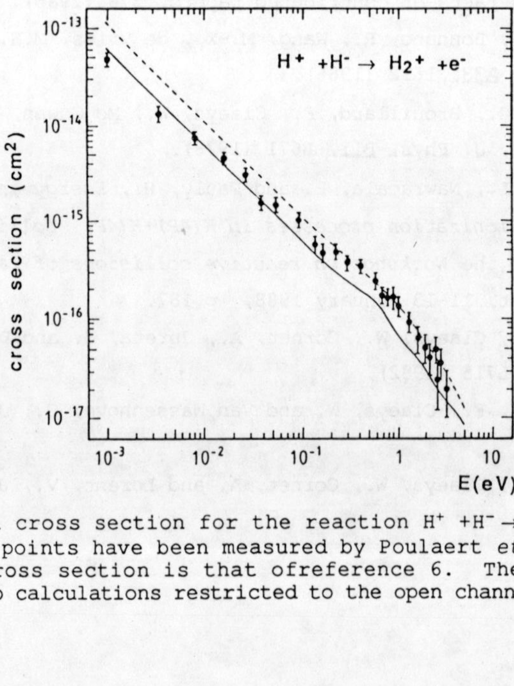

Fig.2 : Total cross section for the reaction $H^+ + H^- \rightarrow H_2^+ + e^-$. The experimental points have been measured by Poulaert *et al* [15]. The calculated cross section is that ofreference 6. The dashed curve corresponds to calculations restricted to the open channels.

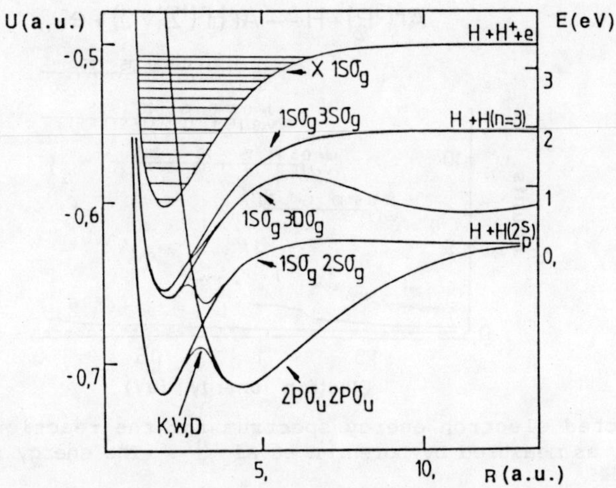

Fig.3 : Potential energy curves relevant to the H(1s) + H(2s) → H_2^+ $+e^-$ associative ionization in the $^1\Sigma_g$ symmetry. The adiabatic curves are those of Kolos, Wolniewicz and Dressler, J.Chem. Phys. 82, 3292 (1985).

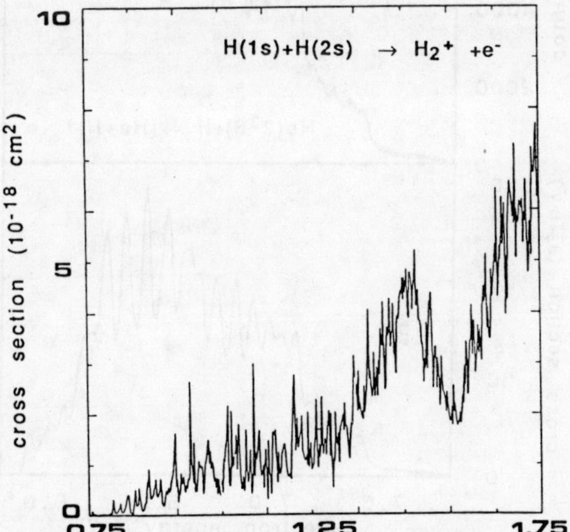

Fig.4 : Total cross section for the reaction H(1s) + H(2s) → H_2^+ $+e^-$. Coupled equations are solved for the two entrance channels.

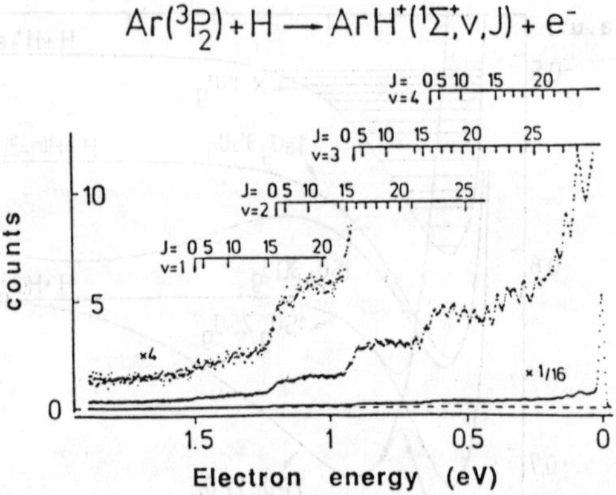

Fig.5: Ejected electron energy spectrum for the reaction $Ar(^3P_2)+H \rightarrow ArH^+ + e^-$, as measured by Lorenzen *et al* [12]. The energy resolution is 20 meV FWHM.

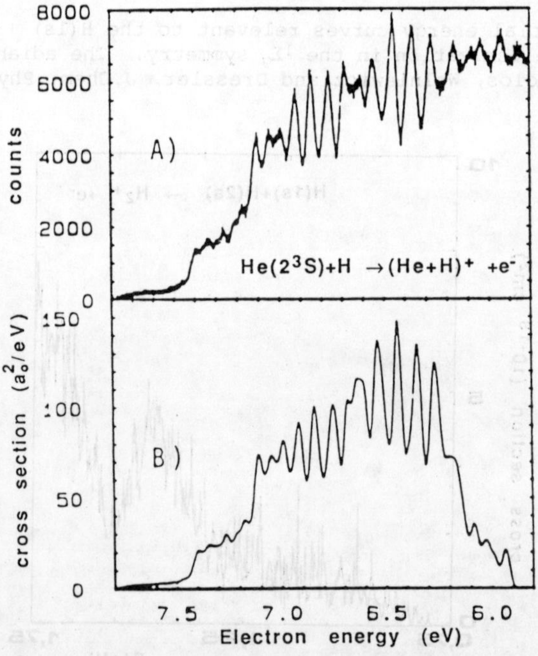

Fig.6 : Comparison between measured high energy part of the electron energy spectrum of the reaction $He(2\,^3S)+H \rightarrow (He + H)^+ + e^-$ at 300 K (A) and quantum mechanical calculation (B) of the spectrum due to associative ionization only, as given in reference 18. Electrons with an energy lower than 6.26 eV come from Penning ionization.

POTENTIAL ENERGY CURVES FOR DISSOCIATIVE RECOMBINATION OF HeH^+

H. H. Michels

United Technologies Research Center
East Hartford, CT 06108

ABSTRACT

The potential energy curves pertinent to the process of dissociative recombination (DR) of e + HeH^+ have been calculated using an *ab initio* valence configuration interaction (VCI) method. Recombination into both the singlet and triplet ion channels has been considered through analysis of the low-lying doublet and quartet states of HeH. With the exception of the ground state, all of the doublet states up to the HeH^+ limit are known to be Rydberg in character. Similarly, we find weak binding for the lowest quartet states of HeH, which appear to be mainly Rydberg to the $a^3\Sigma^+$ state of HeH^+. The nature of a second quartet series, which should be Rydberg to the lowest $^3\Pi$ and first excited $^3\Sigma^+$ states of HeH^+, is uncertain, owing to possible strong perturbations with the ion-pair states of HeH.

INTRODUCTION

The dissociative recombination (DR) of electrons with atmospheric molecular ions was first proposed by Bates and Massey[1] to explain the observed electron density variation in the earth's ionosphere. The calculated and observed rate coefficients for DR of atmospheric molecular ions are large (10^{-6} - 10^{-7} cm^3/sec), owing to the multitude of curve crossings of neutral dissociating molecular states with the ground vibrational state of the corresponding molecular ion. Adams, et al[2] have studied DR in more complex (HCO^+, N_2H^+) molecular ions and McGowan and Mitchell[3] report recombination rates for over thirty molecular species. For all of these studies, a reasonably fast recombination rate was observed.

In contrast, very slow DR rates have been reported for He_2^+ recombination[4], where both theoretical and spectroscopic studies[5,6] have shown that the ground $X^2\Sigma_u^+$ state of He_2^+ does not exhibit curve crossings with either the $[1\sigma_g^2 1\sigma_u]\phi$ Rydberg manifold of neutral He_2 states or the $[1\sigma_u^2 1\sigma_g]\phi$ repulsive manifold of states. More recently, very slow DR rates have been reported for recombination of vibrationally cold H_3^+ ions[7]. These results have been shown by Michels and Hobbs[8] to again be a consequence of the lack of effective curve crossings with the $^1A_1$ ground state of H_3^+ near its potential minimum. Indirect DR of e + H_3^+, via capture into vibrationally excited Rydberg states of H_3, also appears to have a low probability[8], since no effective curve crossings of these Rydberg states with the lower valence states dissociating to H + H + H or H + H_2 were found.

The molecular HeH^+ ion offers a third case where DR may be very slow. Recombination of electrons with the ground $X^1\Sigma^+$ state of HeH^+ would appear to be similar to the e + He_2^+ case, since it has been known for some time[9] that the low-lying excited doublet states of HeH form a Rydberg manifold. Very recent studies by Smith and Adams[10], using the FALP apparatus, appear to indicate a very slow DR rate for HeH^+. The FALP technique permits close to thermalized recombination conditions[11] such that this observed slow rate is indicative of DR to vibrationally cold, ground state HeH^+ ions. Yousif and Mitchell[12], using a merged beam technique[13] have also recently reported recombination studies of HeH^+. Their preliminary results indicate that excited HeH^+ states, possibly the long-lived triplet states, may exhibit large DR rates which appear to rapidly decrease as the excited state population is quenched. In this paper, we examine the location of the low-lying potential energy curves of HeH which are relevant to DR of electrons with either the ground $X^1\Sigma^+$ state or the low-lying triplet states of HeH^+.

THEORETICAL CONSIDERATIONS

The calculations of the potential energy curves for HeH were performed using the *ab initio* valence-configuration-interaction (VCI) method[14,15]. The specific form for the electronic wavefunction may be written as

$$\Psi(R) = \sum_{\mu} c_{\mu} \Psi_{\mu}(R) \qquad (1)$$

where R specifies the internuclear separation of HeH. Each $\Psi_{\mu}(R)$ is a different spatial configuration built from atomic orbitals, properly antisymmetrized and projected over the two-dimensional spin space for this three-electron system. Two different basis sets were examined. The first was the 24 ellipsoidal orbital basis chosen by Green, et al[16] to represent the electronic states of the HeH^+ molecular ion arising from atomic states with $n \le 3$. Preliminary studies of the low-lying doublet states of HeH indicated that this basis was too diffuse for an accurate representation of the ground state of the He atom. Since a balanced description of the low-lying doublet and quartet states of HeH is needed, a second 23 orbital STO basis was developed consisting of a 4s2p1d basis for helium and a 3s2p basis for hydrogen. This basis is given in Table 1, where the calculated and experimental atomic energies are given for comparison.

A molecular correlation diagram for HeH is given in Table 2. It is seen that two doublet Rydberg series arise, the first from He + H* and the second from He* + H. Only the lower of these two Rydberg series is computationally tractable, although pseudo-variational calculations have been reported[17] for the second autoionizing doublet series. A similar analysis of the structurally similar LiHe molecule has also been reported by Schneiderman and Michels[18]. United atom correlations for the low-lying doublet series of HeH have been given by Kimura and Lane[19], but the curve crossing indicated in their work between the ground and first excited $^2\Sigma^+$ states of HeH is in question, in light of the united atom state identifications reported by Green, et al[16].

Several calculations of the low-lying members of the first HeH Rydberg doublet series have been reported[9,17,20-22]. Since there are no repulsive excited states of HeH below the HeH^+ [$X^1\Sigma^+$] ionization limit, DR cannot occur to the ground state of HeH^+ with probability higher than the relatively slow radiative recombination rate (10-40 ns). In Tables 3 and 4 and in Figure 1, we give the first calculations of the low-lying quartet states of HeH. These HeH excited quartet states appear mainly to be Rydberg to the $a^3\Sigma^+$ excited state[16] of HeH^+. However, several additional features are present.

The lowest $^4\Pi$ state is slightly attractive with a dissociation energy of ~130 meV at R_e~4.5 bohrs. This state is therefore the first member of the quartet

Rydberg series with the limit of HeH^+ [$a^3\Sigma^+$]. This lowest triplet state of the HeH^+ ion is bound by 105 meV at R=4.47 bohrs[23]. This picture is qualitatively similar to the behavior of the corresponding states in LiHe[18], except that in the LiHe system, the binding energy of the $a^4\Pi$ state is nearly 1.0 eV, owing to the much larger binding exhibited by the $a^3\Sigma^+$ state of $LiHe^+$. This increased binding of the parent ion state in $LiHe^+$, over that calculated for HeH^+, is the result of the much larger polarization attraction of Li + He^+, as compared to H + He^+. The $^4\Sigma^+$ II and $^4\Sigma^+$ V states of HeH, which diabatically correlate to He $^3P(1s2p)$ + H $^2S(1s)$ and He $^3P(1s3p)$ + H $^2S(1s)$, respectively, are strongly repulsive since the outer MO is mainly anti-bonding. These repulsive $^4\Sigma^+$ states give rise to homogeneous perturbations in the more purely Rydberg $^4\Sigma^+$ III and $^4\Sigma^+$ IV states, both of which exhibit weak binding at R~4.5 bohrs.

Higher manifolds of quartet Rydberg states of HeH must arise which have the He* + H^+ ion states as series limits. These states cannot be variationally treated, since they are autoionizing relative to the lowest $a^3\Sigma^+$ state of HeH^+, but their general behavior and location can be approximately determined. On Figure 1, we indicate the lowest $^3\Pi$ I and the excited $^3\Sigma^+$ II and $^3\Sigma^+$ III states of HeH^+, each of which dissociates to He* + H^+. These excited triplet states of HeH^+ have been shown by Green, et al[16] to closely follow the corresponding H_2^+ states. The corresponding three possible quartet Rydberg series can be constructed by displacements from these ion limits in accordance with the Rydberg structure of the H atom. Some confidence can be given to this method for estimating the behavior of this second manifold of quartet Rydberg states, based on the study reported by Michels[24] of Rydberg-Rydberg interactions in the Li_2 system. In that study, long range interactions were calculated for molecular states connecting up to Li*(n=3) + Li*(n=3). The resultant potential energy curves showed many avoided crossings at large internuclear separations but were clearly Rydberg-like to the Li_2^+ ion cores at shorter (R<10 bohrs) separations.

There is one, presently unknown, complication to this picture of these higher manifolds of quartet Rydberg states. The limit He^- $^4P(2s2p)$ + H^+ gives rise to ion-pair states of $^4\Sigma^+$ and $^4\Pi$ symmetry that will exhibit a long range Coulomb attraction and thus curve-cross many of the states in this upper quartet Rydberg series. The lowest crossing, between the attractive ion-pair curve and the lowest member of the quartet Rydberg series with $^3\Sigma^+$ II as the series limit, is shown in Figure 1. It is not clear that the perturbations caused by the ion-pair states can distort this Rydberg series to the point where these states exhibit curve crossings with the excited HeH^+ triplet states. The character of this interaction, however, could be very non-localized since the interaction involves a redistribution of very diffuse electron density.

DISCUSSION

The potential energy curves shown on Figure 1 for the low-lying quartet states of HeH do not seem to offer a mechanism for DR to the $a^3\Sigma^+$ state of HeH^+, other than a relatively slow radiative recombination rate. Most of the quartet states are nearly Rydberg to the $a^3\Sigma^+$ state of HeH^+ in the region where this ion state exhibits a shallow minimum. The suitability of the $a^3\Sigma^+$ state of HeH^+ as an important source for DR is also in question, since this state is weakly bound[16] and easily dissociated through collisions.

This leaves the second quartet Rydberg manifold of HeH as the most likely series that could exhibit curve crossings with the more strongly bound $^3\Sigma^+$ II or $^3\Pi$ I states of HeH^+. If these series are not strongly perturbed through the mixing of the He^- + H^+ ion-pair states, DR of e + HeH^+ will be very slow for both the ground and excited states of this molecular ion. It is not clear how to access the true nature of these states of HeH which energetically lie between the limits of He^+ + H(1s) and He*(n=2) + H^+. Miller and Schaefer[17] attempted to analyze the first excited doublet manifold of HeH states, which correlate with He*(n=2) + H(1s), in a treatment of Penning ionization in this system. Their excited state potentials are uncertain, however, since the calculated states are not variationally stable with respect to expansion of their chosen basis set. The second quartet series of HeH should be more computationally tractable and CI studies are in progress within the spirit of the stabilization method[25].

Acknowledgements

We thank Professors D. Smith and J. B. A. Mitchell and Drs. T. Green and J. R. Peterson for helpful discussions, and J. B. Addison for carrying out the calculations. This work was supported in part by the Air Force Office of Scientific Research under Contract F49620-85-C-0095. Use of the computational facilities at the Air Force Supercomputer Center - Kirtland is also acknowledged. This paper is dedicated to the memory of my mother, Hilda D. Michels.

REFERENCES

1. D. R. Bates and H. S. W. Massey. Proc. Roy. Soc. A192, 1 (1947).
2. N. G. Adams, D. Smith and E. Alge, J. Chem. Phys. 81, 1778 (1984).
3. J. W. McGowan and J. B. A. Mitchell, in "Electron-Molecule Interactions and their Applications", Vol. II, ed. L. G. Christophoron, Academic Press, NY (1984).
4. J. N. Bardsley and M. A. Biondi, Adv. At. Mol. Phys. 6, 1 (1970).
5. M. L. Ginter and R. Battino, J. Chem. Phys. 52, 4469 (1970).
6. B. Liu, Phys. Rev. Lett. 27, 1251 (1971).
7. D. Smith and N. G. Adams, Ap. J. 284, L13 (1984).
8. H. H. Michels and R. H. Hobbs, Ap. J. 286, L27 (1984).
9. H. H. Michels and F. E. Harris, J. Chem. Phys. 39, 1464 (1963).
10. D. Smith, private communication.
11. E. Alge, N. G. Adams and D. Smith, J. Phys. B: At. Mol. Phys. 16, 1433 (1983).
12. F. Yousif and J. B. A. Mitchell, Bull. APS 33, 1010 (1988).
13. J. B. A. Mitchell, in "Atomic Processes in Electron-Ion and Ion-Ion Collisions", ed. F. Brouillard, Plenum Pub. NY (1986).
14. F. E. Harris and H. H. Michels, Int. J. Quant. Chem. 1S, 329 (1967).
15. A. C. Wahl, P. S. Bertoncini, G. Das and T. L. Gilbert, Int. J. Quant. Chem. 1S, 123 (1967).
16. T. A. Green, H. H. Michels, J. C. Browne and M. M. Madsen, J. Chem. Phys. 61, 5186 (1974).
17. W. H. Miller and H. F. Schaefer III, J. Chem. Phys. 53, 1421 (1970).
18. S. B. Schneiderman and H. H. Michels, J. Chem. Phys. 42, 3706 (1965).
19. M. Kimura and N. F. Lane, Phys. Rev. A37, 2900 (1988).
20. R. Polak, J. Vojtik, I. Paidarova and F. Schneider, Chem. Phys. 55, 183 (1981).
21. I. Paidarova, J. Vojtik and R. Polak, Chem. Phys. 72, 119 (1982).
22. G. Theodorakopoulos, S. C. Farantos, R. J. Buenker and S. D. Peyerimhoff, J. Phys. B: At. Mol. Phys. 17, 1452 (1984).

23. H. H. Michels, J. Chem. Phys. 44, 3834 (1966).
24. H. H. Michels, in "Production and Neutralization of Negative Ions and Beams", ed. J. G. Alessi, AIP Conf. Proc. 158 (1986).
25. A. U. Hazi and H. S. Taylor, Phys. Rev. A1, 1109 (1970).

Table 1. Slater-Type Orbital (STO) Basis and Calculated Atomic Energies.

STO parameters - (He): 1s = 2.000, 1s' = 1.129, 2s = 0.623, 2s' = 0.337, 2p = 0.526, 2p' = 0.208, 3d = 0.413; (H): 1s = 1.063, 1s' = 0.374, 2s = 1.320, 2p = 1.240, 2p' = 1.010. All atomic calculations are full CI with STO parameters optimized for the He (1s2s) and (1s2p) excited states.

State	E_{calc} (hartrees)	E_{exp} (hartrees)
He [$^1S(1s^2)$]	-2.873883*	-2.903724
He [$^3S(1s2s)$]	-2.174234	-2.175229
He [$^3P(1s2p)$]	-2.130633	-2.133164
He [$^3S(1s3s)$]	-2.065381	-2.068674
He [$^3P(1s3p)$]	-2.056535	-2.058065
He [$^3D(1s3d)$]	-2.052454	-2.055618
H^- [$^1S(1s^2)$]	-0.526123	-0.527716

*Optimization of the STO basis for the ground state of He gives E= -2.899779

Table 2. HeH Correlation Diagram

Separated Atoms	E_{∞}(hartrees)	Molecular States
He $^1S(1s^2)$ + H(1s)	-3.403724	$^2\Sigma^+$
He $^1S(1s^2)$ + (n=2)	-3.028724	$^2\Sigma^+$, $^2\Pi$
He $^1S(1s^2)$ + (n=3)	-2.959280	$^2\Sigma^+$, $^2\Pi$, $^2\Delta$
.		
.		
He $^1S(1s^2)$ + H^+	-2.903727	$^1\Sigma^+$
He $^3S(1s2s)$ + H (1s)	-2.675229	$^2\Sigma^+$, $^4\Sigma^+$
He $^1S(1s2s)$ + H (1s)	-2.645964	$^2\Sigma^+$
He $^3P(1s2p)$ + H (1s)	-2.633164	$^2\Sigma^+$, $^4\Sigma^+$, $^2\Pi$, $^4\Pi$
He $^1P(1s2p)$ + H (1s)	-2.623827	$^2\Sigma^+$, $^2\Pi$
He $^3S(1s3s)$ + H (1s)	-2.568674	$^2\Sigma^+$, $^4\Sigma^+$
He $^1S(1s3s)$ + H (1s)	-2.561255	$^2\Sigma^+$
He $^3P(1s3p)$ + H (1s)	-2.558065	$^2\Sigma^+$, $^4\Sigma^+$, $^2\Pi$, $^4\Pi$
He $^3D(1s3d)$ + H(1s)	-2.555618	$^2\Sigma^+$, $^4\Sigma^+$, $^2\Pi$, $^4\Pi$, $^2\Delta$, $^4\Delta$
He $^1D(1s3d)$ + H(1s)	-2.555603	$^2\Sigma^+$, $^2\Pi$, $^2\Delta$
He $^1P(1s3p)$ + H (1s)	-2.555127	$^2\Sigma^+$, $^2\Pi$
.		
.		
He (n=4) + H (1s)	-2.536493	------
.		
.		
He^+ $^2S(1s)$ + H^- $(1s^2)$	-2.527716	$^2\Sigma^+$
.		
.		
He (n=5) + H (1s)	-2.522598	------
.		
.		
He^- $^4P^\circ(1s2s2p)$ + H^+	-2.178172	$^4\Sigma^+$, $^4\Pi$

Table 3. Calculated Electronic Energies of Low-Lying $^4\Sigma^+$ States of HeH

R (bohrs)	Total Energy (hartrees)				
	$^4\Sigma^+$ I	$^4\Sigma^+$ II	$^4\Sigma^+$ III	$^4\Sigma^+$ IV	$^4\Sigma^+$ V
1.0	-1.91836	-1.79955	-1.78896	-1.76888	-1.74858
2.0	-2.54006	-2.45765	-2.43606	-2.42647	-2.39981
3.0	-2.64415	-2.56556	-2.54615	-2.54080	-2.49433
4.0	-2.66264	-2.58533	-2.56991	-2.56008	-2.51546
5.0	-2.66740	-2.60358	-2.57079	-2.56114	-2.53312
6.0	-2.67030	-2.61687	-2.56773	-2.55961	-2.53674
7.0	-2.67227	-2.62380	-2.56573	-2.55809	-2.53728
8.0	-2.67335	-2.62724	-2.56476	-2.55696	-2.53788
9.0	-2.67384	-2.62894	-2.56437	-2.55622	-2.53957
10.0	-2.67404	-2.62976	-2.56430	-2.55576	-2.54210
11.0	-2.67411	-2.63014	-2.56439	-2.55548	-2.54479
12.0	-2.67412	-2.63031	-2.56455	-2.55535	-2.54715
13.0	-2.67412	-2.63038	-2.56471	-2.55532	-2.54898
14.0	-2.67412	-2.63040	-2.56486	-2.55538	-2.55028
15.0	-2.67411	-2.63040	-2.56498	-2.55550	-2.55114
16.0	-2.67410	-2.63040	-2.56506	-2.55567	-2.55166
18.0	-2.67409	-2.63039	-2.56516	-2.55599	-2.55213
20.0	-2.67409	-2.63039	-2.56519	-2.55621	-2.55227
30.0	-2.67408	-2.63038	-2.56520	-2.55638	-2.55231
∞ (calc)	-2.67423	-2.63063	-2.56538	-2.55654	-2.55245

Table 4. Calculated Electronic Energies of Low-Lying $^4\Pi$ States of HeH

R (bohrs)	Total Energy (hartrees)		
	$^4\Pi$ I	$^4\Pi$ II	$^4\Pi$ III
1.0	-1.87963	-1.77739	-1.76889
2.0	-2.51790	-2.43102	-2.42561
3.0	-2.62181	-2.54136	-2.53809
4.0	-2.63523	-2.55860	-2.55705
5.0	-2.63404	-2.55888	-2.55871
6.0	-2.63218	-2.55800	-2.55698
7.0	-2.63120	-2.55738	-2.55537
8.0	-2.63076	-2.55701	-2.55427
9.0	-2.63058	-2.55680	-2.55354
10.0	-2.63050	-2.55667	-2.55308
11.0	-2.63046	-2.55658	-2.55278
12.0	-2.63043	-2.55652	-2.55260
13.0	-2.63042	-2.55648	-2.55249
14.0	-2.63041	-2.55645	-2.55242
15.0	-2.63040	-2.55643	-2.55238
16.0	-2.63040	-2.55641	-2.55236
18.0	-2.63039	-2.55639	-2.55233
20.0	-2.63039	-2.55638	-2.55232
30.0	-2.63038	-2.55637	-2.55231
∞ (calc)	-2.63063	-2.55654	-2.55245

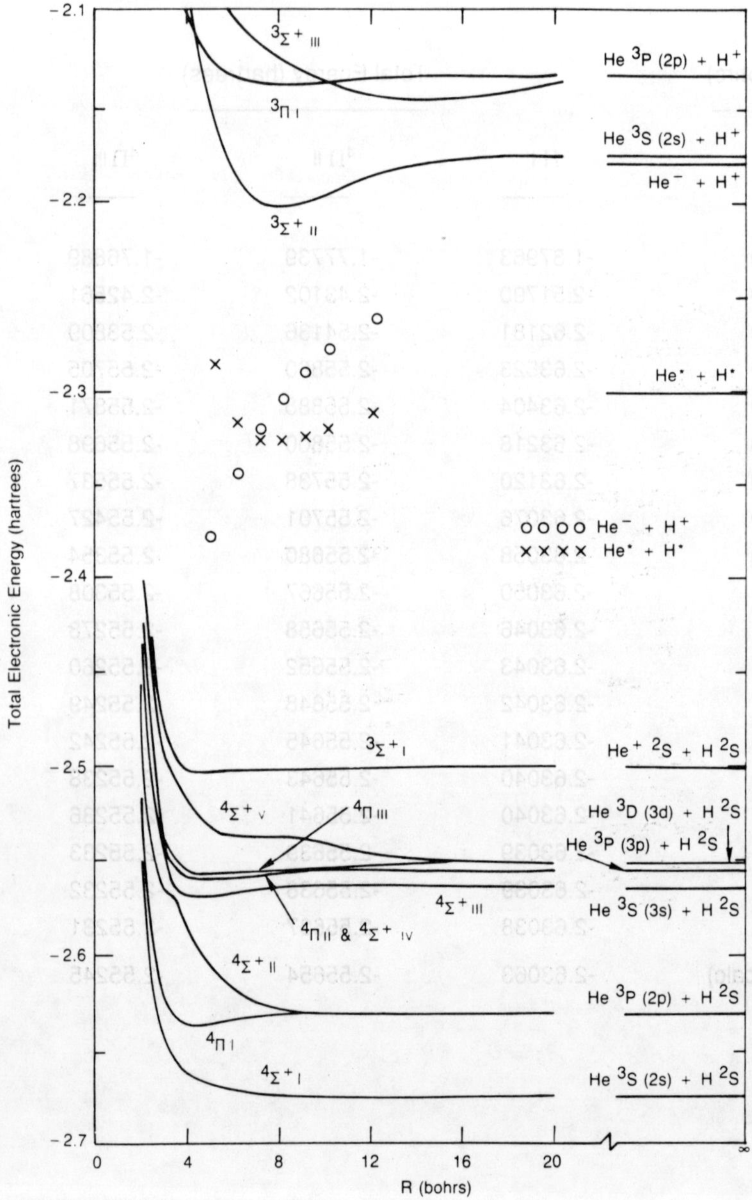

Fig. 1 Comparison of Quartet States of HeH and Triplet States of HeH$^+$

Merged Beam Studies of Dissociative Recombination
Recent Results

J.B.A. Mitchell and F.B. Yousif
Department of Physics
The University of Western Ontario
London, Ontario, Canada. N6A 3K7

ABSTRACT

The merged beam technique as currently practiced allows recombination cross sections for chemically reactive ions to be determined over a wide energy range with high energy resolution. Internal energies of the ions can be controlled and measured. Product branching ratios can be studied. A review of recent results is presented.

The first application of the merged beam technique to the study of electron ion recombination was the experiment of Davis and Barnes[1)] who attempted to measure the capture of electrons by alpha particles. Their experiment produced spectacular but unfortunately irreproducible results which nevertheless prompted the scientific community of the day to take a hard look at recombination. Much of our basic understanding of radiative recombination derives from these studies.

A renewal of interest in merged beam experimentation for electron ion research began in the sixties with a number of rather unsuccessful experiments.

The main problem encountered by the proponents of this method was the difficulty experienced in controlling the intersection angle between the beams as they interacted.

The first successful application of the technique followed the suggestion of D.J. Auerbach of using a trochoidal analyzer to achieve merging of the two beams. Since then the experiment at The University of Western Ontario has been used for numerous measurements of dissociative recombination[2)] and has also been applied to the study of dielectronic recombination[3)]. Currently, plans are underway for measurements of laser stimulated radiative recombination of atomic ions using this apparatus[4)].

In principle the merged beam approach offers a number of advantages for the study of low energy collision processes. The ions are prepared and mass selected prior to electron collision and the products are separated and detected afterwards. Very low collision energies can be achieved and the energy can be varied over a wide range. At low energies the technique can yield a very good energy resolution allowing spectroscopic measurements of the collision process to be made. Since the reagents and products have high velocities in the laboratory frame they are easily detected using nuclear single particle counting techniques and this allows very small cross sections to be measured.

KINEMATICS

If the ion and electron beams have energies , velocities and masses E_i, v_i, m_i and E_e, v_e, m_e respectively and intersect at angle Θ, then the centre of mass collision energy E_{cm} is given in terms of the relative velocity v_r by

$$E_{cm} = \frac{1}{2}\mu v_r^2 \approx \frac{1}{2} m_e (v_i^2 + v_e^2 - 2v_i v_e \cos\Theta)$$

where $\mu = m_i m_e/(m_i + m_e) \approx m_e$

One can define the reduced ion energy, E+, which is the energy that the ions would have if they had the same mass as the electrons, i.e. $E_+ = E_i(m_e/m_i)$. If Θ is small then cos

$\Theta \sim 1 - \frac{\Theta^2}{2}$ and so E_{cm} can be expressed in the form: -

$$E_{cm} = \left[E_+^{1/2} - E_e^{1/2}\right] + \left[E_+ E_e\right]^{1/2} \Theta^2$$

Differentiating E_{cm} with respect to E_+, E_e and Θ yields an expression for the spread in the measured centre of mass energy as a function of the spreads in the reduced ion energy, ΔE+, the electron energy ΔEe, and the intersection angle ΔΘ. Assuming gaussian distributions for these spreads, then: -

$$\Delta E_{cm} = \left[\left\{\left[1 - (E_+/E_e)^{1/2}\right] \Delta E_e\right\}^2 + \left\{\left[1 - (E_e/E_+)^{1/2}\right] \Delta E_+\right\}^2 + \left[2(E_e E_+)^{1/2}\, \Theta\Delta\Theta\right]^2\right]$$

When E+ ≈ Ee the first two terms tend to zero and the energy resolution is dominated by the angular term.

With suitable tuning of the apparatus, the dominant contribution to Θ and $\Delta\Theta$ comes from electrons emitted with significant transverse energies from the thermionic cathode. It is however possible with suitable collimation to reduce θ and $\Delta\theta$ and so achieve high energy resolution.

EXPERIMENTAL TECHNIQUE

The merged beam apparatus at U.W.O. has been described in detail elsewhere[5). The ion beam is formed in the terminal of a 0.5 MeV Van de Graaff accelerator and following mass analysis is injected into the collision chamber. Electrostatic double deflection is used to remove neutrals before the beam enters the interaction region. The electron beam is formed using a barium oxide indirectly heated cathode. Initially parallel to the ion beam, it is shifted laterally using a combination of an axial magnetic field and a transverse electric field in a device known as a trochoidal analyzer, so that it overlaps the ion beam. Within the interaction region, the degree of beam overlap is determined using three equispaced scanners which allow the beam profiles and therefore the effective collision area F, to be measured[6). Following the interaction region, the electrons are demerged using a second trochoidal analyzer and are collected in a faraday cup. The ion beam is electro-statically deflected into a second faraday cup and the neutrals pass straight on to be detected using a surface barrier detector.

The cross section for the process under study is determined using the expression: -

$$\sigma = \frac{C_i e^2}{I_i I_e L} \left| \frac{v_i v_e}{v_i - v_e} \right| F$$

where $I_i I_e$, v_i and v_e are the ion and electron currents and velocities respectively, e is the electronic charge, L the length of the collision region, F the effective collision area and C_i the signal count rate.

True signals due to electron-ion collisions are separated from those due to ion beam-background gas collisions by modulating the electron beam and using two gated scalers to count the detector pulses when the beam is on, (signal + background measured) and when the beam is turned off, (background only). The true signal is then derived from the difference in the scaler readings.

VIBRATIONAL STATE CONTROL

The merged beam technique has been used for measurements of the dissociative recombination cross sections for more than 30 molecular ions, produced in a radiofrequency (rf) ion source[2)]. The ion residence time in this source (~ 35 μs) is insufficient to allow for complete vibrational relaxation and so in general it can be said that these measurements refer to excited ions.

Recently a new source has been installed in the Van de Graaff injector and this employs a radiofrequency field to trap ions, formed by electron impact, for up to several milliseconds within a U-shaped channel. The ions diffuse along the channel before being extracted through an octupole lens. An extraction voltage can be applied to an aperture in front of this lens and it is found that the amount of internal energy in the extracted ions is a function of both source pressure and this extraction voltage. The latter presumably can alter the source residence time although this statement has not been verified experimentally.

Measurement of the internal energy of the ions can in principle be made by measuring the threshold for an excitation process in which electron impact induces a transition to a higher energy state which dissociates. Application of this method to individual species is described below. For the case of a transition to a repulsive upper state the method yields fairly unambiguous results although the shape of the threshold may be distorted by the presence of resonances. For cases such as N_2^+ excitation however, where the upper state is likely to be a bound state which is predissociated by a third state, the analysis is more complex.

An alternative method of determining the degree of vibrational relaxation has become apparent for measurements of simple molecules such as H_2^+ and N_2^+. It has been found that resonances due to rydberg state interactions are visible when the ions are de-excited but these resonances are masked when other vibrational states are present. This effect can be seen in fig. (1).

H_2^+ Recombination

Extensive calculations of H_2^+ dissociative recombination cross sections have been performed in recent years and details concerning these are given elsewhere in this proceedings. At low energies the $1\Sigma_g^+$ resonance is the primary state through which the recombination proceeds. This state intersects the H_2^+ ion state in the vicinity of the v = 1 level and so direct recombination of H_2^+ (v = 0) ions is predicted to be considerably suppressed. Calculations[7)] have shown that the cross sections for individual vibrational states should exhibit very rich structure due to interference between direct capture and indirect capture through vibrationally excited Rydberg states of the neutral molecule.

Earlier merged beam measurements[5)] of H_2^+ recombination did indicate the presence of these resonances, however subsequent attempts to reproduce the structure were unsuccessful. The problem with studying H_2^+ is that when formed via electron impact, all 19 available vibrational states are populated and since the molecule lacks a dipole moment the radiative lifetimes for these states is ~ 10^6 seconds. It is possible however in principle to chemically react away the higher vibrational states via the following endothermic processes.

$$H_2^+ (v > 2) + He \rightarrow HeH^+ + H$$

$$H_2^+ (v > 1) + Ne \rightarrow NeH^+ + H.$$

This was tried in the earlier measurements but the short residence time of the ions in the rf source used at that time, prevented these reactions from removing all the excited ions so that the structure became smeared out.

With the trap source mentioned earlier, the residence time is much longer and so all the higher states can be removed. Crossed beam measurements of the collisional dissociation reaction: -

$H_2^+(v) + He \rightarrow H^+ + H + He$

were performed for various source conditions and by measuring the threshold for this reaction while using a hydrogen/neon source gas mixture, it was shown that H_2^+ ions with only v = 0 and 1 populated could indeed be produced[8].

Preliminary measurements[9] of e - H_2^+ recombination have been performed using ions from this source and the results are shown in fig. 1. Deep narrow resonances were formed in these measurements at 0.045, 0.07, 0.1, 0.2, 0.4, 0.6 and 1.0 eV. It is obvious from the figure that much more work is required before these resonances can be characterized completely but the points where the dips occurred were remeasured several times with the results falling within each others errors each time. The resonance at 0.1 eV is less than 20 meV wide and this illustrates the high resolution capability of the merged beam technique. In future measurements computerized scanning of the energy across the resonances will be performed in order to obtain the shape more accurately.

Measurements were also made of the dissociative excitation of H_2^+ via the transition, $X\ ^1\Sigma_g^+ \rightarrow 2p\sigma u$ and these displayed a threshold indicating that only low vibrational states were present.

H_3^+

This ion has caused quite a sensation in the last few years. The recombination rate had been measured using a number of different techniques which yielded quite good mutual agreement[10]. The generally accepted value for the rate was $2.3 \times 10^{-7}\ (300/Te)^{0.5}$. In 1983 however, Adams et al[11] published results taken using the Flowing Afterglow Langmuir Probe (FALP) method which were at least an order of magnitude below this value. More recently[12] they have revised their value down further to $< 1 \times 10^{-11}\ cm^3\ s^{-1}$ which is characteristic of radiative recombination in atomic ions. The low value apparently confirmed theoretical predictions[13] which indicated that an unfavourable curve crossing between the ion ground state and the resonance through which the direct recombination should proceed, would lead to a reduced rate.

The early merged beam results of Auerbach et al used H_3^+ ions produced in a conventional rf ion source and were known to be vibrationally excited. New measurements[14)] have been performed using ions prepared in the trap ion source and these are illustrated in fig. 2. Concurrently measured dissociative excitation results showed that as the source pressure and ion residence time are increased, the internal energies of the ions decreased. It can be seen that for H_3^+ ions with only v = 0 and 1 populated, the cross section for H_3^+ recombination attains a value of 3×10^{-15} cm^2 at 0.01 eV. (This corresponds to a rate coefficient of ~ 2×10^{-8} cm^3 s^{-1} at 100 K.) Lower values could not be obtained despite further increases in source pressure and residence time. These measurements therefore are in agreement with the earlier, but not the later, FALP results.*

The situation has been complicated further by a recent plasma spectroscopy experiment of Amano[15)] who found a value of (1.8 ± 0.2) × $10^{-7} cm^3 s^{-1}$ for the H_3^+ (v=0) recombination rate at 300K!

It has been suggested, (Johnsen 1987 - private communication) that the source of some of the confusion surrounding H_3^+ recombination may lie in the existence of an alternative decay channel leading to relatively long lived H_3^* molecules.

In other words H_3^+ recombination may proceed as follows:

$$e + H_3^+ \rightarrow H + H + H \qquad \text{-- I}$$

$$\rightarrow H_2 + H \qquad \text{-- II}$$

$$\rightarrow H_3^* \rightarrow H_2 + H + h\nu \qquad \text{-- III}$$

* The microwave afterglow measurements of Biondi's group have been shown to have been in error due to plasma contamination by CH_5^+ ions. The reasons for the disagreements between the FALP results and those of Peart and Dolder (crossed beams) and Mathur et al (Electron Beam Ion Trap) are not so obvious.

Back (Private communication 1988) has suggested that a rydberg state of H_3 formed via electron capture in Amano's apparatus could be collisionally stabilized to a lower, non-autoionizing state and this would explain the rapid disappearance of H_3^+ (v = 0) observed by Amano.

In the FALP apparatus, the decrease in the electron concentration is detected using a langmuir probe. It is possible that an H_3^* ion arriving at the probe would reionize to give an H_3^+ ion and an electron. A lower recombination rate would therefore be measured.

In an effort to substantiate this suggestion an experimental program aimed at measuring the branching ratios for the decay channels of H_3^+ recombination is underway at U.W.O. The partial cross sections for recombinations leading to channels I and II have been measured previously[16] for $H_3^\dagger$ ions with a substantial, (~60%), population of vibrationally excited states. The new experiment uses vibrationally cool H_3^+ ions from the trap source. The detector used to measure the neutral recombination products in the merged beam apparatus is energy sensitive but normally it cannot distinguish between channels I, II and III since for a given channel the fragments arrive almost simultaneously. They therefore appear as one full energy particle. By placing a grid with known transmission t in front of the detector however, the probability is reduced to t^3 while for channel II the probability is reduced to t^2. When one or more of the fragment partners fails to arrive then the recombination event is signalled by a particle with 1/3 or 2/3 of the total kinetic energy. If N_I, N_{II} and N_{III} are the number of recombination events leading to channels I, II and III per second respectively, then the numbers of signals at 1/3, 2/3 and the full beam energy are given by: -

$$N_1 = 3t\,(1-t)^2\,N_I + t(1-t)\,N_{II}$$

$$N_2 = 3t^2\,(1-t)\,N_I + t(1-t)\,N_{II}$$

$$N_3 = t^3\,N_I + t^2\,N_{II} + tN_{III}$$

This takes into account the probability that signals will be measured at 1/3 and 2/3 energy only if some particles are not transmitted to the detector, hence the (1-t) factors. By using single channel analyzers to selectively count detector pulses with heights corresponding to 1/3, 2/3 and full beam energy, N_1, N_2 and N_3 can be measured and hence NI, NII and NIII determined. Further details on the technique and analyzers for isotopic triatomic species are given in references 16 and 17. In practice this is a very difficult experiment since the lower energy signal channels are very susceptible to noise and many checks must be constantly performed to ensure that only true signals are counted. The problem is greatly aggravated for the H_3^+ (v = 0,1) measurement because of the low available ion beam currents ($\sim 1 \times 10^{-11}$ A).

Preliminary findings from this experiment indicate that for H_3^+ (v = 0, 1) channel III, leading to an H_3^* molecule does indeed play an important role in the recombination process. In order to reach the detector the lifetime of the state must be at least 1×10^{-7} seconds. Since the initial capture must proceed into an autoionizing state some form of stabilization must be occurring for such a long lived H_3^* to be seen. Collisional stabilization can be ruled out since beam and background gas densities are very low in this apparatus. It may be that electric fields may play a role in this process as they certainly do in dielectronic recombination[18].

Much further work remains to be performed before this process can be fully characterized but it appears that high lying rydberg states are a major factor in the recombination of low vibrational state ions.

It is interesting to note that the emission spectrum of H_3^* has been measured in a plasma[19] and that the mechanism for the formation of H_3^* has been postulated to be electron ion recombination*.

* Up to now the spectra of HD_2^* and D_3^* but not H_2D^* have also been identified. A measurement of the recombination of H_2D^+ might prove interesting indeed!

HeH^+

The merged beam experiment has been used to study the recombination of this ion and once again very surprising results have been found. Since the first excited state of He has such a high potential energy (19.82 eV) and since the ground state of He^- is unbound, there is no suitable curve crossing with the ground state of HeH^+ to allow dissociative recombination to proceed directly. The recombination rate therefore has always been expected to be very low. Fig. 3 shows results of the recombination cross sections for HeH^+ measured recently at U.W.O. Measurements were made with ions produced using both high and low source extraction potentials. Experience has shown that the former case produces ions with considerable internal excitation energy while the latter results in ions which are much more relaxed. The results are very preliminary in nature but it can be seen that in fact the cross section is far from small. It also displays very striking resonant structure.

Two possible explanations are offered to explain these results. It may be that the HeH^+ ion beam contains a significant fraction of metastable $a^3\Sigma$ or $A^1\Sigma$ states which are predicted[20)] to be weakly bound. Alternatively it may be that the recombination leads to an HeH^* state* which is sufficiently long lived to be detected. Confirmation of either of these explanations awaits measurements of the dissociative excitation of HeH^+ and of the branching ratio of the decay channels. Both these experiments are planned.

N_2^+

A remeasurement of N_2^+ recombination has recently been completed and cross section values lower by a factor of 4 than previously accepted have been found. Excitation measurements indicate that only N_2^+ (v = 0) ions were present in the beam. Johnsen[21)] has pointed out that former N_2^+ measurements may have been made on plasmas containing an unexpected excited state population.

* Metastable HeH^* has been observed in a number of experiments and its emission spectrum has recently been measured by Brooks et al[20)].

FUTURE MEASUREMENTS

The ability of the trap source to produce beams of vibrationally cool ions opens up a wide range of exciting measurements. These will be pursued when scheduling permits and include measurements of the recombination of O_2^+, HCO^+, CH^+ and $C_3H_3^+$. It is also our intention to extend our measurements of the dissociative excitation of H_2^+ and H_3^+, the branching ratio of the recombination of isotopic variants of H_3^+ and to investigate the final product energy distribution from H_2^+ recombination.

ACKNOWLEDGEMENTS

The authors would like to express their gratitude to the U.S. Air Force Office of Scientific Research and the Canadian National Science and Engineering Research Council for continuing support for this research.

REFERENCES

1. Davis, B. and A.H. Barnes, Phys. Rev. 34, 152, 1929.
2. Mitchell, J.B.A. and J.Wm. McGowan in Phys. of Ion Ion and Electron Ion Collisions (ed. F. Brouillard and J.Wm. McGowan) NATO ASI, Baddeck, Nova Scotia. Plenum Pub. N.Y. 1983, p. 279.
3. Mitchell, J.B.A., C.T. Ng, J.L. Forand, D.P. Levac, R.E. Mitchell, A. Sen, D.B. Miko and J.Wm. McGowan, Phys. Rev. Lett. 50, 335, 1983.
4. Mitchell, J.B.A. in Intersections of Nuclear and Particle Physics (editor G. Bunce) AIP Conf. Proceedings 1988.
5. Auerbach, D.J., R. Cacak, R. Caudano, T.D. Gaily, C.J. Keyser, J.Wm. McGowan, J.B.A. Mitchell and S.F.J. Wilk, J. Phys. B10, 3797, 1977.
6. Keyser, C.J., H.R. Froelich, J.B.A. Mitchell and J.Wm. McGowan, J. Phys. E12, 316, 1979.
7. Nakashima, K., H. Takagi and H. Nakamura, J. Chem. Phys. 86, 726, 1987. Giusti-Suzor, A., J.N. Bardsley and C. Derkits, Phys. Rev. A28, 682, 1983.
8. Sen, A., J.Wm. McGowan and J.B.A. Mitchell, J. Phys. B20, 1509, 1987.

9. Hus, H., F.B. Yousif, C. Noren and J.B.A. Mitchell, Phys. Rev. Lett. 60, 1006, 1988.

10. Leu, M.T., M.A. Biondi, and R. Johnsen, Phys. Rev. A8, 413, 1973. MacDonald, J.A., M.A. Biondi and R. Johnsen, Planet Space Sci. 32, 651, 1984. Peart, B. and K.T. Dolder, J. Phys. B7, 1948, 1974. Mathur, D., S.U. Khan, J.B. Hasted, J. Phys. B11, 3615, 1978. Mitchell, J.B.A., C.T. Ng, L. Forand, R. Janssen and J.Wm. McGowan, J. Phys. B17, L909, 1984.

11. Adams, N.G., D. Smith and E. Alge, J. Chem. Phys. 81, 1778, 1984.

12. Smith, D. and N.G. Adams, J. Chem. Soc. Faraday Trans. 83, 149, 1987.

13. Michels, H.H. and Hobbs, Ap. J. 286, L27, 1984.

14. Hus, H., F.B. Youssif, A. Sen and J.B.A. Mitchell, Phys. Rev. A38, 1988.

15. Amano, T., Ap. J. Lett. (in press) 1988.

16. Mitchell, J.B.A., J.L. Forand, C.T. Ng, D.P. Levac, R.E. Mitchell, P.M. Mul, W. Claeys, A. Sen and J.Wm. McGowan, Phys. Rev. Lett. 51, 885, 1983.

17. Forand, J.L., J.B.A. Mitchell and J.Wm. McGowan, J. Phys. E18, 623, 1985.

18. Davis, J. and V.L. Jacobs, Phys. Rev. 12A, 2017, 1975.

19. Herzberg, G., J. Chem. Phys. 70, 4806, 1979.

20. Brooks, R.L., J.L. Hunt and J.J. Miller, Phys. Rev. Lett. 58, 199 1987.

21. Johnsen, R., Int. J. Mass. Spec. Ion Phys. 81, 67, 1987.

22. Noren, C., F.B. Yousif and J.B.A. Mitchell (in preparation).

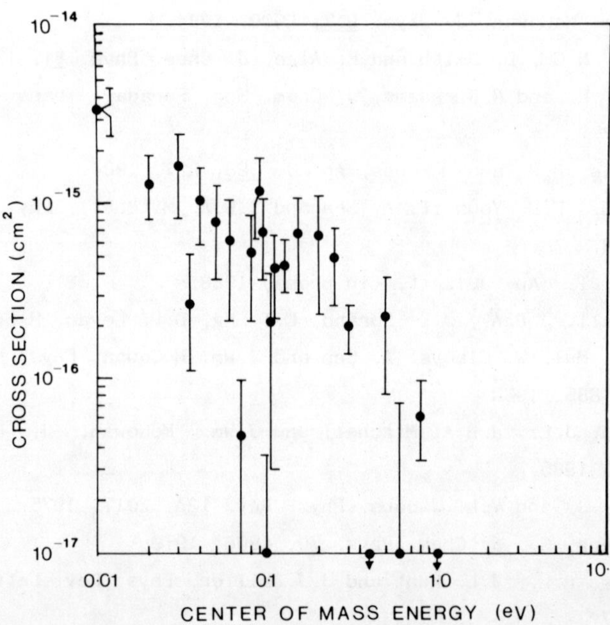

Figure 1 Dissociative recombination cross sections H_2^+ in low vibrational states.

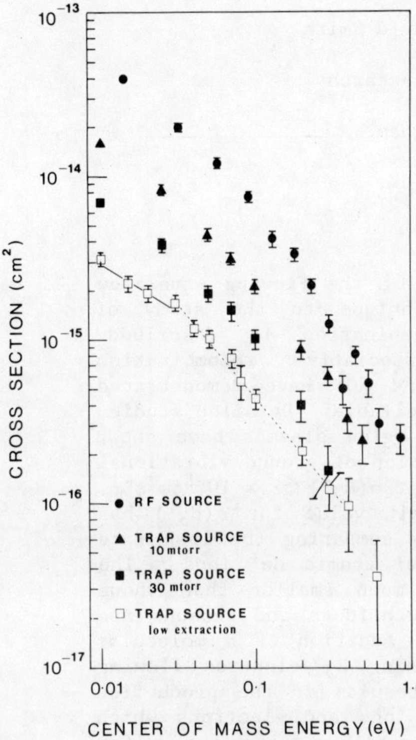

Figure 2 Dissociative recombination cross sections for H_3^+ ions

(a) with states up to v > 4.

(b) with states v ≤ 3

(c) with states v ≤ 2

(d) with v = 0 and 1 only

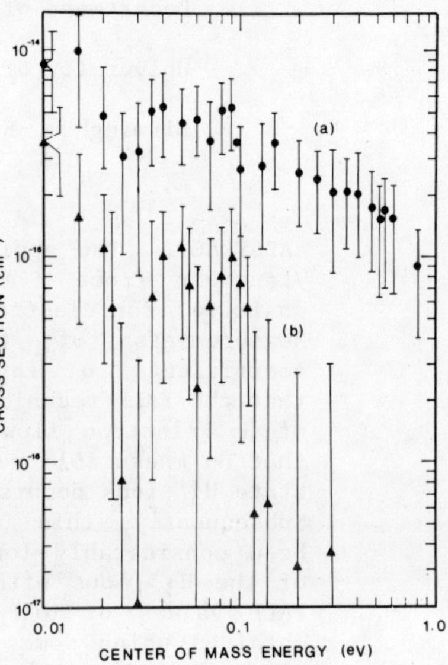

Figure 3 Recombination cross sections for HeH^+ ions produced

(a) with high potential extraction.

(b) with low extraction potential.

FALP STUDIES OF POSITIVE ION/ELECTRON RECOMBINATION

Nigel G. Adams and David Smith

Department of Space Research

University of Birmingham

Birmingham, B15 2TT, UK.

ABSTRACT. The application of the Flowing Afterglow /Langmuir Probe (FALP) technique to the study of positive ion/electron recombination is described. Measurements of the dissociative recombination coefficients, α, for O_2^+ and NO^+ have demonstrated that the FALP technique is reliable. Detailed studies of H_3^+/electron flowing afterglow plasmas have shown that no measurable recombination of ground vibrational state H_3^+ ions occurs and that $\alpha(H_3^+) \lesssim 2 \times 10^{-8} cm^3 s^{-1}$. Subsequently, this upper-limit value for $\alpha(H_3^+)$ has been considerably lowered by comparing the loss rate of the H_3^+ ions with that of atomic He^+ ions. The FALP values of $\alpha(H_3^+)$ are much smaller than those obtained using some other techniques and reasons are given for the differences. Addition of a molecular gas, X, to the non-recombining H_3^+/electron flowing afterglow plasma generally results in the production of a plasma comprising XH^+ ions and electrons which invariably undergoes recombination. Thus the α for several XH^+ ion species have been determined, some over appreciable temperature ranges, including species for which X = CO, N_2, H_2O, CH_4 and CH_3OH. Work is now in progress to identify the neutral products of some dissociative recombination reactions. With this objective in mind, both vacuum ultraviolet and laser induced fluorescence spectroscopy are being applied to the determination of the neutral fragments of recombination reactions. Preliminary results concerning H_3O^+ recombination are reported.

INTRODUCTION

Positive ion/electron recombination processes are important in controlling the degree of ionization in most types of ionized gases and plasmas. Radiative, dielectronic, collisional-radiative, collisional and dissociative recombination are well-characterized processes, and which of these processes occurs or dominates the ionization loss in a particular plasma depends on the physical and chemical properties of the plasma, i.e. the number densities of ions and electrons, the kinetic temperatures of the charged species and the nature of the ions. Thorough discussions of these various recombination processes have been given in some authoritative texts[1-4]. Plainly, dissociative recombination can only occur in ionized media containing molecular ions. However, this encompasses most laboratory plasmas (except the very high temperature fusion plasmas in which all primary molecular gases are fully dissociated and ionized) and many naturally-occurring plasmas such as the terrestrial ionosphere and interstellar gas clouds. A proper understanding of the ion chemistry of such media and the equilibrium degree of ionization (controlled by the balance between the ionization rate and recombination rate) demands a knowledge of the rate coefficients for dissociative recombination reactions, α, the variations of the α with temperature, and the identity of the neutral products of the reactions, often for a considerable number of different positive ion species.

The physical conditions under which recombination occurs have a controlling influence on the rate of recombination, and thus experiments designed to study recombination, in order to provide data to be used to facilitate ion chemical models of ionized media, should attempt to simulate the conditions pertaining to the ionized media of interest as far as possible. Accurate laboratory simulation of the physical conditions is not practically possible for media such as the ionosphere (low ionization density, unequal electron and ion temperatures, solar radiation field which influences the internal states of the recombining ions, etc.)[5,6] and interstellar gas clouds (very low ionization density, very low temperatures)[7,8]. However, using afterglow plasma techniques, recombination can be studied at temperatures not greatly different from those appropriate to the above media, although the ionization density and the neutral gas density are necessarily much greater in order to enhance the recombination loss relative to diffusion loss[2,9] and this difference must be considered when utilizing such laboratory data in ion-chemical models of naturally-occurring plasmas. A definite advantage in studying dissociative recombination at relatively high gas number densities is that the recombining molecular ions undergo multiple collisions with ambient atoms and/or molecules prior to their recombination with the electrons which assists in removing any nascent internal excitation. Then it can usually be assumed with confidence that the ions under study are in their ground rovibronic states.

On the other hand, a disadvantage of a relatively high gas pressure is that it can catalyse the production of "cluster ions" (i.e. the bonding of ligand molecules to the "core ions" which are to be studied) especially at low temperatures. Since the recombination coefficients for "cluster ions" are often much greater than for the simple "core ions"[10] then the presence of even small fractions of cluster ions can enhance the loss rate of electrons from the plasma and lead to erroneous values of the α for the "core ions". The low pressure, essentially collisionless merged beam method for studying recombination processes effectively circumvents this problem and can provide data on the cross sections, σ, for the dissociative recombination of a very wide variety of mass analysed positive ions[11]. However, although such data are of great value to the understanding of the fundamental nature of the recombination process, they are not directly applicable to models of low temperature media because the interaction energy of the ions and electrons, and hence the rate coefficients for recombination, cannot be ascertained sufficiently accurately at the very low interactions energies necessary. Also the degree of internal excitation of the ions is often uncertain in merged beam experiments, although this is presumably not a fundamental limitation since ion source developments should enable such problems to be minimised[12].

The strengths and limitations of merged beam methods and afterglow methods have been discussed in previous research papers and reviews and so they will not be given here[4,9,13]. In this paper, we are entirely concerned with the application of the flowing afterglow/Langmuir probe (FALP) method developed in our laboratory to the study of positive ion/electron recombination reactions, especially dissociative recombination reactions, of ions considered to be involved in atmospheric and interstellar ion chemistry.

THE FALP EXPERIMENT

In a continuous electrical discharge through a gas, the mean energies or "temperatures" of the electrons and ions, especially the electrons, are inevitably greater than the neutral gas temperatures by virtue of the energy that the charged particles gain from the electric field sustaining the discharge. When the electric field is removed the temperatures of the charged particles relax to that of the ambient neutral gas and an afterglow plasma may be formed in which the temperatures of charged and neutral particles are equal. Reactions between the composite particles of the afterglow plasma can then be studied under thermalised conditions. This is the basis of the time-resolved or stationary afterglow (SA) technique which has been exploited for many years by Biondi and his colleagues to determine dissociative recombination coefficients for a wide variety of molecular positive ion species. This pioneering work has been described in several reviews (see, for example, references 2) and 10)). The SA has also been used to great effect to study diffusion processes[14] and ion-neutral reactions [15]. However, the SA method is limited in its scope because reactant ions must be formed by subjecting a pure gas or, more generally, a gas mixture to an electrical discharge. This usually generates more than one ionic species and sometimes internally excites the neutral gas (or gases) which creates great problems of data interpretation and can lead to erroneous results.

With the advent of the flowing afterglow (FA) method, these problems have been circumvented[16,17]. In this method, ionization is created by a continuous electrical discharge or some other form of ion source in the upstream region of a carrier gas (usually pure helium at about 1 torr) which is constrained to flow rapidly along a flow tube (about 1m long and

10cm diameter) by the action of a large mechanical (Roots-type) pump. A flowing afterglow plasma is thus created along the flow tube in which the temperatures of the charged particles have relaxed to the carrier gas temperature by the time they reach the downstream region. Gases can now be introduced into this thermalised plasma to create plasmas of different ionic composition as a result of the reactions of the added gases with the ambient ions and/or helium metastable atoms. Thus, for example, the addition of sufficient O_2 to the helium afterglow (comprised of He^+, He_2^+ ions, triplet He metastable atoms and electrons at number densities of the order of 10^8 - $10^{11} cm^{-3}$, and ground state helium atom number densities of about $(2\text{-}4) \times 10^{16} cm^{-3}$ at 300K) results in the production of an O_2^+/electron afterglow plasma. By controlling the flow rate of the reactant gas (O_2 in this case) into the afterglow and monitoring the number density of the reactant ions (e.g. He^+) in the plasma utilizing a downstream mass-spectrometer/detection system, the rate coefficient for the ion-molecule reaction and the ion products of the reaction can be determined. The critical point is that the reactant gas is not subjected to the electrical discharge (as it is in the SA) and so it is not significantly internally excited. Sequential additions of gases along the flowing afterglow produce sequential ion-neutral reactions and thus the technique has great chemical versatility.

The FA was originally used to determine the rate coefficients for a large number of ion-neutral reactions, some over wide temperature ranges, and this has led to the characterization of several ion-neutral reaction processes and also to the elucidation of the ion chemistry of the ionosphere[18]. However, in recent years it has been largely replaced by the selected ion flow tube (SIFT) method for the study of ion-neutral reactions[17,19,20].

A new application of the FA arose following the development of the single Langmuir probe technique to measure electron and ion number densities and electron temperatures in stationary afterglow plasmas[21]. It was a natural development to unite the probe diagnostic method and the FA and this created the flowing afterglow/Langmuir probe (FALP) apparatus with which several different types of plasma reaction processes can be studied. The FALP technique has been described in detail previously[9,22] and therefore will only be alluded to here. A small cylindrical Langmuir probe (a tungsten wire approximately 0.5 cm long and 25 μm diameter) can be positioned at any point along the afterglow plasma axis. The potential of the probe is swept by ± 2 volts relative to the stainless steel walls of the flow tube and the current flowing to the probe from the plasma is measured. In this way, a current/voltage characteristic is obtained for the probe/plasma system which is analysed by a microcomputer and thus the electron number density (n_e) , ion number density (n_+) and electron temperature (T_e) can be readily obtained at any position (z) along the axis of the thermalised plasma.

In the pure helium plasma, diffusion of electrons and ions is the only loss process and from a measurement of n_e versus z, the ambipolar diffusion coefficient can be determined. The addition of small amounts of argon to the afterglow plasma results in the destruction of helium metastable atoms and the production of Ar^+ ions by Penning reactions. This is a convenient way of removing metastables, which could otherwise be a source of ionization in the downstream afterglow following the addition of reactant gases.

To determine the dissociative recombination coefficient, α, for a particular positive ion species, it is necessary to create an afterglow plasma ideally comprising only that ionic species and electrons. This is possible for a large variety of species because of the chemical versatility of the FA, although small fractions of unwanted ionic species are sometimes unavoidably present and then these species have to be accounted for in the analysis of data and the derivation of the α. The first α to be determined using the FALP method was that for O_2^+, i.e. $\alpha(O_2^+)$[23]. The O_2^+/electron plasma was created in the downstream region by the addition of O_2 to the helium afterglow as mentioned above. In the region of the afterglow upstream of the O_2 addition point, the gradient of n_e with z was small and, as expected, characteristic of diffusive loss of ionization, but downstream of the O_2 addition point, the gradient of n_e was much steeper indicating the occurrence of the additional loss process of dissociative recombination of O_2^+ ions with electrons (see

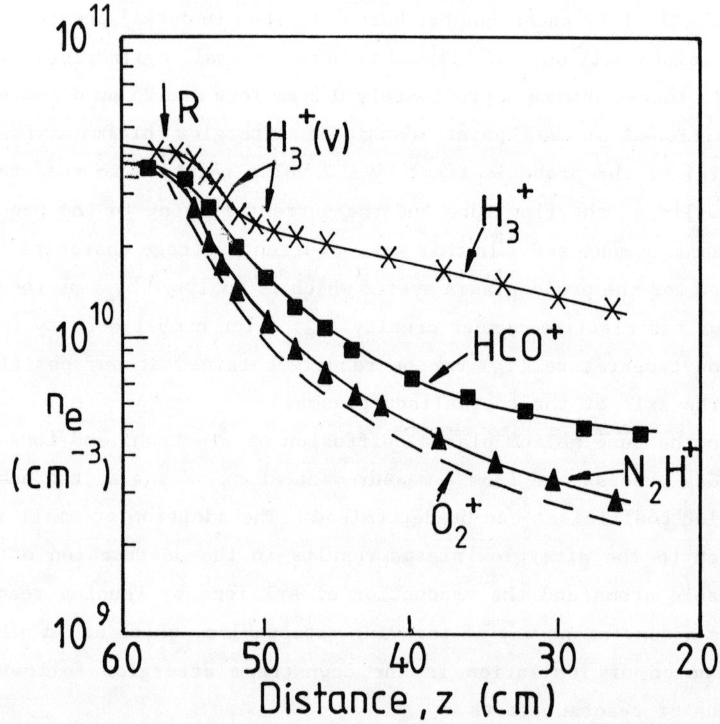

Fig.1. A sample of data obtained from the FALP experiment. The reduction of the electron number density, n_e, with distance along the plasma column, z, is very gradual upstream of the entry port, R, through which gases are added to form molecular ions in the afterglow plasma. On addition of H_2 at R, an obvious increase in the loss rate of electrons from the plasma occurs which is due to recombination of vibrationally excited H_3^+ ions. Subsequently, the linearity of the log n_e versus z plot is characteristic of electron and ground state H_3^+ ion loss by ambipolar diffusion and not by recombination. The addition of small amounts of CO or N_2 to the H_3^+ plasma creates HCO^+ and N_2H^+ plasmas. The relatively rapid reduction of n_e with z in these plasmas is a manifestation of dissociative recombination of HCO^+ and N_2H^+ ions with electrons. Electron loss by recombination in an O_2^+ plasma is also shown (see text).

Fig. 1). Under conditions for which diffusive loss is small relative to recombination loss, i.e. high n_e to enhance recombination and high carrier gas pressure to inhibit diffusion, then a plot of n_e^{-1} versus z is linear and the slope of the plot is equal to α/v_p , where v_p is the flow velocity of the plasma (a parameter which is readily determined[24]). The $\alpha(O_2^+)$ at 300K determined in this way is in excellent agreement with those obtained previously by several other workers using different techniques [25]. (A measurement of $\alpha(O_2^+)$, and indeed $\alpha(NO^+)$ which is also now well known[23,26], is an excellent way of calibrating any system to be used to determine dissociative recombination coefficients.) The temperature of the FALP apparatus can be varied over the approximate temperature range of 90 to 600K, and so α values can be determined over this temperature range for a wide variety of positive ion species (Note that problems can arise due to condensation of vapours at the lower temperatures and this restricts the temperature range over which the α can be determined for some ions). The first paper reporting FALP measurements of α included those for O_2^+, NO^+, and NH_4^+[23]. These data illustrated that when the α are relatively small at 300K (i.e. between 10^{-7} and $10^{-6}cm^3s^{-1}$) then they varied significantly with T (e.g. $\alpha(O_2^+)$ $\sim T^{-0.7}$; $\alpha(NO^+)$ $\sim T^{-0.9}$) whereas when α is large at 300K (i.e. $>10^{-6}cm^3s^{-1}$) then the variation with T is less marked, as is the case for both H_3O^+ and NH_4^+ [23,27]. This feature of recombination reactions had been revealed previously by SA studies[28,29].

STUDIES OF H_3^+ RECOMBINATION

H_3^+ ions are formed in ionized gases rich in H_2 via the reaction of H_2^+ ions with H_2.

$$H_2^+ + H_2 \longrightarrow H_3^+(\upsilon) + H \tag{1}$$

It has been established experimentally that the H_3^+ ions formed in this reaction are vibrationally excited (the vibrational distribution peaking at υ=3 or 4)[30,31]. It is also known that the $H_3^+(\upsilon)$ ions are relaxed to their ground vibrational states in collisions with H_2 (via the process of proton transfer)[30-32].

$$H_3^+(\upsilon) + H_2 \longrightarrow H_3^+ + H_2^*. \tag{2}$$

An appreciation of these reactions facilitates the discussion of H_3^+ recombination given below. H_3^+ ions are important precursor ions in interstellar cloud ion chemistry; they are formed by reaction (1) following the production of H_2^+ by cosmic ray ionization of H_2. Hence a knowledge of the rate of recombination of H_3^+ ions in interstellar clouds is of great importance to the understanding of the physics and chemistry of interstellar clouds[7,8] (and similarly of the atmospheres of the outer planets[33]). Therefore a good deal of experimental effort has been expended in recent years to determine the recombination coefficient for H_3^+ ions.

Very early SA studies of $\alpha(H_3^+)$ in the 1950's gave conflicting results, although significantly (following the relatively recent FALP studies described below) one SA study by Persson and Brown[34] indicated that $\alpha(H_3^+)$ was small ($<3 \times 10^{-8}cm^3s^{-1}$) at 300K. Subsequent SA studies at 300K by Leu et al[35] and Macdonald et al [36] indicated that $\alpha(H_3^+)$ was "typical" for most molecular ions at about $2 \times 10^{-7}cm^3s^{-1}$, and this was given credence by the merged beam studies of Mitchell et al[37] who derived a similar $\alpha(H_3^+)$ for a temperature of 300K from their cross section data. Hence $\alpha(H_3^+)$ values of this magnitude were naturally invoked in models of the ion chemistry of interstellar clouds.

To study the recombination of H_3^+ ions using the FALP apparatus with the preferred helium carrier gas, a potential problem arises viz: He^+ ions react only very slowly with H_2[38]. Fortunately, however, H_2 reacts relatively rapidly with He_2^+ [39] and helium metastable atoms[40]. Therefore, it was important to ensure that the majority of the He^+ ions formed in the discharge were converted to He_2^+ by the three-body reaction of He^+ with two He atoms and this is the case at the normal helium operating pressures of 1 torr. On addition of H_2 to the afterglow (at a concentration of about one percent of the helium concentration), the He_2^+ ions were rapidly converted to HeH^+, He_2H^+ and H_2^+ ions and the metastables to H_2^+ ions and these product ions all react rapidly with H_2 to form H_3^+ ions[41]. Thus an afterglow plasma can be created at 300K consisting mainly of H_3^+ ions and electrons but with a small percentage of He^+ ions as determined by the downstream mass spectrometer detection system. However, at the lower temperature of 95K at which recombination of H_3^+ ions was also studied, the three-body association reaction which converts He^+ to He_2^+ is much faster and the plasma flow velocity is much smaller so that the conversion of He^+ to He_2^+ is complete prior to the downstream reaction zone and no significant He^+ was observed in the plasma mass spectrum. (Interestingly, a small signal of He_3^+ was observed.) So on the addition of sufficient H_2 to the plasma an H_3^+/electron plasma was formed. On the addition of large flows of H_2 at 95K, H_5^+ appeared in the mass spectrum as a result of the three-body association reaction of H_3^+ with H_2 (with helium as a third-body). The appearance of H_5^+ correlated with a rapid reduction in the electron density in the afterglow plasma (indicating the onset of H_5^+ recombination with electrons).

Having created the H_3^+/electron plasma, the n_e was measured as a function of z using the movable Langmuir probe[41]. A typical result of the many data sets obtained at 300K is shown in Fig. 1 where it can be seen that following an initial relatively rapid reduction in n_e then the log n_e versus z plot is linear with a small slope which is indicative of slow diffusive loss of electrons. This plot should be compared with that also obtained at 300K for the O_2^+/electron plasma given in Fig. 1 from which an $\alpha(O_2^+)$ of $2 \times 10^{-7} cm^3 s^{-1}$ is derived. Clearly, the H_3^+/electron plasma is decaying principally by ambipolar diffusion and not recombination and certainly $\alpha(H_3^+) \ll \alpha(O_2^+) = 2 \times 10^{-7} cm^3 s^{-1}$. An analysis of the FALP data indicates that $\alpha(H_3^+)$ must be less than $2 \times 10^{-8} cm^3 s^{-1}$ at 300K and a similar upper limit value is also obtained at 95K. Similar results were obtained in a D_3^+/electron plasma. These values for $\alpha(H_3^+)$ are clearly in conflict with the values derived from both SA and merged beam experiments. However, after obtaining the FALP data and prior to its publication, we became aware of the theoretical work of Michels and Hobbs[42] which predicted that $\alpha(H_3^+)$ should be very small for ground vibrational state H_3^+ ions but appreciable for vibrationally-excited ions. This work not only supports the FALP results but also provides an explanation for the initial relatively rapid reduction in n_e with z in the H_3^+/electron plasma. Since the H_3^+ ions are formed via reaction (1) then they are initially vibrationally excited and a fraction of them can recombine immediately after formation, although the majority will be relaxed in collisions with H_2 (i.e. via reaction (2)).

These FALP results prompted further SA afterglow experiments and it now appears that the SA values for $\alpha(H_3^+)$ were anomolously high due to impurities of CH_5^+ in the afterglows[26]. Since $\alpha(CH_5^+)$ is $1.1 \times 10^{-6} cm^3 s^{-1}$ at 300K only a small percentage of CH_5^+ is required as a contaminant ion to confuse the determination of $\alpha(H_3^+)$. Also, recent merged beam studies of $\sigma(H_3^+)$ have shown that the σ decreases markedly as the fraction of $H_3^+(\nu)$ in the ion beam is decreased, and an upper limit estimate of $\alpha(H_3^+)$ for $H_3^+(\nu \leq 1)$ ions of about $1 \times 10^{-8} cm^3 s^{-1}$ has been derived from the data[12], in agreement with the FALP data.

Subsequently, we have developed a method by which we can compare the rates of recombination of various ions in the FALP plasma by observing the relative mass spectrometer signals of those ions for greatly differing electron number densities, n_e, in the plasma (remembering that a large n_e enhances recombination and would therefore reveal ions that have significant recombination coefficients). We have found no measurable differences between the loss of He^+, HeH^+ and H_3^+ ions for very large changes in n_e and therefore conclude that $\alpha(H_3^+) \simeq \alpha(HeH^+) \simeq \alpha(He^+)$ within the limits of accuracy of measuring relative peak heights of the mass spectrometer signals.

What then is the actual value of $\alpha(H_3^+)$? If $\alpha(H_3^+)$ is equal to $\alpha(He^+)$ then at the low values of n_e at which three-body effects are insignificant (which is certainly the situation in interstellar gas clouds) then we may assume that He^+ can only radiatively recombine with electrons and that the the radiative recombination coefficient of He^+, $\alpha_r(He^+)$, at plasma temperature of 300K is $\lesssim 10^{-11} cm^3 s^{-1}$ [1]. Presumably therefore $\alpha_r(H_3^+)$ will be similar. These α_r increases with decreasing temperature and will thus somewhat exceed $10^{-11} cm^3 s^{-1}$ in cold interstellar clouds.

For large n_e and especially at low temperatures, the collisional-radiative recombination process can occur. According to the published table[1]), the collisional-radiative recombination coefficient, α_{cr}, for H^+ ions at a temperature of 250K and an n_e of $10^{10}cm^3s^{-1}$, i.e. approximating to the physical conditions in the FALP plasma, is about $10^{-9}cm^3s^{-1}$ (note that α_{cr} is not expected to depend significantly on the nature of the ion as long as it is singly-charged). At a fixed temperature, α_{cr} increases approximately linearly with n_e and thus at 250K and $10^{13}cm^{-3}$ then α_{cr} is about $10^{-6}cm^3s^{-1}$. Clearly under such conditions the simple reciprocal density recombination law (i.e. $n_e^{-1} \propto z$) will not hold since it assumes that α is independent of n_e. This cannot be put to the test in the FALP since n_e in excess of a few times $10^{10}cm^{-3}$ cannot be attained. However, very recently, measurements have been made of the temporal decay of ground vibrational state H_3^+ ions in the afterglow of an electrical discharge through H_2 using an infrared absorption technique to detect and monitor the number density of ground state H_3^+ ions[43]). A value for $\alpha(H_3^+)$ of $1.8 \times 10^{-7}cm^3s^{-1}$ at a temperature of about 210K was reported which again cast some doubt on the much lower value deduced from the FALP data. However, it must be said that this experiment involved a number of uncertainties which leads one to doubt the result. The electron density was not determined in this experiment and so n_e was equated to the ground state H_3^+ number density, and the likely presence of significant number densities of H^+ ions was ignored. Also the electron temperature was uncertain and could have been in excess of the gas temperature (rather than equal to the gas temperature as was assumed) in the very early ($\sim 10^{-3}$ to 10^{-4}s) afterglow following a heavy current discharge through the H_2. The presence of $H_3^+(\nu)$ and also H_5^+, both of which are known to recombine rapidly with electrons, cannot be ruled out entirely, and the inevitable vibrational excitation and dissociation of the H_2 (generating reactive H atoms) introduces uncertainties into the ion chemistry which are difficult to quantify. So it seems possible that the loss rate of ground state H_3^+ ions from the plasma may be enhanced since they are coupled via the interaction processes to the rapidly recombining $H_3^+(\nu)$ and H_5^+. What seems even more likely is that, at the high n_e in the plasma ($\gtrsim 10^{13}cm^{-3}$), the collisional-radiative process mentioned above is contributing to, and is probably the major loss process for H_3^+ ions. Under such circumstances, also noted above, a linear plot of the reciprocal number density of H_3^+ against time in the decaying afterglow would not be expected, yet this was apparently obtained and resulted in the large value of $\alpha(H_3^+)$. This is a perplexing situation. Clearly, further experiments are needed, if possible at much smaller n_e which would avoid many of the problems alluded to above. Meanwhile, the $\alpha(H_3^+)$ value derived from this experiment should be treated with some caution. This is surely not the last word on this problem !

DISSOCIATIVE RECOMBINATION OF PROTONATED MOLECULES

The addition of a wide variety of molecular gases, X, to an H_3^+/electron plasma initiates the proton transfer reactions:

$$H_3^+ + X \longrightarrow XH^+ + H_2 \qquad (3)$$

Such reactions are rapid, essentially occurring on each collision between H_3^+ and X, when the proton affinity (PA) of X exceeds $PA(H_2)$. This then provides the means of determining the α for many protonated molecules, i.e. XH^+ ions. Fig.1 shows the n_e versus z data obtained when CO and N_2 are added to an H_3^+/electron plasma in the FALP apparatus[41]. It clearly shows that HCO^+ and N_2H^+ plasmas decay more rapidly than H_3^+ plasmas, and it is also evident that $\alpha(N_2H^+) > \alpha(HCO^+)$ (the values are 1.7×10^{-7} and 1.1×10^{-7} in cm^3s^{-1} at 300K respectively). These α values were obtained from plots of $n_e{}^{-1}$ versus z as mentioned previously. These measurements together with measurements at 95K indicated that $\alpha(N_2H^+)$ and $\alpha(HCO^+)$ varied approximately as T^{-1}. The larger recombination coefficients (e.g., $\alpha(CH_5^+)$) increased more slowly with decreasing temperature in accordance with the results of other studies for different ion species[27].

More recently, a FALP study of the α for several more protonated species has been carried out at 300K, varying in complexity from H_2F^+ to the clustered species $H^+(C_2H_5OH)_3$[44]. The results of this study revealed a trend of increasing α with increasing complexity of the recombining ion. Thus $\alpha(H_2F^+) = 1.1 \times 10^{-7} cm^3s^{-1}$ and $\alpha(H^+(C_2H_5OH)_3) \simeq 2 \times 10^{-6} cm^3s^{-1}$, again a not unexpected result in the light of previous studies[10]. These data find valuable application to the ion-chemical modelling of interstellar molecular clouds in which it is certain that many of these protonated molecules exist.

THE NEUTRAL PRODUCTS OF DISSOCIATIVE RECOMBINATION REACTIONS

It is obviously desirable to know the products of dissociative recombination reactions of polyatomic ions in order to obtain a better understanding of both the fundamentals of the reaction process and of the ion chemistry of real plasmas such as interstellar gas clouds. Theoretical effort has been directed to this problem with some conflicting results[1, 46-48]. Yet no experimental data are available relating to ground state polyatomic ions (although some information is available on the products of recombination of a few internally excited ionic species[13,45]). However, experiments are underway at the Universities of Birmingham, U.K. and Rennes, France to determine the products of dissociative recombination of some important interstellar species including H_3O^+, NH_4^+ and H_2CN^+ . To this end, vacuum ultraviolet (VUV) absorption spectroscopy (to detect atomic radicals such as H, O and N) and laser induced fluorescence (LIF) spectroscopy (to detect molecular radicals such as OH, NH and CN) are being used to probe recombining plasmas in FALP apparatuses.

Preliminary results have been obtained using the Birmingham FALP/LIF facility which indicate that OH radicals are generated in the dissociative recombination reaction of H_3O^+ with electrons. Although the experiment is not yet fully quantitative the time is not far off when the percentage of the recombination events which lead to OH radicals will be determined. Also, when the VUV absorption experiments at the Lyman α wavelength reveal the percentage of H atoms which are released following recombination, then a good deal will be known about the products of recombination of H_3O^+ ions.

ACKNOWLEDGEMENTS

Our thanks are due to the United States Air Force for financial support of this work. We also acknowledge the involvement of Charles R. Herd in the spectroscopy studies.

REFERENCES

1) Bates, D.R. and Dalgarno, A., in "Atomic and Molecular Processes", Pure and Applied Physics Vol.13, Ed. D.R. Bates (Academic:New York, 1962); Bates, D.R., in "Recent Studies in Atomic and Molecular Processes", Ed. A.E. Kingston (Plenum:New York, 1987) p.1.

2) Bardsley, J.N. and Biondi, M.A., Adv. Atom. Mol. Phys. 6, 1 (1970).

3) Massey, H.S.W. and Gilbody, H.B., "Electronic and Ionic Impact Phenomena" Vol IV, (Clarendon:Oxford, 1974).

4) Mitchell, J.B.A., in "Atomic Processes in Electron-Ion and Ion-Ion Collisions", Ed. F. Brouillard (Plenum:New York, 1986)p. 185.

5) Thomas, L., Radio Sci. 9, 121 (1974).

6) Smith, D. and Adams, N.G., Topics in Current Chemistry 89, Plasma Chemistry I, 1 (1980).

7) Dalgarno, A. and Black, J.H., Rept. Prog. Phys. 39, 573 (1976).

8) Smith, D. and Adams, N.G., Int. Rev. Phys. Chem. 1, 271 (1981).

9) Smith, D. and Adams, N.G. in "Physics of Ion-Ion and Electron-Ion Collisions" Ed. F. Brouillard and J.W. McGowan (Plenum:New York, 1983) p.501.

10)Biondi, M.A., Comm. Atom. Mol. Phys. 4, 85 (1973).

11)Mitchell, J.B.A. and McGowan, J.W. in "Physics of Ion-Ion and Electron-Ion Collisions" Ed. F. Brouillard and J.W. McGowan (Plenum:New York, 1983)p. 279.

12)Hus, H., Youssif, F., Sen, A. and Mitchell, J.B.A., Proceedings 5th International Swarm Seminar, Birmingham, England 29-31 July 1987, p.82.

13)Adams, N.G. and Smith, D., in "Rate Coefficients in Astrochemistry" Ed. T.J. Millar and D.A. Williams (Reidel:Dordrecht, 1988) in press.

14)Smith, D., Dean, A.G. and Adams, N.G., Z.Physik 253, 191 (1972).

15)Adams, N.G., Dean, A.G. and Smith, D., Int. J. Mass Spectrom. Ion Phys. 10, 63 (1972/73).

16)Ferguson, E.E., Fehsenfeld, F.C. and Schmeltekopf, A.L., Adv. Atom. Mol. Phys. 5, 1 (1969).

17)Smith, D. and Adams, N.G., in "Gas Phase Ion Chemistry" Vol 1, Ed. M.T. Bowers (Academic:New York, 1979) p.1.

18)Ferguson, E.E., Fehsenfeld, F.C. and Albritton, D.L., in "Gas Phase Ion Chemistry" Vol 1, Ed. M.T. Bowers (Academic:New York, 1979) p.45.

19)Smith, D. and Adams, N.G., Adv. Atom. Mol. Phys. 24, (1988) in press.

20)Adams, N.G. and Smith, D., in "Techniques for the Study of Gas-Phase Ion-Molecule Reactions", Ed. J.M. Farrar and W.H. Saunders Jr. (Wiley-Interscience: New York, 1988) in press.

21)Smith, D. and Plumb I.C., J. Phys. D 5, 1226 (1972).

22)Smith, D., Adams, N.G., Dean, A.G. and Church, M.J., J.Phys. D. 8, 141 (1975).

23)Alge, E., Adams, N.G. and Smith, D., J. Phys. B 16, 1433 (1983).

24)Adams, N.G., Church, M.J. and Smith, D., J.Phys. D 8, 1409 (1975).

25)Walls, F.L. and Dunn G.H., J. Geophys. Res. 79, 1911 (1974).

26)Johnsen, R., Int. J. Mass Spectrom. Ion Processes 81, 67 (1987).

27)Smith, D. and Adams, N.G. in "Swarms of Ions and Electrons in Gases", Ed. W. Lindinger, T.D. Mark and F. Howorka, (Springer-Verlag:Vienna, 1984) p.284.

28)Huang, C-M., Biondi, M.A. and Johnsen, R., Phys. Rev. A14, 984 (1976).

29)Huang, C-M., Whitaker, M., Biondi, M.A. and Johnsen, R., Phys. Rev. A18, 64 (1978).

30)Leventhal, J.J. and Friedman, L., J. Chem. Phys. 50, 2928 (1969).

31)Bowers, M.T., Chesnavich, W.J. and Huntress, W.T. Jr., Int. J. Mass Spectrom. Ion Phys. 12, 357 (1973).

32)Kim, J.K., Theard, L.P. and Huntress, W.T. Jr., Int. J. Mass Spectrom Ion Phys. 15, 223 (1974).

33)Dalgarno, A., in "Rate Coefficients in Astrochemistry", Ed. T.J. Millar and D.A. Williams (Reidel:Dordrecht, 1988) in press.

34)Persson, K.B. and Brown S.C., Phys. Rev. 100, 729 (1955).

35)Leu, M.T., Biondi, M.A. and Johnsen, R., Phys. Rev. A8, 413 (1973).

36)Macdonald, J.A., Biondi, M.A. and Johnsen, R., "Planet. Space Sci.

32, 651 (1984).

37)Mitchell, J.B.A., Ng, C.T., Forard, L., Janssen, R. and McGowan, J.W., J. Phys. B. 17, L909 (1984).

38)Johnsen, R. and Biondi, M.A., J. Chem. Phys. 61, 2127 (1974); 74, 6996 (1981).

39)Adams, N.G., Bohme, D.K. and Ferguson, E.E., J. Chem. Phys. 52, 5101 (1970).

40)Lindinger, W., Schmeltekopf, A.L. and Fehsenfeld, F.C., J. Chem. Phys. 61, 2890 (1974).

41)Adams, N.G., Smith, D. and Alge, E., J.Chem. Phys. 81, 1778 (1984).

42)Michels, H.H. and Hobbs, R.H. Ap. J. (Letters) 286, L27 (1984).

43)Amano, T., Ap. J. (Letters) submitted.

44)Adams, N.G. and Smith, D., Chem. Phys. Letts. 144, 11 (1988).

45)Smith, D., Phil. Trans. Roy. Soc. Lond. A324, 257 (1988).

46)Herbst, E., Ap. J. 222, 508 (1978).

47)Bates, D.R., Ap. J. (Letters) 306, L45(1986).

48)Millar, T.J., DeFrees, D.J., McLean, A.D. and Herbst, E., Astron. Astrophys. 194, 250 (1988).

RECOMBINATION MEASUREMENTS IN MICROWAVE PLASMA AFTERGLOWS

Rainer Johnsen
Department of Physics and Astronomy
University of Pittsburgh

I. INTRODUCTION

The term "electron-ion recombination" means different things to different people. Some of us, both theorists and experimentalists, view recombination as an elementary collision process while others think of recombination as a process leading to deionization of a plasma. The first group may view the multitude of processes occurring in a recombining plasma as "complications" on the way to understanding recombination. One may also take the view, as I will do in my talk, that understanding plasmas is one of the principal motivations for studying recombination and that the "complications" are of interest in their own right.

An extensive review of our recombination measurements in microwave afterglows has been given only recently [1]. There, I discussed the problem of microwave heating of electrons in afterglow plasmas. Building on earlier work by Phelps[2] and also by Penetrante and Bardsley[3], we concluded that inelastic collisions of electrons with molecular additives may lead to a reduction of the electron temperature below the value that would be obtained in a pure atomic gas. I will briefly review the present status of this problem and will discuss the consequences for earlier data obtained by the microwave afterglow technique.

Vibrational excitation of recombining ions may affect their recombination rates and product distributions, but only few afterglow experiments in which the vibrational states of the ions were known have been performed and their interpretation may not be as straightforward as was originally assumed by the experimenters. The recombination studies by Zipf[4] on N_2^+ recombination will be given an interpretation which differs from that proposed originally.

The third topic concerns the role of the ambient neutral gas in the recombination process. In plasmas at high neutral density (near atmospheric), electron-ion recombination may be enhanced by three-body collisions. We have developed a technique to study the decay of photoionized plasmas in the pressure range from about 200 Torr to 2 atmospheres. This work is still in progress, but the results indicate that recombination of fast-recombining ions occurs with about the same rate as at low pressures. Exceptions may exist in the case of slowly recombining ions, where the effect of three-body stabilization is more pronounced.

Although recombination with electrons is the principal subject of this meeting, I will briefly touch on our concurrent work on ion-ion recombination in atmospheric-density plasmas. We find that the dependence of ion-ion recombination rates on ambient gas density is strongly affected by the effect of the ambient gas on mutual neutralization. The results corroborate earlier experimental work and theoretical suggestions.

II. MICROWAVE HEATING OF ELECTRONS IN AFTERGLOW PLASMAS

The microwave afterglow technique, as developed by Biondi and coworkers[5], has contributed a large body of recombination data for a wide range of ions. One of its principal advantages is the ability to elevate the electron temperature during the afterglow by application of microwave power. This allows us to measure recombination rates as a function of electron temperature in the range from 77 K to a maximum of about 5000 K, the range of greatest interest for applications in ionospheres, the interstellar medium and laboratory discharges. It has been realized only recently that the traditional use of the Margenau[6] theory to calculate electron temperatures is not always justified when the plasma contains molecular additives, even when those are present in only minute concentrations. Inelastic collisions of electrons with molecules (rotational and vibrational excitation) constitute a non-negligible sink for the electron energy which must be included in the electron power balance. The purpose of the additives, of course, is to produce the desired ion species by Penning ionization and subsequent ion-molecule reactions and thus the problem cannot be eliminated by dispensing with the additives.

A theoretical treatment of electron heating in microwave afterglows is made more difficult by the fact that the microwave heating field is not spatially uniform. Contrary to earlier assumptions, heat conductance within the electron gas is not sufficient to smooth out spatial nonuniformities. The most thorough analysis of microwave heating of electrons has been carried out by Dulaney et al.[7] in the course of the analysis of their data on recombination of nitric oxide ions. As a result of this work, previously existing discrepancies between the results of different recombination measurements have now been resolved.

In my recent review article[1] I assessed the degree to which results of earlier afterglow measurements were affected by inelastic collions of electrons with molecules and nonuniform electron temperatures. In some cases, the inferred dependence of the recombination coefficients on electron temperature may have been too weak. The effects were found to be significant in the case of the recombination studies of nitric oxide ions[7]. In many other cases, such as in most studies of other diatomic ionx, no revisions are required. The measurements of dimer-ion (e.g. N_4^+, O_4^+)

recombination coefficients required addition of rather large concentrations of molecular gases with the result that the electron temperature was affected more strongly. Direct experimental evidence showing the effect of molecular additives on the electron temperature has been obtained by Dulaney et al.[8]. The data most in doubt are those pertaining to ammonia- and water-cluster ions. There is a strong suspicion that microwave heating did not substantially increase the electron temperature above that of the gas. As a result, the inferred lack of a dependence of the recombination coefficients on electron temperature may be only apparent. In view of the importance of cluster ions to many applications, this work should be repeated. A compilation of recombination coefficients measured by the microwave afterglow technique and other methods is given in Tables 1 and 2. Comments on the reliability of the data and other information are given in the column labelled "remarks". A more detailed discussion of the microwave heating problem is given in the review article[1].

Our studies of recombination of O_4^+ and HCO^+ have now been published[8,9]. The factor-of-two discrepancy between the microwave afterglow results for the recombination coefficient of HCO^+ and the FALP measurements still presents a problem. Since the afterglow results are higher one might suspect that the result was affected by ion clustering, but no cluster ions were observed in that experiments and corrections were applied for the presence of impurity ions.

III. THE ROLE OF VIBRATIONALLY EXCITED IONS IN AFTERGLOWS

The absence of information concerning the vibrational states of recombining ions has been a problem for many years. Often, aeronomers who needed particular values for recombination coefficients to explain their ionospheric observations felt free to propose recombination coefficients far in excess of the laboratory values, pointing to the possibility that vibrationally excited ions might recombine faster than those in ground states. The recombination of N_2^+ ions will be discussed here as an example.

The microwave afterglow measurements by Zipf[4] on N_2^+ recombination are sometimes quoted as providing definitive proof that, at least in the case of this ion, vibrational excitation is ineffective in enhancing dissociative recombination. This conclusion needs to be reexamined for the following reason: Zipf determined the relative abundance of N_2^+ ions in v=0, 1,2 using laser induced fluorescence in the B-X band. His observations that these ions decayed at nearly the same rate then was interpreted by Zipf as indicating nearly identical recombination rates for ions in these states.

Table 1: (see Table 2 for "remarks and abbreviations)

Ion	α(300K)	[cm^3/s]	x	Temp. [K]	method	remarks	Reference
Ne_2^+	1.75±0.2	[-7]	0.49	300-4600	μwa-ms	A	10
Ar_2^+	9.1 ±1.0	[-7]	0.61	300-8500	μwa	A	11
Kr_2^+	1.6±0.2	[-6]	0.55	300-8400	μwa	A	12
Xe_2^+	2.3±0.2	[-6]	0.6	300-7500	μwa	A	13
$N0^+$	4.3±0.3	[-7]	0.38	380-5470	μwa-ms	wrong x	25
$N0^+$	4.2	[-7]	0.9	200-600	FALP		15
$N0^+$	2.3	[-7]	0.5	>10,000	MB		17
$N0^+$	4.3	[-7]	0.85	5000	trap		14
$N0^+$	4.4	[-7]	0.75	300-4500	μwa-ms	A	7
N_2^+	1.8±0.3	[-7]	0.39	300-5000	μwa-ms	A	28
N_2^+	3.5	[-7]	0.5	300->10000	MB	E	17
N_2^+	2.2	[-7]	--	300	μwa-ms	F	4
N_2^+	1.78	[-7]	0.37	700-2700	shock		19
0_2^+	1.95±0.2	[-7]	0.7	300-1200	μwa-ms	A	28
			0.56	1200-5000			
0_2^+	1.95	[-7]	0.7	200-600	FALP		15
0_2^+	1.95	[-7]	0.66	200-5000	trap		18
0_2^+	2.3	[-7]	0.5	300-10000	MB		17
H_3^+	2.3	[-7]		300	μwa-ms	D	20
H_3^+	1.7	[-7]	1.0	500-3000	μwa-ms	D	21
H_3^+	~3.5	[-7]	~0.5	(0.01-0.7eV)	MB	F	22
H_3^+	<2	[-8]	---	95K, 300K	FALP	G	16
HCO^+	2.0	[-7]	1	300, 205	μwa-ms	A	23
HCO^+	1.1	[-7]	~0.8	300, 95	FALP		16
HCO^+	2.4±0.4	[-7]	0.69	293-5500	μwa-ms	A	9

Table 2

Ion	α(300K)	[cm³/s]	x	Temp. [K]	method	remarks	Reference
$H_3O^+(H_2O)$	2.5±0.5	[-6]	0.08	300-8000	μwa-ms	C	24
$H_3O^+(H_2O)_2$	3.0±0.6	[-6]	0.08	300-8000	μwa-ms	C	24
$H_3O^+(H_2O)_3$	3.6±0.7	[-6]	0	300-8000	μwa-ms	C	24
$H_3O^+(H_2O)_4$	5	[-6]	0	300-8000	μwa-ms	C	24
$H_3O^+(H_2O)_5$	˜5	[-6]	0	300-8000	μwa-ms	C	24
$NH_4^+(NH_3)$	2.82	[-6]	0.147	300-3000	μwa-ms	C	25
$NH_4^+(NH_3)_2$	2.68	[-6]	0.05	300-3000	μwa-ms	C	25
$NH_4^+(NH_3)_3$	˜3	[-6]	0	200	μwa-ms	C	25
$NH_4^+(NH_3)_4$	˜3	[-6]	0	200	μwa-ms	C	25
$N_2^+N_2$	1.4±0.2	[-6]	0.41	300-6000	μwa-ms	B	26
CO^+CO	1.3±0.3	[-6]	0.34	300-4000	μwa-ms	B	27
$CO^+(CO)_2$	1.9±0.4	[-6]	˜0	300-1000	μwa-ms	B	27
			0.33	1000-9000	μwa-ms		
$O_2^+O_2$	4.2±0.4	[-6]	0.48	143-5500	μwa-ms	A	8

Explanations: x defined by

$$\alpha(T) - \alpha\,(300)\,(T_e/300)^{-x}$$

where T_e is the electron temperature. The cross section data obtained by the beam and trap methods have been converted to rate coefficients. Abbreviations: μwa-ms: microwave afterglow-mass spectrometer. μwa: microwave afterglow. MB: merged beam. FALP: flowing afterglow/Langmuir probe. shock: shock tube. trap: ion trap

Remarks: A: no revisions required. B: x likely to be ≈0.5. C: x≈0 may be due to lack of heating (see text). D. Absolute value possibly affected by impurities. E. Absolute value should be reduced by factor of two (Mitchell, private comm.). F. Refers to vibr. excited ions. G. Refers to ions in v=0

We know now, from the work of Lindinger et al[29] and Kemper and Bowers[30], that quenching of vibrationally excited nitrogen ions in collisions with neutral nitrogen and also with the neon buffergas present in that experiment are fast processes. Under the condition of Zipf's experiment vibrationally excited ions should have decayed in a few microseconds. How then did Zipf's ions in v=1, 2 manage to survive for times on the order of many milliseconds? The answer seems to be that the neutral nitrogen in that experiment was also vibrationally excited and that the ions were kept vibrationally "hot" by exchanging energy with the reservoir of vibrationally hot neutrals. Zipf's observations then should be interpreted as yielding an effective recombination coefficient for a population of excited ions at a vibrational temperature of about 1500 K. This is still a valuable piece of information, but no rates for ions in particular vibrational states can be inferred.

Preliminary merged beam measurements by Mitchell et al.[31] seem to indicate that N_2^+ ions in v=0 recombine more slowly (by a factor of about four) than the ions in afterglow experiments. Even though this result may not be in direct conflict with the microwave afterglow results, it still creates a problem:
If we accept lower values for the recombination coefficients for v=0 ions the coefficients for v=1 ions must be drastically increased to yield the same effective recombination coefficient for a distribution of vibrationally excited ions at 1500 K (roughly 90% in v=0, 9% in v=1, 1% in v=2 in Zipf's work). For example, if we take $\alpha(v=0)$ as one fourth of the afterglow value of 2(-7) cm^3/s, we are forced to conclude that $\alpha(v=1) \approx 1.5\ (-6)\ cm^3/s$, which may be hard to accept unless theoretical calculations show that recombination by v=1 ions is unusually rapid. New FALP data[32] provide a more direct argument against lower values for $\alpha(v=0)$. In that experiment only v=0 ions should have been present, but the recombination coefficients were found to be close to the microwave afterglow results.

The presence of vibrationally hot neutral molecules in afterglow plasmas might be expected to have some effect on the electron temperature as a result of superelastic collisions of electrons with vibrationally excited molecules. If, as is indicated by Zipf's work, the vibrational temperature is near 1500 K under typical afterglow conditions, the electrons in a "room temperature" afterglow would not be at room temperature and the measured recombination coefficients would be far too low. The question of coupling of electron and vibrational temperatures has been discussed by Hurle[33] in the case of pure nitrogen afterglows, where the effect is important. Using his numbers, one estimates that afterglows in mixtures of rare gases and small amounts of nitrogen will not exhibit a significant enhancement of the electron temperature. Thus, this problem does not arise under the conditions of typical afterglow experiments.

One might also ask whether microwave heating of electrons would indirectly lead to vibrational excitation of the ions. The route of energy transfer would be from the electrons to vibrational energy in the neutrals and from there to the ions. The answer to the question appears to be negative. The cross sections for vibrational excitation of nitrogen molecules are small below 2 eV and the estimated rate of energy transfer should be negligible under the conditions of microwave afterglow experiments.

IV. ELECTRON-ION RECOMBINATION AT HIGH AMBIENT GAS DENSITIES

Most experiments on electron-ion recombination have been carried out at low ambient gas densities. Under these conditions, two-body recombination between an ion and an electron of positive energy is the only mechanism of importance. In the presence of an ambient gas, electrons may also be captured into states of slightly negative energy (Rydberg states) which subsequently decay by dissociation, collisional ionization or collisional stabilization. The effect of the ambient gas is to enhance the rate of recombination over that at zero gas density. The degree of enhancement is of some interest for applications in high-pressure discharges, such as those used in gas lasers.

We have carried out a series of recombination measurements in ambient helium at atmospheric densities. The experimental method, which has been described earlier[34], makes use of uv-photoionized afterglow plasmas. The electron density is determined by radio-frequency conductivity probes rather than by the microwave frequency shift method. Mass analysis of the plasma ions is performed by ion sampling from the plasma. It is difficult to produce atmospheric density plasmas with the desired simple ion composition, but we have succeeded in producing plasmas containing mainly water cluster ions, diatomic argon ions, or HCO^+ ions. The measured recombination rates for water cluster and argon ions were found to be within about twenty percent of those measured in low-pressure experiments. The large two-body recombination coefficients of these ions mask the three-body contribution. However, the measured rate coefficient for HCO^+, a slowly recombining ion, was significantly larger (by factors of 4 to 5 at 1 atm of helium) than the low-pressure values. A three-body recombination coefficient of about 2×10^{-26} cm^6/s would be inferred from a comparison of low- and high-pressure data. The numbers seem reasonable, but there is little experimental data for comparison and more data at different pressures need to be taken to improve the accuracy.

V. ION-ION RECOMBINATION

The subject of ion-ion recombination is somewhat removed from the main topic of this meeting, but many of my colleagues here have worked on ion-ion recombination and may be interested in some new experimental results. The technique that we used is similar to that used in the high pressure electron-ion recombination studies. An electron-ion plasma is produced by uv-photoionization; attachment of electrons to oxygen molecules and subsequent ion molecule reactions then produces the desired ion-ion plasma. We were able to produce conditions where the recombination process

$$H_3O^+ (H_2O)_n + NO_3^- + M \dashrightarrow \text{neutrals}$$

with n=2,3 and M=He or M=Ar dominated the afterglow decay. Both positive and negative ions were identified by mass spectrometry. Neutral products were not observed, but proton transfer to form nitric acid and water molecules is expected to be the dominant channel.

We find that the recombination rates increase with density, as expected from theory. However, an extrapolation of the pressure dependence to zero density yields values far in excess (about 1(-6) cm^3/s) of those observed at very low densities. A sharp rise of the recombination coefficient must be present at pressures below that covered by our experiment (200 Torr to 2 at). To some extent, our data corroborate the FALP results by Adams and Smith[35] who observed a strong positive pressure dependence of the recombination coefficient in the 1 to 8 Torr range. The observed pressure dependence can be reproduced by a model which includes both three-body recombination (i.e. collisional energy transfer from the orbiting ion pair to the gas) and mutual neutralization by proton transfer. The results will be published in the near future.

Acknowledgment:
This work was, in part, supported by the NASA Planetary Atmospheres Program under Grant NGL 39-0110137, by the ARMY Research Office under Grant DAAG 29-84-K-0010, and by Air Force Cambridge/DNA under contract No. F19628-83-0037.

References:

1. R. Johnsen, Int. J. Mass Spectr. and Ion Proc., 81, 67 (1987).
2. A. V. Phelps, private communication.
3. B. M. Penetrante and J. N. Bardsley, Phys. Rev. A 34, 3252 (1986).
4. E. C. Zipf, Geophys. Res. Lett. 7, 645 (1980).
5. M. A. Biondi, "Electron-ion recombination in gas lasers", in Applied Atomic Collision Physics, Vol. 3, (Academic Press, 1982).
6. H. Margenau, Phys. Rev. 69, 508 (1946).
7. J. L. Dulaney, M. A. Biondi, and R. Johnsen, Phys. Rev. A, 36, 1342 (1987).
8. J. L. Dulaney, M. A. Biondi, and R. Johnsen, Phys. Rev. A, 37, 2539 (1988).

9. B. Ganguli, M. A. Biondi, R. Johnsen, and J. L. Dulaney, Phys. Rev. A, 37, 2543 (1988).
10. L. Frommhold, M. A. Biondi, and F. J. Mehr, Phys. Rev. A165, 44 (1968).
11. Y. S. Shiu and M. A. Biondi, Phys. Rev. A 16, 1817 (1977).
12. Y. S. Shiu and M. A. Biondi, Phys. Rev. A 17, 868 (1978).
13. Y. S. Shiu, M. A. Biondi, and D. P. Sipler, Phys. Rev. A15, 494 (1977).
14. C. Huang, M. A. Biondi, and R. Johnsen, Phys. Rev. A11, 901 (1977).
15. E. Alge, N. G. Adams, and D. Smith, J. Phys. B 16, 1433 (1983).
16. N. G. Adams, D. Smith, and E. Alge, J. Chem. Phys. 81 1778 (1984).
17. P. M. Mul and J. W. McGowan, J. Phys. B 12, 1591 (1979).
18. F. L. Walls and G. H. Dunn, J. Geophys. Res. 79, 1911 (1974).
19. A. J. Cunningham and Hobson, J. Phys. B. 5, 2328 (1972).
20. M. T. Leu, M. A. Biondi, and R. Johnsen, Phys. Rev. A8, 413 (1973).
21. J. A. MacDonald, M. A. Biondi, and R. Johnsen, Planet. Space Sci. 132, 651 (1984).
22. J. B. A. Mitchell, C. T. Ng, L. Forand, R. Janssen, and J. W. McGowan, J. Phys. B17, 1909 (1984).
23. M. T. Leu, M. A. Biondi, and R. Johnsen, Phys. Rev. A8, 420 (1973).
24. C. M. Huang, M. Whitaker, M. A. Biondi, and R. Johnsen, Phys. Rev. A18, 64 (1978).
25. C. M. Huang, M. A. Biondi, and R. Johnsen, Phys. Rev. A14, 984 (1976).
26. M. Whitaker, M. A. Biondi, and R. Johnsen, Phys. Rev. A24, 743 (1981).
27. M. Whitaker, M. A. Biondi, and R. Johnsen, Phys. Rev. A 23, 1481 (1981).
28. F. J. Mehr and M. A. Biondi, Phys. Rev., 181, 264 (1969).
29. W. Lindinger, F. Howorka, P. Lukac, S. Kuhn, H. Villinger, E. Alge, and H. Ramler, Phys. Rev. 23, 2319 (1981).
30. P. R. Kemper and M. T. Bowers, J. Chem. Phys. 81, 2634 (1984).
31. J. B. A. Mitchell, Paper presented at this Conference.
32. D. Smith, private communication.

33. I. R. Hurle, J. Chem. Phys. 41, 3592 (1964).
34. R. Johnsen, Rev. Sci. Instr. 57, 428 (1986).
35. D. Smith and N. G. Adams, in "Physics of ion-ion and electron-ion collisions", F. Brouillard and J. W. McGowan eds., Plenum Publishing Co., page 501.

EXPERIMENTAL STUDIES OF DISSOCIATIVE RECOMBINATION BRANCHING RATIOS

B.R. ROWE*, J.L. QUEFFELEC, Département de Physique Atomique et Moléculaire, Université de Rennes 1, 35042 RENNES (FRANCE) and *Laboratoire d'Aérothermique du CNRS, 4Ter route des Gardes, 92190 MEUDON (FRANCE).

ABSTRACT: a new technique is reviewed which allows the branching ratios of dissociative recombination to be studied. The molecular ions are formed by charge exchange reaction with a monoatomic ion in a highly ionized plasma flow. After recombination, the products are detected using V.U.V. absorption methods.

1 INTRODUCTION

Dissociative recombination (hereafter DR) of molecular ions with electrons is an important process in ionized gases and consequently in many fields of physics and chemistry such as astrochemistry, planetary atmosphere studies and plasma physics.

The importance of DR in these various applications has been recognized for a long time and has stimulated theoretical and experimental works. However the story of DR is full of controversy and of contradictory results. Even concerning the thermal DR rate coefficient, which has been extensively studied in different laboratories by a variety of techniques [1] [2] [3], there remain some disagreements and surprises as shown by the recent findings on the H_3^+ DR [4] [5] [6]. However for numerous ions the measured rate coefficients can be used with confidence.

In contrast with this situation, there is a sparcity of data and much confusion concerning the nature and the state of DR products. Results are available mainly for diatomic ions. The excited radiative states formed in DR of rare gas diatomic ions have been studied at Pittsburgh [7]. In pioneering experiments, Kley [8] and Zipf [9] [10] have studied respectively the production of metastable $N(^2D)$ from NO^+ and of $O(^1D,^1S)$ from O_2^+ DR. In this last case, rigorous theoretical calculations of the branching ratios were performed [11] [12].

Concerning polyatomic ions, some spectra induced by DR have been observed as the ultraviolet emission of CO in CO_2^+ DR [13] [14], but to our knowledge only two complete direct determinations of the neutral products have been performed: H_3^+ and its deuterated analogs [15] and H_2O^+ [16]. A rigorous calculation of the branching ratios of polyatomic ions DR would be a considerable task. Very recently, Bates, on the basis of qualitative considerations has inferred the nature of the DR products for several polyatomic ions [17] but his results are in contradiction with a pioneering treatment of Herbst [18] based on statistical phase space theory.

In this paper, studies of DR branching ratios made in Rennes University using a plasma flow tube apparatus are reviewed. They concern metastable production in diatomic ion DR (O_2^+, N_2^+), the influence of the vibrational excitation of CO_2^+ on the vibrational state of the $CO(A^1\Pi)$ product and the determination of the yield of oxygen and hydrogen atoms through H_2O^+ DR.

2 EXPERIMENT

The technique used in Rennes Laboratory has been described in detail elsewhere [16] [14] [19]. The fundamentals of the method lie in creating a highly ionized plasma flow where only one species of monoatomic ion is present (Ar^+,He^+, or N^+ usually). As long as the plasma ion is monoatomic, the electron density can decay only through collisional-radiative recombination, which is a very slow process in the present experiment. The studied diatomic or polyatomic ion is created by ion-molecule reaction with this primary parent monoatomic ion. Its destruction through DR happens then prior to any other reaction. This leads to a sharp drop of the electron density and the neutral atomic products are detected using V.U.V. absorption techniques.

The plasma flow is created in a water cooled tube which is pumped back by a roots blower. At the upstream end of the tube an arc plasma generator is set. The plasma generator exit is the throat of a copper nozzle which constitutes the anode and is at ground potential. Consequently, the plasma is a flowing afterglow with thermalized electrons. The parent gas is introduced near the exit of the tube using a circular injector directed against the flow. This design yields a faster conversion of the primary ions.

At the downstream part of the apparatus a flange holds either a langmuir probe or a quadrupole mass spectrometer. The langmuir probe is used here to measure electron temperature and density [20] and the mass spectrometer checks the nature of the ions involved in the chemistry.

The neutral temperature is determined from the rotational structure of the (0,0) band of the second positive system of N_2, (added for this purpose in very small amount to the flow). It is generally found to be very close to the electronic temperature prior to recombination, the high electron density ensuring thermalization between electrons and ions.

The product atom densities are deduced from the absorptions of the convenient lines emitted by a V.U.V. source [20]. For a quantitative analysis the knowledge of the line profile is required.

Measurements of the product atom densities for various initial electron densities and several plasma conditions allowed the determination of the branching ratios of several DR reactions as presented below.

3 RESULTS AND DISCUSSION

3.1 Metastable Formation In N_2^+(v=0,1) DR.

DR of N_2^+ ions is thought to be a major source of kinetically excited atoms in the martian atmosphere, therefore leading to nitrogen escape from Mars [21]. However atoms formed by DR have enough energy to escape only if the pair formed is (^{4}S, ^{2}D) for v=0.

This reaction was the first study conducted using the present technique [19]. N_2^+ ions are formed in an argon buffer gas by charge exchange from Ar^+ ions:

$$Ar^+ + N_2 \longrightarrow N_2^+ + Ar$$

These N_2^+ ions are then destroyed by dissociative recombination:

$$N_2^+ + e \longrightarrow N + N(^4S,^2D\, or\, ^2P)$$

In the plasma flow there is no sufficiently fast process to excite the nitrogen vibrational level. Absorption measurements using ArII lines have shown that the fine structure state of Ar^+ ions are in Boltzmann equilibrium at the plasma electron temperature. Consequently, the $P_{1/2}$ density represents at most 15% of the $P_{3/2}$ one in this experiment; moreover, the $P_{3/2}$ state reacts about twice as fast with nitrogen than $P_{1/2}$ [22] . Therefore N_2^+ ions are formed mainly by charge exchange between the $P_{3/2}$ state and N_2(v=0). This reaction has been extensively studied and is known to yield mainly N_2^+(v=1), [23] [24] [25] which can react back with Ar to give N_2.

N_2^+(v=1) is also known to be quickly relaxed on the ground vibrational level by resonant charge exchange reaction with nitrogen. However, in the present experiment, N_2^+ usually recombines prior to any other reaction except for a very large flow of parent nitrogen.

For a good analysis of the the experimental results, it is necessary to include the quenching of $N(^2P)$ and $N(^2D)$ by superelastic collisions. DR of nitrogen ions on the ground vibrational level can only produce the atom pairs ^{4}S+^{2}D, ^{4}S+^{2}P or ^{2}D+^{2}D, while the ^{2}P+^{2}D channel is very sligthly exothermic for N_2^+(v=1). It was found that in both cases the main channel is ^{2}D+^{2}D since the branching ratio is always more than 0.85 toward N(^{2}D). For vibrationally excited ions, the results seem indicate that a minor ^{2}P+^{2}D channel is open. These results support the thermospheric model of Frederick and Rusch [26] for the F-region, which leads to a very large branching ratio toward $N(^2D)$ in N_2^+ DR. They imply that the escape flux of nitrogen from Mars is less than previously calculated, but in better agreement with the observations (see the paper of J. Fox in these proceedings).

3.2 Metastable Formation In $O_2^+(v)$ DR.

O_2^+ is the major ionised species in a large part of the terrestrial, Martian and Venusian upper atmospheres. More than forty years ago, Bates and Massey [27] suggested that O_2^+ DR could explain the high rates of electron recombination observed in the ionospheric E and F1 regions. Since these early times, it has also been recognized that O_2^+ DR is responsible of the green and red lines yielded by metastable atomic oxygen $O(^1S)$ and $O(^1D)$ in the Earth airglow [28]. There is a lot of literature concerning atmospheric chemistry of metastable atomic oxygen and its link with O_2^+ DR. However, laboratory data are quite confusing [9] [10] because of the problem of the vibrational state of the ions. Recently, rigorous calculations of the branching ratios became available [11] [12].

In Rennes Laboratory, the DR products of O_2^+ ions have been studied [29] for various states of the ions. In a very complete set of experiments, Ar^+ was used as the primary ion. O_2^+ is then obtained by charge transfert and it is known that a quite peaked distibution around v=9 is obtained [30]. The situation is more complicated for oxygen than for nitrogen, since $O(^1D)$ is quenched not only in superelastic collisions with electrons but also in collisions with O_2 and Ar. Numerical simulations of various experimental results for different plasma conditions were necessary to obtain reliable branching ratios. A typical comparison between experimental values and numerical simulation is shown in figure 1. The different rate coefficients implicated in the numerical calculations are carefully chosen in the literature and values of the branching ratios are adjusted to give the best fit to the experimental data. The influence of the choice of the rate coefficient was also studied. The numerical model includes the vibrational quenching of O_2^+ by molecular oxygen but ions recombine largely before being quenched. The results are the following:

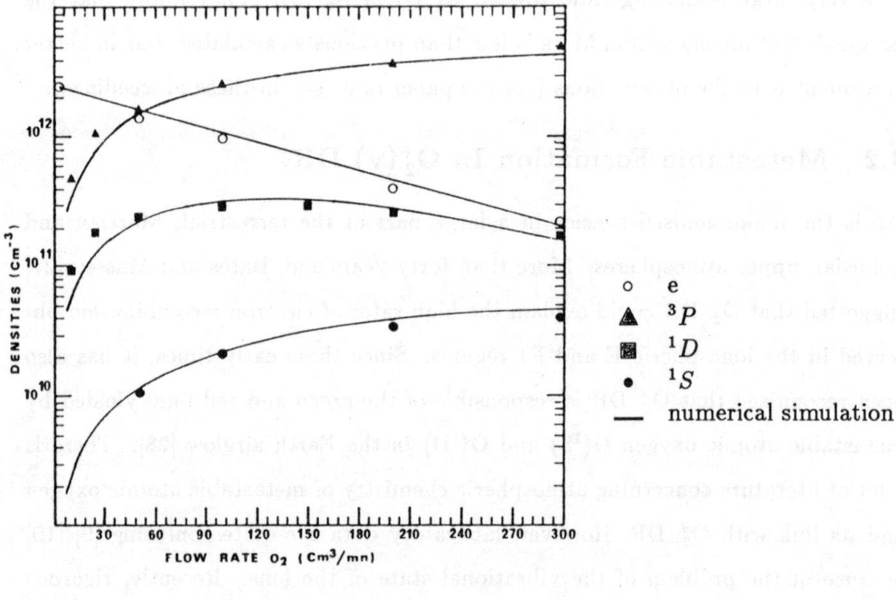

Figure 1: experimental values and numerical simulation.

- Concerning O(^1D): a branching ratio of 0.5 is found as giving the best fit. This result is quite independent of the choice of the quenching rates, with the exception of molecular oxygen. However the value of this rate coefficient is reliable [31] and a different value does not reproduce correctly the shape of the O(^{1}D) curve. This branching ratio is close to the previous determination of Zipf [9] for lower vibrational states supporting the idea that it is not strongly influenced by vibrational excitation.

- Concerning O(^{1}S): the deduced branching ratio is 0.22. However this value is proportional to the retained value for the quenching rate coefficient by electrons which results from a theoretical calculation [32].

In another set of experiments, Xe^+ was used as the primary ion. It was formed by charge exchange from He^+ in a helium buffer. Due to the plasma temperature it is possible to populate the v=0, 1, 2 states in these experiments, despite the endothermicity of the charge exchange between Xe^+ and O_2 for v=1, 2 formation.

Only the production of O(^{1}S) was studied, O(^{1}D) being quenched by Xe with a very large rate. Concerning O(^{1}S) an upper limit of 0.025 is found for the branching ratio from O_2^+(v=0) DR that is much lower than the widely used 0.1 value. It was not possible to obtain an unique solution for the three branching ratios from v=0,1,2 but it is rather interesting to note that the recent theoretical values of Guberman [12] reproduce the experimental data within the stated uncertainties.

3.3 Fourth Positive System Of CO In DR Of Vibrationally Excited CO_2^+

CO_2^+ is one of the major ions of both Venusian and Martian atmospheres [21] [13]. Its DR can lead to the $A^1\Pi$ state of CO, yielding the fourth positive system, a prominent emission feature of Venusian and Martian airglows.

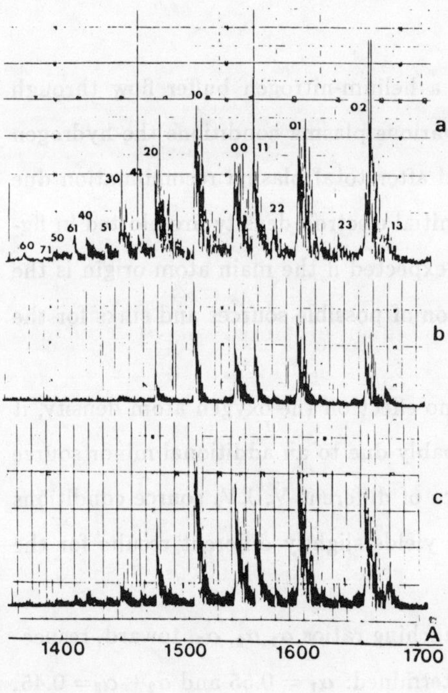

Figure 2:the fourth positive system of CO under various plasma conditions.

This DR problem was studied by Gutcheck and Zipf [13] but again the results were quite confusing since an analysis of their experimental conditions shows that the ions are vibrationally relaxed prior to DR while they attributed their observed spectrum to vibrational excitation.

Using the present apparatus the CO fourth positive spectrum obtained from CO_2^+ DR for various conditions of vibrational excitation has been observed [29]. The spectra are shown in figure 2. Under plasma conditions (a) CO_2^+ ions are formed by charge exchange from Ar^+ ions in an argon buffer with high electron density (3.7×10^{11} cm^{-3}). They are vibrationally excited and no quenching occurs prior to DR. In (c) conditions the ions are formed in the same way but some quenching occurs due to the lower electron density (1.8×10^{10} cm^{-3}).

In (b) conditions N^+ ions are used instead of Ar^+ as the primary ion and the initial vibrational excitation of CO_2^+ is less important due to a smaller exothermicity of the charge exchange. In this (b) case quenching prior to DR can also occur and it is clear from figure 2 that no levels of $CO(A^1\Pi)$ above v'=2 are significantly populated. In contrast, for (a) conditions levels up to v'=7 are observed. This shows that the $CO(A^1\Pi)$ vibrational distribution created in CO_2^+ DR is directly linked to CO_2^+ vibrational excitation, as has been suggested previously [13].

3.4 The Yield Of Oxygen And Hydrogen Atoms Through DR Of H_2O^+ Ions

In this case, H_2O^+ ions were formed in a helium-nitrogen buffer flow through charge exchange with N^+ ions [16]. For various plasma conditions the hydrogen and oxygen atom densities were measured after total plasma recombination due to H_2O^+ DR. These densities versus the initial electron density are plotted in figure 3, displaying a very good linearity as expected if the main atom origin is the plasma recombination. Careful examination of possible sources and sinks for the atoms confirm this conclusion [16].

As a variation of the water flow rate has no effect on the oxygen atom density, it has a little one for hydrogen. This is probably due to an additional minor source of H atoms in this experiment. The choice of different V.U.V. source conditions (i.e. of the line profile) for the Lyman α yields slightly different results for the densities too.

These measurements allowed the three branching ratios α_1,α_2, α_3, toward, respectively, OH+H, H_2+O and O+2H to be determined: $\alpha_1 = 0.55$ and $\alpha_2 + \alpha_3 = 0.45$, with $\alpha_2 < 0.21$ and $\alpha_3 > 0.24$. The limits are due to the uncertainties in the atoms measurements.

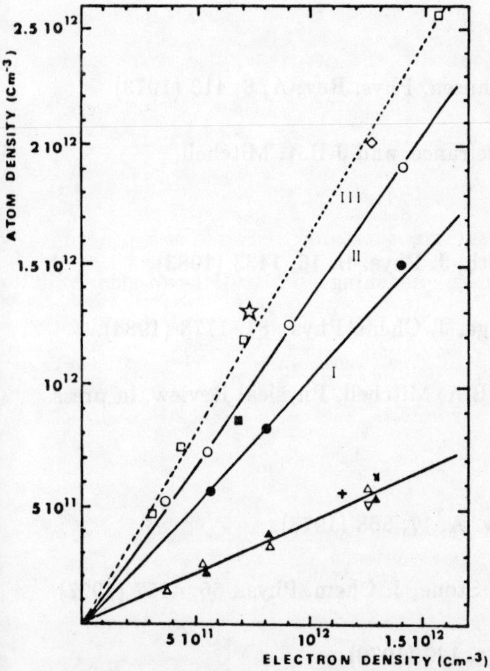

Because of the formation process, these results refer to vibrationally excited ions. They support Herbst's conclusions [18] that several channels are usually accessible in DR, in contradiction with the recent theoretical work of Bates [17].

Figure 3:Oxygen and hydrogen atom densities vs initial electron density. Hydrogen atoms □ ▩ ◇ ☆ O ●, oxygen atoms △▲+×. Unfilled:large flow rate (1.2 10^{-3}g/s); filled: little flow rate (1.4 10^{-4}g/s); □ O (II and III) two different line profiles for Lyman-α ∇ 1306.04 Å O1 line.

4 Conclusion

In this short review a technique has been presented which allows for the determination of product branching ratios for DR reactions. However, in its present state, it is usually limited to vibrationally excited ions.

In many applications, as astrochemistry, the ions are deexcited. New measurements using a FALP apparatus are in progress, in order to obtain results for ground state ions.

References

[1] M.T. Leu, M.A. Biondi and R. Johnsen, Phys. Rev. A, 8, 413 (1973).

[2] P.M. Mul, J.Wm. McGowan, P. Defrance, and J.B.A. Mitchell, J. Phys. B, 16, 3099 (1983).

[3] E. Alge, N.G. Adams, and D. Smith, J. Phys. B, 16, 1433 (1983).

[4] N.G. Adams, D. Smith, and E. Alge, J. Chem. Phys., 81, 1778 (1984).

[5] H. Hus, F. Youssif, A. Sen, and J.B.A. Mitchell, Physical Review, in press.

[6] T. Amano, Astrophys. J., in press.

[7] Y.J. Shiu, M.A. Biondi, Phys. Rev. A, 17, 868 (1978).

[8] D. Kley, G.M. Lawrence, and E.J. Stone, J. Chem. Phys., 66, 4157 (1977).

[9] E.C. Zipf, Bull. Am. Phys. Soc., 24, 129 (1970).

[10] E.C. Zipf, J. Geophys. Res., 85, 4232 (1980).

[11] S.L. Guberman, Planet. Space Sci., 36, 47 (1988).

[12] S.L. Guberman, Nature, 327, 408 (1987)

[13] R.A. Gutcheck and E.C. Zipf, J. Geophys. Res., 78, 5429 (1973).

[14] F. Vallée, B.R. Rowe, J.C. Gomet, J.L. Queffelec and M. Morlais, Chem. Phys. Lett., 124, 317 (1986).

[15] J.B.A. Mitchell, J.L. Forand, C.T. Ng, D.P. Levac, R.E. Mitchell, P.M. Mul, W. Claeys, A. Sen and J.Wm. McGowan, Phys. Rev. Lett., 51, 885 (1983).

[16] B.R. Rowe, F. Vallée, J.L. Queffelec, J.C. Gomet and M. Morlais, J. Chem. Phys. 88, 845 (1988).

[17] D.R. Bates, Astrophys. J., 306, L45 (1986).

[18] E. Herbst, Astrophys. J., 222, 508 (1978).

[19] J.L. Queffelec, B.R. Rowe, M. Morlais, J.C. Gomet and F. Vallée, Planet. Space Sci., 33, 263 (1985).

[20] M. Dudeck, G. Poissant, B.R. Rowe, J.L. Queffelec and M. Morlais, J. Phys. D Appl. Phys., 16, 995 (1983).

[21] J.L. Fox and A. Dalgarno, J. Geophys. Res., 88, 9027 (1983).

[22] T. Kato, K. Tanaka and I. Koyano, J. Chem. Phys. 77, 337 (1982).

[23] W. Lindinger, F. Howorka, P. Lukac, S. Kuhen, H. Villinger, E. Alge and H. Ramler, Phys. Rev. A, 23, 2319 (1981).

[24] D. Smith and N.G. Adams, Phys. Rev. A, 23, 2327 (1981).

[25] T. Kato, K. Tanaka and I. Koyano, J. Chem. Phys., 77, 834 (1982).

[26] J.E. Frederick and D.W. Rusch, J. Geophys. Res., 82, 3509 (1977).

[27] D.R. Bates and H.S.W. Massey, Proc. R. Soc., 192, 1 (1947).

[28] M. Nicolet, Phys. Rev., 93, 633 (1954).

[29] F. Vallée, J.L. Queffelec, B.R. Rowe, J.C. Gomet and M. Morlais, J. Chem. Phys. to be submitted.

[30] G. Mauclaire, R. Derai, S. Fenistein and R. Marx, J. Chem. Phys., 70, 4017 (1979).

[31] J.A. Davidson, H.I. Schiff, T.J. Brown, G.E. Streit and C.J. Howard, J. Chem. Phys., 69, 1213 (1978).

[32] K.A. Berrington and P.G. Burke, Planet. Space Sci., 29, 377 (1981).

PREDISSOCIATION OF 3s $^{1,3}\Pi_g$ AND $^{3,5}\Pi_u$ RYDBERG STATES OF O_2: MECHANISMS AND PATHWAYS

W. J. van der Zande, W. Koot, and J. Los
FOM Institute for Atomic and Molecular Physics
1098 SJ Amsterdam, The Netherlands

J. R. Peterson
Molecular Physics Department
SRI International, Menlo Park, CA 94025, USA

ABSTRACT

High-resolution translational spectroscopy is used to examine the dissociating states of O_2 produced by electron capture by keV O_2^+ ions from Cs. $^{1,3}\Pi_g$ vibrational levels up to v = 8, resulting from 3s capture by the O_2^+ $X^2\Pi_g$ ground state are predissociated by electronic configuration mixing or by spin-orbit coupling, and yield products at three different dissociation limits, direct evidence of strongly diabatic behavior at the avoided crossings of both singlet and triplet Π_g valence state potentials of the dissociating channels. Capture by the metastable $a^4\Pi_u$ ions in the beam form 3s $^{3,5}\Pi_u$ Rydberg states. The $^5\Pi_u$ is observed for the first time. The $^3\Pi_u$ decays mostly by autoionization.

1. INTRODUCTION

There are two important aspects of dissociative recombination (DR). The first and most thoroughly studied experimentally is the reaction rate, or cross section, which is important in determining the electron density and conductivity, and the relative populations of ion species in planetary ionospheres and molecular plasmas. The second is the distribution of ground and excited electronic final states of the dissociation products. These products are very important in controling the neutral chemistry and species population in the excited atmospheres following ionization and in causing optical emissions in airglows. A directly related aspect is the kinetic energy released by the dissociation to the final products, which can induce chemical reactions and also the escape of the products from planetary atmospheres.

Very little experimental work has yet been carried out to determine the final states of DR reactions because of the difficulty of identifying the neutral products, which often have very slow radiative decay rates. Most of the DR rates for atmospheric species have been determined in decaying plasmas, drift tubes and flow tubes at near-thermal temperatures where collisional quenching interferes with optical detection of the final states of the dissociations. However, two translational spectroscopic techniques have recently been used in fast molecular beams to determine the final states and products of dissociative electron capture from Cs by molecular ions. The decay of bound Rydberg states in these initially formed reactions resembles that in DR, but lower excited electronic states are initially formed. The research described here does not pertain directly to DR, but demonstrates a powerful technique for determining the final products, and provides an unusually detailed picture of some of the processes that control the predissociating mechanisms responsible for DR and also the interactions among possible competing channels of the outgoing "half-collisions," which determine the final product states.

The dissociative processes discussed here results from the capture of 3.9 eV-bound electrons by a fast molecular ion beam in the transit of a Cs vapor oven. Translational spectroscopy is used to determine the final product states by measuring the distribution of kinetic energy released to the fragments. The first such experiments on O_2 were performed by Peterson and Bae,[1] who concluded that excited valence states about 4 eV below both the $X^2\Pi_g$ ground state and the $a^4\Pi_u$ metastable state of O_2^+ were produced by near-resonant, Franck-Condon-favored electron capture transitions. Both of these ion states are present in most O_2^+ beams. The dissociations were found to yield ^3P and ^1D products with three main KER distributions peaking near 1, 3, and 7 eV. The work described here used a much higher resolution technique to examine the same reaction.

2. EXPERIMENTAL

The experimental technique used is a fairly recent development by the FOM Institute for Atomic and Molecular Physics in Amsterdam. It permits high resolution translational spectroscopic measurements of the barycentric kinetic energy released (KER) to the dissociating products, and is described by de Bruijn and Los.[2] It employs a localized source of excitation and dissociation of the molecule (in this case a charge-transfer cell, length ~0.1 cm), followed by a relatively long (>60 cm) drift path in which the dissociated fragments separate from the beam axis at angles that depend on the KER and the dissociation angles in the c.m. frame. After traversing the drift space, the two dissociation fragment atoms are detected by a position- and time-sensitive detector employing multichannel plates. The output signals are processed by a computer, which converts them to the radial positions and the difference in the arrival times of the two fragments, obtains a KER value for each event, and stores the data in memory.

The KER is the difference between the total internal energy (electronic, vibrational and rotational) of the excited molecule formed by the electron capture event just prior to dissociation, and the total electronic energy of the final dissociated products. The latter energy can usually be deduced, hence the former can be determined and the reaction pathways that connect the two can be identified.

3. RESULTS

The results of the initial experiment are shown in Figure 1. The KER spectrum has three major distributions peaking near 1, 3, and 7 eV as found by Peterson and Bae[1], but all are highly structured! Clearly, predissociating Rydberg states rather than valence states are produced in the electron capture from Cs. In addition, a smaller structure appears near 5 eV, which was unobservable earlier.

Discussion of the results is facilitated by reference to Fig. 2, which shows most relevant O_2 and O_2^+ potentials. Analysis of the spectra has shown that the electron capture from Cs yields predominantly Rydberg states formed by the addition of a 3s electron states, to the initial $X^2\Pi_g$ or $a^4\Pi_u$ ion core, thus forming $^1\Pi_g$, $^3\Pi_g$, $^3\Pi_u$, and $^5\Pi_u$, which are the lowest Rydberg states leading to each ionic limit. The vibrational structures in each group in Figure 1 consist of unresolved rotational states belonging to each vibrational level. Their spacings are similar to the respective ion cores, demonstrating their Rydberg characteristics. The potentials of only the $^3\Pi_g$ and $^5\Pi_u$ Rydberg states are shown in Fig. 2; their spin partners are slightly higher. These curves were obtained by shifting the parent ion curves downward to agree with the experimental energies. We will discuss the results in greater detail below.

3.1 Predissociation of the $d^1\Pi_g$ and $C^3\Pi_g$ Rydberg States.

During the last several years the first four vibrational levels of both of these states have been observed in various resonance-enhanced multiphoton ionization (REMPI) experiments,[3-6] which clarified some earlier observations in electron scattering experiments. Observation of v > 3 in REMPI is prevented by the vanishing Franck-Condon factors from the ground state. These experiments disclosed some effects of perturbations in both states, and varying linewidths indicated the effects of predissociation, However, there was no knowledge of the final products.

In our KER results, the 1 and 3 eV groups result from predissociation of the 3s $^{1,3}\Pi_g$ states to the $O(^3P)+O(^1D)$ and $O(^3P)+O(^3P)$ limits respectively, as concluded by Peterson and Bae;[1] however the details are much more complex than they perceived. An enlargement of the KER spectrum below 5 eV is shown in Fig. 3 (note the 3x amplification above 2.2 eV). These Rydberg states are predissociated by the $1^1\Pi_g$ and $1^3\Pi_g$ valence states. In the calculated potentials of Saxon and Liu,[7] shown in Fig. 2, each of these $1\Pi_g$ valence states has an avoided crossing with the corresponding $2\Pi_g$ valence state at R ~ 1.6Å. The crossings allow a distribution of final products to all of the three dissociation limits shown in Fig. 2, and in fact all are produced in varying degrees and by different mechanisms. We now discuss the predissociation of each Rydberg state.

3.1.1 The $C^3\Pi_g$ State: We first note in Fig. 3 that most KERs are in the 1 eV group. Most of these peaks are due to the $C^3\Pi_g$ state in vibrational levels v = 0 - 8, indicating that most of the "half-collisions" pass diabatically from the $1^3\Pi_g$ valence state to $2^3\Pi_g$ at the avoided crossing. Only 8% of the products follow the adiabatic channel, even though the radial kinetic energy is only about 1 eV at the crossing. From this probability and the slopes of the diabatic curves estimated from Ref. 3, we used the Landau-Zener formula to obtain an electrostatic coupling of 59 meV at the crossing.

The observed energy levels of the $C^3\Pi_g$ vibrational levels deduced from the KER values are in good agreement with those obtained from the REMPI experiments for v = 0 - 3. Levels up to v = 8 are seen to contribute to the KER spectrum in Figure 3. From these and other similar data taken in higher resolution, the absolute energies of $C^3\Pi_g$ v = 0 - 8 have been obtained from this KER group.[8] In addition, we have been able to deduce the coupling between the Rydberg state and the $1^3\Pi_g$ valence state to be 79 meV.[9] This result was achieved by applying the Fermi "golden rule" to a high-resolution measurement of the KER linewidth of 19 meV for v = 1.[9] As part of this analysis, the $1^3\Pi_g$ valence state potential of Saxon and Liu was shifted to shorter R by 0.025Å, in accordance with the results of Friedman and Dalgarno,[10] who optimized the location of the calculated potential in calculations of the decay rates to fit the linewidths from REMPI data.[3,4] In predissociations involving potentials that intersect on the outer radius of the bound state, the decay rates can vary greatly with each vibrational level, as they depend on the Franck-Condon factors between each $^3\Pi_g$ vibrational level and the continuum of the dissociating state.

The first two vibrational levels of $^3\Pi_g$ are also seen in the 3 eV group, which dissociates to the $O(^3P)+O(^3P)$ limit. As noted above, they are only 8% of the amounts in the 1 eV group. Above v = 1, they should also contribute, but are buried by the contributions from $d^1\Pi_g$, which will be discussed.

Finally, in the spectrum just above 5 eV (not shown) a peak is identified as resulting from $C^3\Pi_g$ v = 9, which is evidently predissociated by the higher $2^3\Pi_g$ valence state and exits diabatically to ^3P + ^3P. This observation is consistent with the Saxon-Liu potentials shifted by ΔR = 0.024 Å, as suggested by Friedman and Dalgarno.[10]

3.1.2 The $d^1\Pi_g$ State: We first were surprised to find that the first two KER peaks above 1 eV are actually double. This fact was confirmed in higher resolution data.[8,9] Contributions are not only by $C^3\Pi_g$ but also by $d^1\Pi_g$, which is about 80 meV higher in energy at v = 0. Dissociation to this limit can only be via the $1^3\Pi_g$ valence state, thus apparently the coupling to the more distant $1^1\Pi_g$ is weaker than the spin-orbit coupling to $C^3\Pi_g$ (and thence to $1^3\Pi_g$) for v = 0 and 1. Actually, some of part v = 1 is predissociated by $1^1\Pi_g$ and appears in the 3eV group, along with all of v = 2 and 3. These states are all energetically lower than the O(^{1}D)+O(^{1}D) limit, and thus are forced to take the adiabatic route at the 1-$2^1\Pi_g$ avoided crossing, to O(^{3}P)+O(^{3}P).

However, in Fig. 3 peaks can be seen from v = 5 to 8 at very low KER values below 1 eV, which indicate that when energetically possible the predissociation by $1^1\Pi_g$ also proceeds at least partly diabatically ^{1}D + ^{1}D at the avoided crossing. By reducing the beam energy and lengthening the flight path before the detector, the low KER data in Fig. 4 were obtained, showing clearly that the conjecture was correct.[8,11] We assume that the predissociation above v = 4 goes predominantly diabatically across the 1-$2^1\Pi_g$ crossing as in the triplet case. Not only does v = 4 of the mass 16 O_2 isotope, which is only 94 meV above O(^{1}D)+O(^{1}D), go to that limit, but measurements with $^{18}O_2$ show that v = 4 dissociates similarly with a KER of only 38 meV.[11] It should be noted that the kinetic energy at the avoided valence crossing is almost 1 eV for both isotopes.

The structure in the high energy tail of the v = 4 peak is real and due to individual rotational states. Figure 5 shows part of a KER spectrum obtained with a 4 keV ion beam. The resolved rotational lines are clearly visible. This spectrum has been modeled[11] using the Gaussian apparatus function and three adjustable parameters which determined energy of J = 1 with respect to $O(^1D)+O(^1D)$, the rotational constant, and the Boltzmann rotational temperature of 550 K. The result is shown as the dashed line in Fig. 5.

For states higher than v = 4, the rotational lines were unresolved because of the instrument width. However, the decay widths for low J could be established in several cases by fitting the low-energy onset. The shape reflected a width exceeding that of the apparatus function. The Fermi golden rule, which approximates the natural decay widths, can be separated into a product of an R-independent electronic Hamiltonian H_{el} and the bound-free Franck-Condon factors. Because the F-C factors are strongly dependent on the relative locations of the repulsive valence potential and Rydberg potentials, the variation of the decay widths from level to level can be used to optimize the location of the valence potential (whose calculated values have some uncertainty). The $1^1\Pi_g$ potential of Saxon and Liu was allowed to shift both vertically and radially in order to optimize the fit of the F-C factors with the variation in decay width. A shift to smaller R by 0.039 Å and downward by 320 meV gave the best fit,[11] and is within the uncertainties estimated by Saxon and Liu.[7] Using this shift, the magnitude of H_{el} = 55 meV was calculated from the observed decay width of 10 cm^{-1} for $^{16}O_2$, v = 4.[11] From this value of H_{el}, the decay widths of all vibrational levels v = 4-8 for both mass 16 and 18 isotopes were calculated. The decay widths determined from mass 16 were found also to hold for the different values of $^{18}O_2$, and even explained an anomously small contribution from $^{18}O_2$, v = 6.[11], whose predicted F-C factor was only 5×10^{-4} of those levels nearby.

The nearly diabatic nature of the $^1\Pi_g$ valence state explains variations in the rotational constant B_v for low v's observed in REMPI experiments.[3,6] While B_0 is 1.68 cm^{-1}, identical to that of the $X^2\Pi_g$ ion and characteristic of a true Rydberg state, the v = 1 and 3 levels have reduced B values of 1.57 and 1.16 cm^{-1} (v = 2 was not identifiable). The reduced B suggests that these levels, which cannot dissociate to ^1D + ^1D, have vibrational wavefunctions with substantial amplitudes at larger R than the pure Rydberg states. For v < 4, these states are effectively bound by a double-well potential formed by the $d^1\Pi_g$ state and the diabatic 1-2 $^1\Pi_g$ valence potential, until they ultimately decay adiabatically via the avoided valence-state crossing. For v = 4, however, our measurements show that B_4 = 1.58 cm^{-1}. Thus the levels above $O(^1D)+O(^1D)$ dissociate to it diabatically and have Rydberg-like B_v's, with decay widths determined only by the Rydberg-valence coupling.

3.2 The $^{3,5}\Pi_u$ States

These states are formed by the addition of a 3sσ orbital to the metastable $a^4\Pi_u$ ion core, which populates up to 50% of ion beams, depending on the ion source conditions.[12] The $H^3\Pi_u$ state has been observed to cause autoionizing resonances in photoionization studies,[13-15] but the 3s $^5\Pi_u$ state is previously unreported.

The KER spectrum from the 3s $^5\Pi_u$ state occupies the 7 - 8.5 eV part of the spectrum in Fig. 1. The relative shape of the spectrum depends on the ion source electron energy. The bimodal shape of the spectrum in Fig. 1 results from both direct ionization, which peaks near v = 4, and a low-v component from $b^4\Sigma_g^- \rightarrow a^4\Pi_u$ cascading in the ion source, this process creates predominantly $O_2^+(^4\Pi_u)$, v = 0 ions. (It is tacitly assumed that the vibrational distributions in the Rydgerg states are closely representative of those in the parent ions in the beam, as has been found to characterize H_2^+ + Cs electron capture.[16]) The vibrational constants ω_e = 1008 ± 40 cm^{-1}, $\omega_e x_e$ = 40 cm^{-1} were determined from the peak positions, and are close to those of $a^4\Pi_u$ (1035 and 10 cm^{-1}).[12] From the KER values it is clear that the state dissociates to $O(^3P)+O(^3P)$ via the $^5\Pi_u$ valence state shown in

Fig. 2. The term value T_0 is 12.13 eV, which is 4.0 eV below $a^4\Pi_u$, so the electron capture is near-resonant (IP(Cs) = 3.89 eV).

The $H^3\Pi_u$ state was found to have T_0 = 12.353 eV,[14] thus it lies 220 ± 100 meV above 3s $^5\Pi_u$. Although this state has been oberved in photoionization spectra,[13-15] their interpretation has been hampered by irregularities. We assign the small structures in the 5 - 6 eV region of Fig. 1 to contributions from the predissociation of $H^3\Pi_u$, v = 1 - 4, by a valence state that correlates to the $O(^3P)+O(^1D)$ limit.[9] In fact, the 2, 3, and 4 $^3\Pi_u$ valence states all penetrate the $H^3\Pi_u$ Rydberg potential close enough to the minimum that any or all could take part. In any case, the term values T_v obtained from a higher resolution KER spectrum agree very well with those from photoionization,[13,14] and the relative KER intensities explain some relative intensities missing from those spectra.[12] The fact that predissociation competes with the usually faster autoionization is interesting, but understanding it will require a detailed analysis.

3.3 Vibrational Excitation in the Ion Beam

As mentioned earlier, the vibrational distributions in the Rydberg states observed here are assumed to be very similar to those in the two parent electronic states in the ion beams. The strong propensity for $\Delta v = 0$ in near-resonant electron capture transitions to these states results both from the near-diagonal nature of the F-C factors and from the "softness" of the collision, which allows the F-C factors to have major influence. This propensity has been demonstrated in H_2 by de Bruijn et al.[16] This characteristic implies that the rovibrational population in the Rydberg levels represents that in the ion beam, and thus can be used to monitor conditions in the ion source. We find that the populations in the $a^4\Pi_u$ state in the ion beam result from F-C factors with the O_2 ground state in direct ionization plus contributions from radiative cascading from $b^1\Sigma_g^-$.[12] For the $X^2\Pi_g$ state, however, the extensive population beyond v = 3 cannot be explained by direct ionization or cascading. From experiments in which the ion source electron energies were varied we conclude that the population of these levels in the ion beam (and thus the 3s $^{1,3}\Pi_g$ states) results from autoionization of ns $^3\Pi_u$ states excited by electron impact in the ion source.[9] The H 3s $^3\Pi_u$ and J 3s $^3\Pi_u$ states have been observed in photoionization spectra.[13,14] The substantial effects of autoionization on the population of higher vibrational levels by electron impact has not been observable before.

4. SUMMARY

This study of the KER spectra from predissociating Rydberg states has provided a wealth of new information on the location and properties of all four 3s Rydberg states in O_2, as well as on their predissociation mechanisms and final products. In addition, new information was obtained on the Rydberg-valence and also the valence-valence couplings, as well as on the extensive vibrational excitation in O_2^+ produced by electron impact. It has also demonstrated an effective technique that can be used for the determination of the dissociative recombination pathways and products, using fast beams.

ACKNOWLEDGEMENTS

This was supported by the Netherlands Foundation for Fundamental Research on Matter (FOM) and was made possible by a financial support of the Dutch Organization for the Advancement of Research (NWO), JRP was supported by the National Science Foundation and the Air Force Office of Scientific Research.

REFERENCES

1. Peterson, J. R., and Bae, Y. K., Phys. Rev. A. 30 2807 (1984).
2. deBruijn, D. P., and Los, J., Rev. Sci. Instrum., 52, 1020 (1982).
3. Sur, A., Ramana, C. V., and Colson, S. D., J. Chem. Phys. 83 3429 (1984).
4. Sur, A., Ramana, C. V., Chupka, W. A., and Colson, S. D., J. Chem. Phys. 84, 69 (1986)
5. Katsumata, S. Sato, K., Achiba, Y., and Kimura, K., J. Electron Spectrosc. 41, 325 (1986).
6. Johnson III, R. D., Long, G. R., and Hudgens, J. W., J. Chem. Phys. 87, 1977 (1987).

7. Saxon, R. P. and Liu, B., J. Chem. Phys. 67 5432 (1977).
8. van der Zande, W. J., Koot, W., Peterson, J. R., and Los, J., Chem. Phys. Lett. 140 283 (1987)
9. van der Zande, W. J., Peterson, J. R. and Los, J., to be submitted to J. Chem. Phys.
10. Friedman, R., and Dalgarno, A., Bull. Am. Phys. Soc. 32, 1254 (1987).
11. van der Zande, W. J., Koot, W., Peterson, J. R., and Los, J., submitted to J. Chem. Phys. (1988).
12. van der Zande, W. J., Koot, W., Peterson, J. R., and Los, J., submitted to Chem. Phys. (1988).
13. Dehmer, P. M., and Chupka, W. A., J. Chem. Phys. 62 4525 (1984).
14. Nishitani, I. Tanaka, Tanaka, K., and Koyano, I., J. Chem. Phys, 81 241 (1975).
15. D.M.P. Holland, and West, J. B., Z. Phys. D, Atoms, Molecules and Clusters 4, 367 (1987).
16. de Bruijn, D. P., Neuteboom, J., and Los, J., Chem Phys. 85 233 (1984).

FIGURES

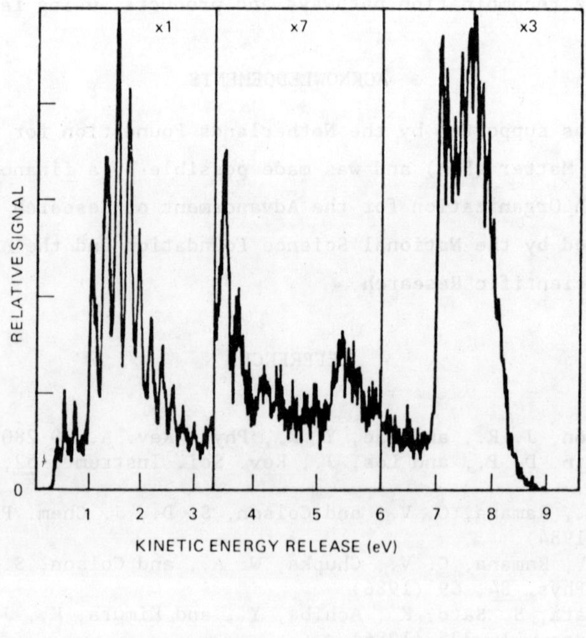

Figure 1. The KER spectrum from 4 keV O_2^+ + Cs beam.

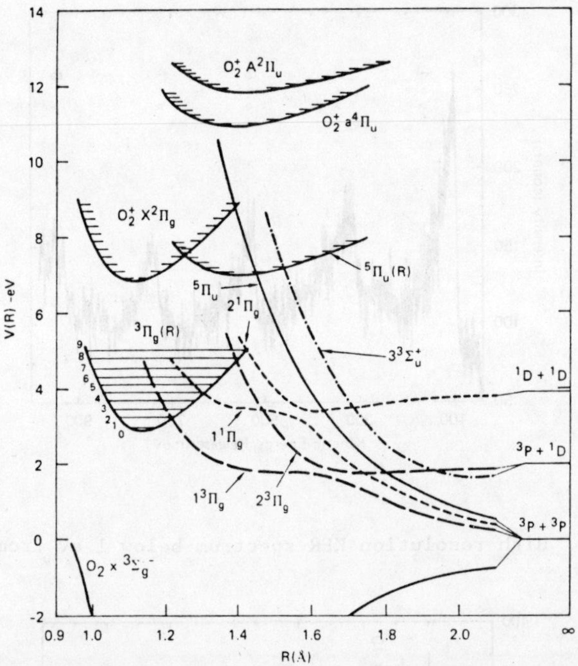

Figure 2. Potentials of selected states. The $^1\Pi_g$ and $^3\Pi_u$ Rydberg states lie just above their partners.

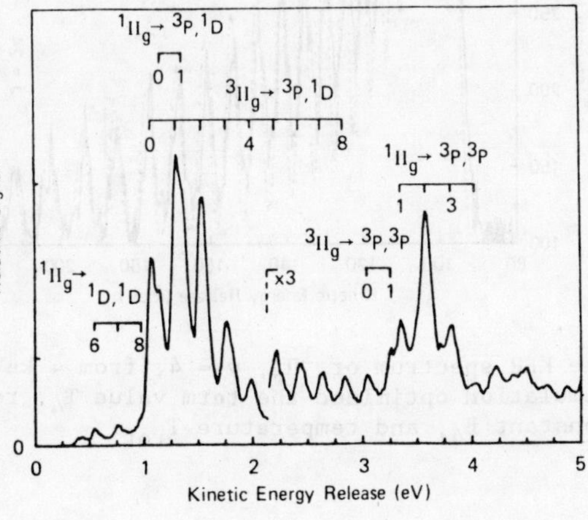

Figure 3. 4 keV O_2^+ KER spectrum below 5 eV.

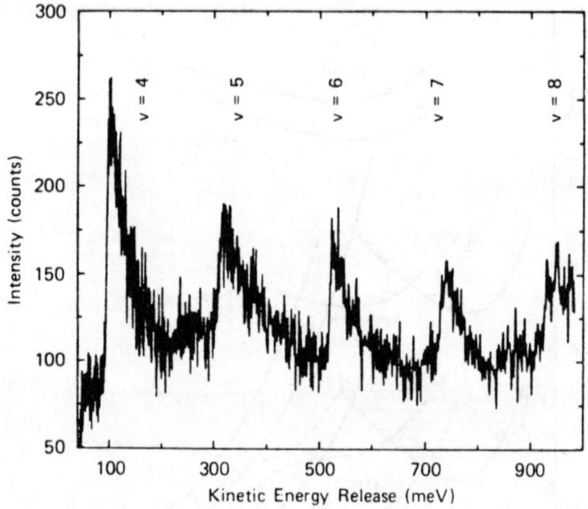

Figure 4. High resolution KER spectrum below 1 eV from 7 keV O_2^+.

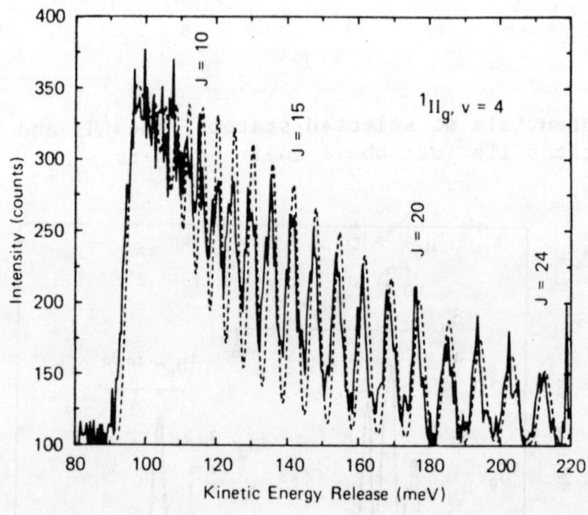

Figure 5. The KER spectrum of $^1\Pi_g$, v = 4, from 4 keV O_2^+. Dashed simulation optimized the term value T_4, rotational constant B_4, and temperature T_{rot}.

LASER STIMULATED RADIATIVE RECOMBINATION: A FIELD IONIZATION DETECTION APPROACH

T.J. Morgan, Department of Physics,
Wesleyan University, Middletown, CT 06457
and
J.B.A. Mitchell, Department of Physics
The University of Western Ontario
London, Ontario, Canada. N6A 3K7

ABSTRACT

An experiment is planned to study electron-proton radiative recombination enhanced by a CO_2 laser which will stimulate capture into the n = 10 or n = 11 states. These high levels are easily ionized by electric fields and this provides a means of specific detection of the recombination products.

INTRODUCTION

In the last few years considerable interest has been expressed in experiments aimed at producing neutral anti-hydrogen by combining positrons with antiprotons from a storage ring such as LEAR. Several approaches to this problem have been presented[1,2,3,4,5] and it seems reasonable to assume that $\overline{H}$ formation rates up to $10^8 s^{-1}$ can be achieved given current technology[6]. The present paper addresses one of these approaches, namely the laser stimulated recombination of antiprotons and positrons in a merging beam configuration. Such an experiment has been proposed for the LEAR facility[3,5] but before resources are allocated to such a project it is desirable to provide a small scale laboratory demonstration of the principle using normal matter (i.e. protons and electrons). The merged beam experiment at U.W.O. is ideally suited for such a demonstration and a proposed experimental program is described in the following.

The cross section for e + p recombination may be expressed in the form[7]:

$$\sigma = A \frac{\nu_0}{\nu} \cdot \frac{h\nu_0}{\frac{1}{2}mv^2} \cdot \frac{g}{n^3} \qquad (1)$$

where $A = 2.11 \times 10^{-22}$ cm^2, $g \sim 1$ and $h\nu = 13.6$ eV. n is the principal quantum number, ν is the frequency of the emitted radiation and m and v are the electron mass and velocity respectively. $h\nu = \frac{1}{2} mv^2 + h\nu_0/n^2$ and so it is seen that $\sigma \sim 1/(nE_e)$ where E_e is the kinetic energy of the incoming electron assuming the proton to be stationary. It can be seen from equation 1 that the cross section for radiative recombination, even for low collision energies, is very small. For example for $E_e = 0.010$ eV, $\sigma(n = 1) = 2.9 \times 10^{-19}$cm^2 while $\sigma(n = 10) = 2.9 \times 10^{-20}$cm^2.

Neumann et al[3] have proposed that the neutralization rate can be enhanced through the process of stimulated radiative recombination, namely: -

$$h\nu + e + p \rightarrow H\ (n) + 2h\nu$$

where the stimulating photon, supplied by a laser, has the same frequency as the emitted photon. They showed that one can define a gain factor g in the form: -

$$g = \frac{\sigma(\text{STIM})}{\sigma(\text{SPON})} = \frac{Pc^2}{F\Delta\nu 8\pi h\nu^3} \qquad \text{- - (2)}$$

where σ(STIM) and σ (SPON) are the cross sections for the stimulated and spontaneous radiative recombination processes, P is the laser power in watts, c the speed of light, F the cross sectional area of the laser beam, h being Planck's constant, ν the laser frequency and $\Delta\nu$ is determined primarily by the velocity spread Δv of the electrons.

$$\Delta\nu = \Delta v \cdot \frac{mv}{h} \qquad \text{- - (3)}$$

For the enhancement to be successful, the velocity spread of the collisions should be as small as possible and in order for σ(SPON) to have a reasonable value, v should also be small. The merged beam technique has the ability to satisfy both these conditions[8].

The enhancement factor scales as ν^{-3} and so it is attractive to consider pumping down from the ionization continuum to a high n state (see fig. 1). Indeed if one considers an electron-proton collision energy of 0.01 ev, then pumping to the n = 10 or n = 11 level of H can be accomplished using a carbon dioxide laser. This is obviously an asset since high powered dc CO_2 lasers are readily available commercially.

Two problems do however arise. Once formed, the excited H atoms are very susceptible to reionization, by photoionization, if too high a laser power is used, and in an actual experiment, to field ionization as the $\overline{H}$'s pass through the magnetic fields of the storage ring.

It can be shown[3)] that the maximum laser power to avoid subsequent photoionization is given by: -

$$P = \frac{Fh\nu}{\tau\sigma_{ph}}$$

where τ is the time during which the excited state is illuminated and σ_{ph} is the photoionization cross section. For H (n = 10) the natural lifetime is ~ 1.9 μs. τ can however be reduced by illuminating the ion beam over a restricted area. Since the ions are moving rapidly in the laboratory frame the exposure time can be made very short.

The field ionization is a more severe problem. The electric field required to ionize a hydrogenic atomic in a given state n may be expressed as[(9)]: -

$$E = \frac{3.2 \times 10^{10}}{n^4}$$

In the context of the antiproton experiment this is the Lorentz field q $\overline{v}$ x $\overline{B}$ encountered by the fast particles passing through the bending magnets of the storage ring. One solution is to reduce the operating energy of the antiproton storage ring although this might lead to a deterioration of stored beam quality. An alternative solution has been proposed by Conti[(10)] and involves using a second laser beam to pump the excited state of $\overline{H}$ down to a lower state which would be immune to field ionization. It is intriguing to envisage using the same laser to both enhance the recombination and to pump to a lower state, (see fig. 1). Since the pump down only occurs when the photon energy corresponds to the energy separation of the two levels, the same laser can only be used for fixed or 'magic' energies. Thus for example stimulation down to the n = 10 level followed by pump down to the n = 7 level can occur for $E_e \sim 6$ meV. If a CO laser was used, stimulation to n = 9 followed by pump down to n = 6 occurs for $E_e = 40$ meV.

An experiment is planned employing electrons and protons as the reactants and with a modest power (10 watt) CO_2 laser to test the stimulated recombination process. The problem with such an experiment lies in the fact that a large background signal of H atoms will be formed via the charge exchange of protons on the residual gas in the collision chamber. A novel technique however can be used to overcome this problem.

The purpose of the experiment is to demonstrate that the use of the CO_2 laser will lead to an enhancement of the production of H(n = 10) atoms. It is sensible therefore to detect only atoms with n ≥ 10 and this may be accomplished by using a field ionizer to reionize the atoms formed via recombination, the resulting protons being detected ,fig. 2. The cross section for charge exchange scales as n^{-3} so the H(n = 10) population formed in this way will be 1/1000 of the total H background formation rate and so can be neglected.

The objectives of the experiment will be: -

(a) To demonstrate the laser enhancement of the radiative recombination.

(b) To investigate how this scales with laser power.

(c) To demonstrate pump down. This should be apparent as a drop in the H(n = 10) signal as the center of mass energy is scanned through the 'magic' energy.

<u>OTHER EXPERIMENTS</u>

The technique of field ionization detection of recombination products should prove of great value for other measurements. In particular the measurement of branching ratios for dissociative recombination decay channels is of great importance and field ionization detection can be used to look for channels leading to highly excited atoms or molecules. Recent studies in our laboratory have shown that for H_3^+ recombination, the channel leading to the formation of metastable rydberg H_3 molecules accounts for about 15% of the total cross section. Field ionization would allow us to identify exactly which states are being formed. Earlier studies[12)] on H_2^+ recombination indicated that reactions such as

$$e + H_2^+ (v) \rightarrow H(n\ell) + H(v)$$

where $n > 4$ accounted for a large percentage of the total recombination process. Whereas the ionizer to be used initially in the proposed equipment can easily ionize states with $n \geq 9$, a new design currently being studied at Wesleyan University should allow states with $n \geq 6$ to be detected. This could ultimately be installed in the apparatus.

In addition to radiative and molecular ion recombination, the field ionization detection technique can be used in the study of dielectronic recombination of multielectron ions[13)] and again this opens up a wide range of interesting experiments.

ACKNOWLEDGEMENTS

Many of the ideas presented in this paper have arisen from discussions with Drs. A. Rich, R. Conti, D. Gidley, and G. Gabrielse . The authors would also like to acknowledge the encouragement received from the U.S. AFOSR, U.S. AFAL and the RAND Corporation.

REFERENCES

1. Humberston, J.W., M. Charlton, F.M. Jacobsen and B.I. Deutch, J. Phys. B. 20, L25, 1987.
2. Gabrielse, G., L. Haarsma, S.L. Ralston and W. Kells, Laser Spectroscopy 8 (ed. W. Persson and S. Svandberg, Springer Verlag, N.Y. 1987, p. 26.
3. Neumann, R., H. Poth, A. Winnacker and A. Wolf, Z. Phys. A., 313, 253, 1983.
4. Rich, A., R. Conti, W. Frieze, D.W. Gidley, M. Skalsey, T. Steiger, J. Van House, H. Griffin, W. Zheng and P.W. Zitzewitz, To be published in Proc. NATO Adv. Research Workshop, "Atomic Physics with Positrons", University College, London. 1987.
5. Poth, H., Appl. Phys. A43, 287, 1987.
6. Mitchell, J.B.A., in Proceedings of RAND Workshop on Antiproton Science and Technology, Santa Monica, CA, 1987, ed. B. Augenstein.

7. Menzel, D.H., App. J. 85, 330, 1937.

8. Auerbach, D., R. Cacak, R. Caudano, T.D. Gaily, C.J. Keyser, J. Wm. McGowan, J.B.A. Mitchell and S.F.J. Wilk, J. Phys. B. 10, 3797, 1977.

9. Brouillard, F., in Atomic & Molecular Processes in Controlled Thermonuclear Fusion, ed. C.J. Joachain and D.E. Post, (Plenum Press, NY 1983).

10. Conti, R.S. (Private Communication) 1988.

11. Hus, H., F. Yousif, A. Sen and J.B.A. Mitchell, Phys. Rev. Lett. 60, 1006, 1988.

12. Phaneuf, R.A., Crandall, D.H. and Dunn, G.H., Phys. Rev. A11, 528, 1975 and Vogler, M. and Dunn, G.H., Phys. Rev. A11, 1983, 1975.

13. Muller, A., Belic, D.S., De Paola, B.D., Dunn, G.H., Mueller, D.W. and Timmer, C., Phys. Rev. Lett. 56, 127, 1986.

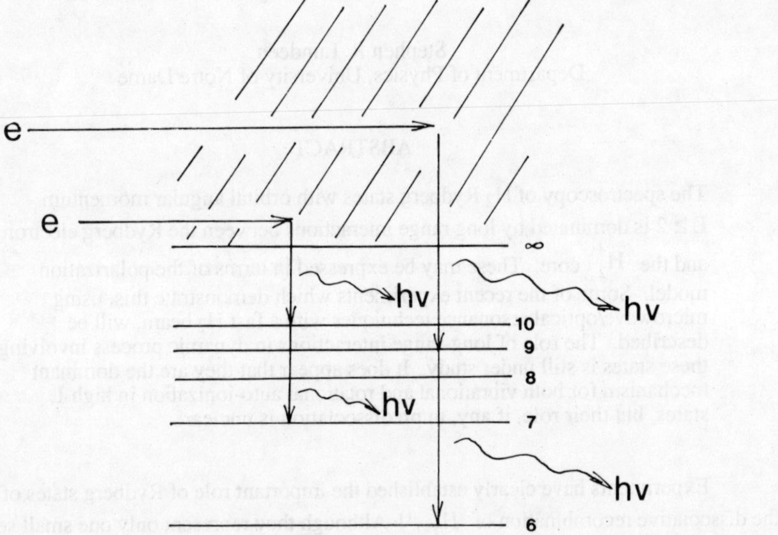

Figure 1 Stimulated recombination of e + H^+ followed by de-excitation to a lower level.

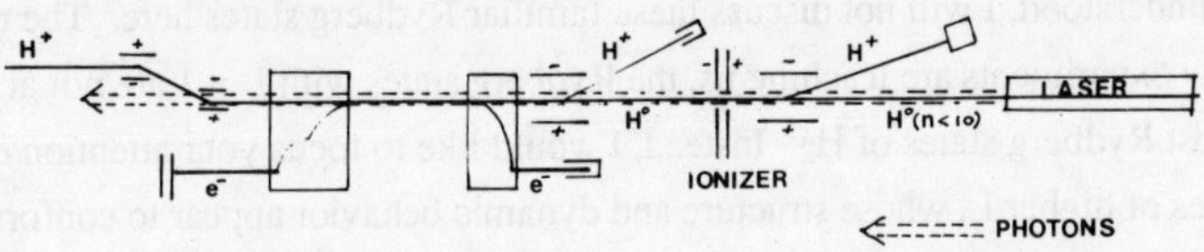

Figure 2 Schematic of merged beam, normal matter simulation experiment at U.W.O.

LONG-RANGE INTERACTIONS IN H_2 RYDBERG STATES

Stephen R. Lundeen
Department of Physics, University of Notre Dame

ABSTRACT

The spectroscopy of H_2 Rydberg states with orbital angular momentum $L \geq 2$ is dominated by long range interactions between the Rydberg electron and the H_2^+ core. These may be expressed in terms of the polarization model. Some of the recent experiments which demonstrate this, using microwave/optical resonance techniques with a fast H_2 beam, will be described. The role of long-range interactions in dynamic process involving these states is still under study. It does appear that they are the dominant mechanism for both vibrational and rotational auto-ionization in high-L states, but their role, if any, in predissociation is unclear.

Experiments have clearly established the important role of Rydberg states of H_2 in the dissociative recombination of H_2^+.[1] Although they represent only one small sector of the configuration space for the four-particle system H_2, the study of these states is itself a rich and challenging field of investigation which is advancing rapidly, primarily due to the application of new laser-based experimental techniques. Until recently, almost all experimental knowledge about H_2 Rydberg states was derived from studies of the nP series, accessible from the ground state in absorption.[2] These experiments have been very fruitful, stimulating the development of Multi-Channel Quantum Defect Theory (MQDT)[3] and the first theories of vibrational auto-ionization.[4] Although these experiments continue to be improved, and there are certainly still aspects of the behavior of the nP series which are not well understood, I will not discuss these familiar Rydberg states here. The reason is that, as new experiments are teaching us, the Rydberg states with L = 1 are not at all typical of most Rydberg states of H_2. Instead, I would like to focus your attention on the Rydberg states of higher L, whose structure and dynamic behavior appear to conform to a pattern with much wider applicability. Thus, if we wish to develop an appreciation of the general features of H_2 Rydberg state behavior, consideration of the higher L states may be instructive.

In particular, I will call your attention to two widely held misconceptions about H_2 Rydberg states which are actually inappropriate generalizations from the studies of nP states.

1) The quantum defects of H_2 Rydberg states are due predominantly, to short-range interactions.
2) MQDT is necessary to understand the behavior of these states because of the importance of inter-series mixing.

In fact, experimental studies now seem to indicate that almost all H_2 Rydberg states, all but those of the lowest L (L= 0, 1) may be understood quite successfully without the use of MQDT and considering only long-range interactions. The picture which emerges from these studies is therefore quite different and complementary to the intuition developed from studies of the P states. I will outline this picture for you this afternoon and briefly describe some of the experiments which have helped to establish it. Then I will consider the implications of this picture for understanding the dynamic behavior of the "general" Rydberg state of H_2, including the processes of auto-ionization and predissociation.

One major qualitative difference in the description of high-L Rydberg states is that they are characterized by different quantum numbers than are usual for molecular states. Rydberg electrons in non-penetrating high-L orbits are coupled so weakly to the orientation of the H_2^+ internuclear axis that it is a good approximation to assume, as illustrated in Figure 1, that the H_2^+ ion rotates freely within the orbit of the Rydberg electron, which in turn sees an almost spherically symmetric potential. In this case the eigenstates of the system would be characterized by the vibrational (ν) and rotational (R) quantum numbers of the core, by the principal quantum number (n) and orbital angular momentum (L) of the Rydberg electron, and by the vector sum of the angular momenta of the two parts of the system $\vec{N} = \vec{R} + \vec{L}$. We denote such a state by:

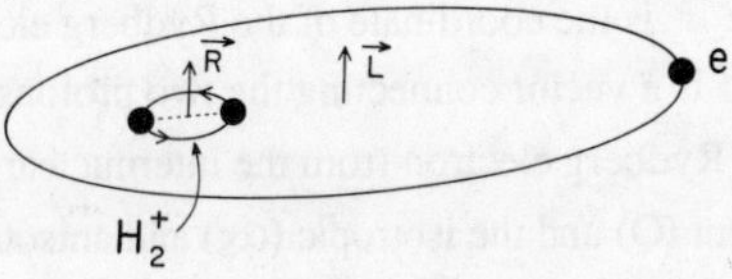

Fig. 1. Diagram illustrating the angular momentum coupling scheme characterizing high-L H_2 Rydberg states. Both the orbital angular momentum of the Rydberg electron (L) and the angular momentum of the H_2^+ ion (R) are good quantum numbers, which add vectorially to give $\vec{N} = \vec{R} + \vec{L}$, the total angular momentum (exclusive of spin).

$$(\nu, R)nL_N$$

The wavefunctions have an "atom-like" character, with the rotational angular momentum of the ion core $(\vec{R})$ playing the role of the "spin" of the atomic nucleus. These states are sometimes referred to as "Hund's case d" states or "L-uncoupled" states. They also describe some high n members of the nP Rydberg series, although series mixing effects usually prevent them from being eigenstates. In contrast, for high-L states, these states form a basis which is very near to the eigenvectors except in the case of accidental very close coincidences between members of the different series.

This pattern may be established formally by writing the Hamiltonian of the system as

$$H = H^o(H_2^+) + H^o(H) + V \qquad (1)$$

where $H^o(H_2^+)$ is the Hamiltonian for the free H_2^+ ion, $H^o(H)$ is the hydrogenic Hamiltonian which describes the Rydberg electron's motion, and V is the remainder, which describes coupling between the two parts of the zeroth-order Hamiltonian. The zeroth-order wavefunctions are products of eigenstates of H_2^+ and H. The structure of these states is largely accounted for by including only long range interactions between the Rydberg electron and the H_2^+ core. The dominant long-range interactions, the quadrupole energies and the adiabatic dipole polarization energies may be represented by an effective potential seen by the Rydberg electron in addition to the dominant Coulomb potential. Thus we take for V in Eq. (1):[5),6)]

$$V_{eff}(\vec{\rho}, \vec{r}) = \frac{-eQ(\rho)}{r^3} P_2(\cos\theta) - \frac{\alpha_S(\rho)}{2r^4} - \frac{\alpha_T(\rho)}{3r^4} P_2(\cos\theta) \qquad (2)$$

where $\vec{r}$ is the coordinate of the Rydberg electron relative to the nuclear center of mass and $\vec{\rho}$ is a vector connecting the two protons. The angle θ gives the angular displacement of the Rydberg electron from the internuclear axis ($\cos\theta = \hat{r}\cdot\hat{\rho}$). The quadrupole moment (Q) and the isotropic (α_S) and anisotropic (α_T) dipole polarizabilities are parameters characterizing the free H_2^+ core, which are functions of the internuclear separation ρ. This potential has no dependence on the coordinate of the core electron; all dependence on the core's electronic structure is contained in the parameters Q, α_S, and α_T. The quantum defects of H_2 Rydberg states may be estimated, in an a-priori manner, by taking the expectation value of V_{eff} in a basis of zeroth order wavefunctions which are products of free H_2^+ rotational and vibrational wavefunctions and hydrogenic Rydberg wavefunctions, combined according to the usual rules of angular momentum addition to be eigenstates of $\vec{N} = \vec{R} + \vec{L}$. :

$$\psi^o_{vRnLNm}(\vec{\rho}, \vec{r}) = \sum_{m_r, m_l} \langle R, m_r, L, m_l \mid R, L, N, m\rangle \frac{g_{vR}(\rho)}{\rho} Y_{Rm_r}(\hat{\rho})\psi_{nLm_l}(\vec{r}) \qquad (3)$$

The expectation value of Eq. (2) with states of the form Eq. (3) is easily calculated to give the first order Rydberg energies:

$$E^{(1)}(\nu RnLN) = -\frac{e^2}{2}<\alpha_S(\rho)>_{\nu R}<r^{-4}>_{nL}$$

$$-\left[e<Q(\rho)>_{\nu R}<r^{-3}>_{nL}+\frac{e^2}{3}<\alpha_T(\rho)>_{\nu R}<r^{-4}>_{nL}\right]<RLN\,|\,P_2(\cos\theta)\,|\,RLN> \qquad (4)$$

where

$$<RLN\,|\,P_2(\cos\theta)\,|\,RLN> = \left[\frac{3S(S-1)-4\,L(L+1)R(R+1)}{2(2R-1)(2R+3)(2L-1)(2L+3)}\right] \qquad (5)$$

with S = R(R+1) + L(L+1) − N(N+1)

$$<r^{-3}>_{nL} = \left[n^3L(L+.5)(L+1)\right]^{-1} \qquad (6a)$$

$$<r^{-4}>_{nL} = \frac{\left[3n^2-L(L+1)\right]}{2n^5(L-.5)(L)(L+5)(L+1)(L+1.5)} \qquad (6b)$$

and $<Q(\rho)>_{\nu R}$, etc. are expectation values of the core parameters in a particular state. Typical values are[7)]

$$<Q(\rho)>_{0,1} = 1.64257 \ ea_o^2$$

$$<\alpha_S(\rho)>_{0,1} = 3.1809 \ a_o^3$$

$$<\alpha_T(\rho)>_{0,1} = 4.0273 \ a_o^3$$

The Rydberg structure predicted by this long-range potential is in good to excellent agreement with observed structures for a wide variety of H_2 Rydberg states. Among the observations are studies of the 4D[5)], 12D[8)], and 12-19F states[9)] by molecular beam laser spectroscopy, studies of the emission spectra of 5G and 4F states[10),11)], and studies of n = 10 F, G, H, I and K states by fast beam laser and microwave spectroscopy.[6),12),13),14)] The effects of inter-series mixing on the spectroscopy of these states is not negligible, but it has been established experimentally that these effects are also dominated by long-range interactions and can be estimated by calculating off-diagonal matrix elements of Eq. (2) between the states in question.[6),12)] Conventional perturbation theory, rather than MQDT may be used since the perturbations in Rydberg state positions by members of other series are generally much smaller than the separation between adjacent members of a given series. This condition is <u>not</u> fulfilled in nP states, where the coupling matrix elements are much larger, leading there to the necessity of MQDT.

The polarization model, represented in its simplest form by Eq. (2) may be derived rigorously using perturbation theory,[5),6)] subject to the basic assumptions:

1) The two electrons in H_2 are distinguishable.
2) There is no penetration of the core by the Rydberg electron.

These conditions are satisfied with increasing precision as L increases. The polarization model has recently been extended to include higher-order contributions, including non-adiabatic corrections to the dipole polarization energies, quadrupole polarization energies, dipole-octupole polarization energies, and other smaller terms.[6] Within this extended polarization model, it is also possible to estimate with some confidence the reliability of the predicted structures based on the convergence of the expectation value of the polarization series. This treatment also makes explicit the model's dependence upon the electronic structure of the H_2^+ ion.

Over the past few years, our group at Notre Dame has developed one of the new experimental techniques which have been used to explore the spectroscopy of high-L H_2 Rydberg states.[6),12),13)] I would like to briefly describe our technique, primarily to show you the clarity with which the high-L structure is revealed by experiment. Our technique is based on formation of the Rydberg states by charge exchange from a fast beam of H_2^+.

Any of the wide variety of vibration-rotation states of H_2^+ present in the fast beam may capture electrons to form Rydberg states in the neutralized beam.Figure 2 illustrates a few of the Rydberg states bound to the lowest four energy states of H_2^+, $v = 0$, $R = 0, 1, 2$, and 3. Sufficiently highly excited Rydberg electrons could be detected by Stark ionization in a strong electric field, but this would be a crude method of detection without any selectivity. However, we can combine Stark ionization with laser excitation to detect particular Rydberg states both selectively and efficiently. Detection of n = 10 states by this method is illustrated by Figure 2. First the natural population of the very highly excited states is removed by Stark ionization. Then a laser selectively excites n = 10 Rydberg electrons,within each series, to a very highly excited state (n = 27 in this illustration). Finally the upper state is Stark ionized and the resulting fast

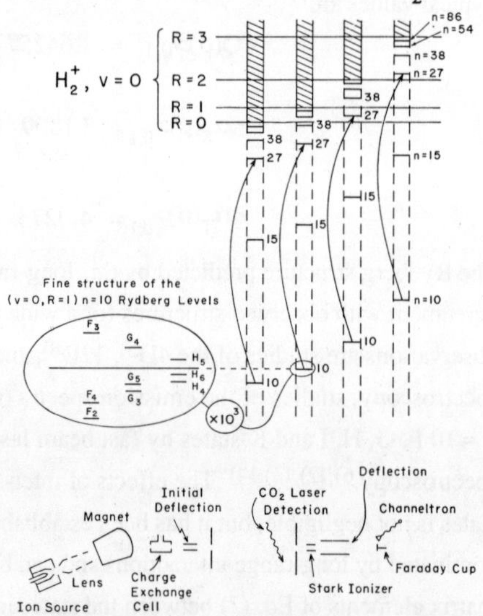

Fig. 2. Diagram illustrating the method of selective detection of individual n = 10 Rydberg states in a fast beam. Rydberg states are formed by charge exchange from a beam of H_2^+ ions. Particular n = 10 states are selectively ionized by laser excitation from 10 to 27 using a Doppler-tuned CO_2 laser followed by Stark ionization of the n = 27 product state.

H_2^+ ions are deflected into a channeltron. Using this scheme, n = 10 Rydberg states, in the fast beam, may be detected with near 100% efficiecy and very low background. This happens, of course, only when the laser is tuned to match the excitation frequency exactly. Actually, however, due to the fine structure of particular Rydberg states, which is given by Eq. (4) and illustrated in Figure 2 for (0, 1) 10F and 10G states, the necessary excitation frequency is slightly different for each particular n = 10 Rydberg state, varying by about 1/1000 of the hydrogenic transition frequency. The laser frequency resolution is sufficiently great that we can distinguish the various n = 10 Rydberg states by their excitation frequencies and therefore selectively detect only a particular Rydberg state. A fragment of the excitation spectrum for 10-26 and 10-27 Rydberg-Rydberg transitions, shown in Figure 3, illustrates the excitation of (0, 1)10F and (0, 1)10G Rydberg states. Shown in the excitation spectrum are three separate peaks for each value of L, corres-

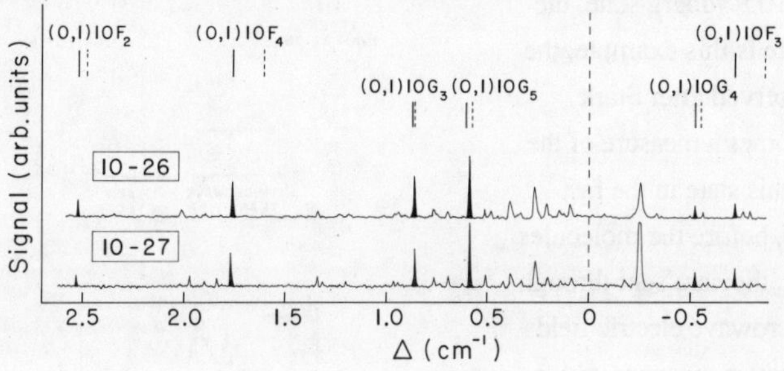

Fig. 3. Curves show experimental spectra of laser induced ion current as a function of the frequency difference Δ between the laser and the hydrogenic transition frequency. The solid peaks represent excitation of (0, 1)10F and (0, 1)10G states to n = 26 and n = 27 final states. The vertical solid bars above the spectra show the predicted positions of these peaks, obtained with the polarization model. The dashed vertical bars indicate the positions which would be predicted in the absence of inter-series perturbations.

ponding to the three eigenstates with different values of $\vec{N} = \vec{R} + \vec{L}$. . Also shown in Figure 3 are the expected excitation frequencies, calculated with the polarization model. A careful study of the excitation spectrum reveals that the quantum defects of these states are predicted to within the experimental accuracy, of about 1%, by this model.[6)] From a more complete spectrum of this type, we have identified more than 20 individual n = 10 Rydberg states in the fast beam.[12)] They are all bound to the lowest four states of H_2^+ ,v = 0, R = 0, 1, 2, and 3 and have L = 3, 4, 5, and 6. The absence of vibrationally excited Rydberg states in the spectrum is extremely clear experimentally since, due to the different quadrupole moment of vibrationally excited H_2^+ , these lines would fall at predictable and well-resolved excitation energies. Their absence is consistent with the expected vibrational auto-ionization of these states with lifetimes much less than the transit time between the charge exchange cell and laser (≅ 1 μsec.).

The reason why we choose to study n = 10 states is experimental convenience; they can be excited with a CO_2 laser. A grating selected single fequency line of the CO_2 laser may be used, with fine tuning of the laser frequency accomplished with the Doppler effect, by varying the angle of intersection between the fast H_2 beam and the laser.

In more recent work,[13),14)] we have carried out ultra-high resolution spectroscopy of these high-L Rydberg states by the use of microwave/optical double resonance. This technique is illustrated in Figure 4. When the laser is tuned to excite a particular n = 10 Rydberg state, the (0, 1)10G_5 state is this example, the ion current observed after Stark ionization becomes a measure of the population of this state in the fast beam. Now if, before the molecules reach the laser, they are sent through a region of microwave electric field whose frequency is resonant with a transition between the (0, 1)10G_5 state and another Rydberg level (the (0, 1)10H_6 state in this illustration), the resonant microwave transitions will change the population of the two states and the detected ion current. By tuning the microwave frequency until this condition is observed, the frequency interval separating the two Rydberg states may be measured directly with high precision. This is illustrated in Figure 4 for the (0, 1)10G_5 - (0, 1)10H_6 transition, which occurs near 10 GHz.[13)] The resonance linewidth, due to the transit time through the microwave field is about 1 MHz, allowing us to easily determine the fine

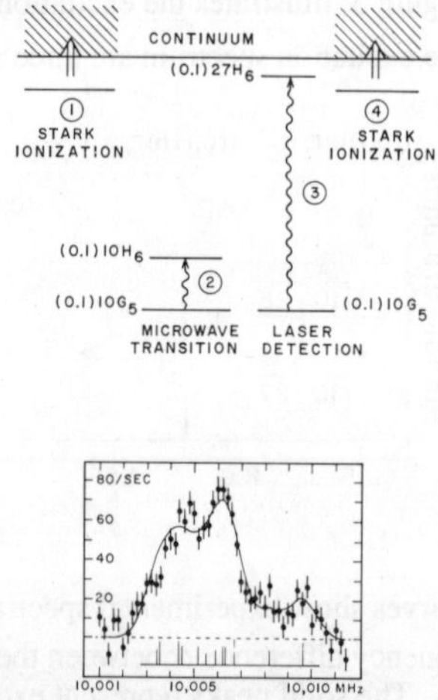

Fig. 4. Diagram illustrating the method of observing microwave-induced resonant transitions between n = 10 H_2 Rydberg states. The laser-induced ion current monitors the population of one of the resonant levels, and is therefore sensitive to microwave-induced population changes. Also shown is a typical microwave resonance spectrum, for the (0, 1)10G_5 to (0, 1)10H_6 transition.

structure interval to ≤ 0.5 MHz, or about 10^{-5} cm^{-1}. This is about three orders of magnitude more precise than measurements derived from studies of the optical spectrum, and corresponds to determination of the quantum defects to about 10^{-5}. We have just completed a systematic study of the spectroscopy of a subset of these n = 10 Rydberg states using this method.[14] Figure 5 shows the microwave transitions which have been studied, corresponding to transitions among (0,1) n = 10 Rydberg states with L = 4, 5, 6, and 7. The results are in good agreement with the predictions of the extended polarization model, to within the estimated precision of that model, except that it appears that the quadrupole moment of H_2^+ (0, 1) may have been slightly under-estimated by existing calculations.

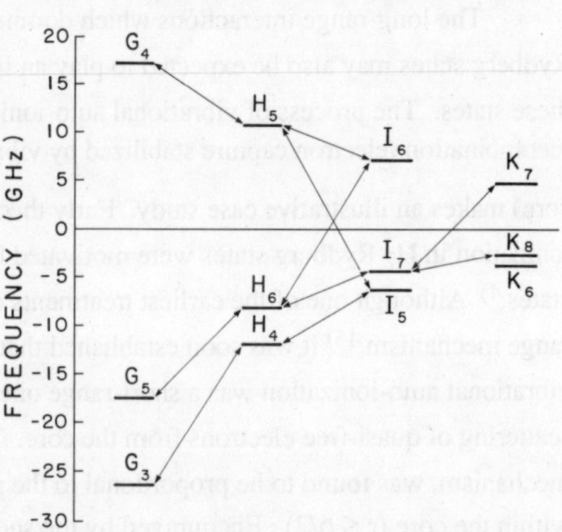

Fig. 5. Level diagram showing the fine structure of (0, 1)10L_N states of H_2 for 4 ≤ L ≤ 7. The arrows indicate microwave resonance transitions which have been observed using the technique illustrated in Fig. 4.

A bi-product of these high resolution microwave studies was the first resolution of the finer level of structure in the states due to magnetic fine and hyperfine structure. This seems to be in good agreement with a relatively simple ad hoc model which we developed.[13] One interesting conclusion which emerged from these studies was that, in general, these high-L Rydberg states are not eigenstates of the total electronic spin, thus it is inappropriate to refer to them as "singlet" or "triplet" states.

These experiments which I have described, and others at lower L carried out by other groups with different techniques, indicate clearly that long-range interactions alone, without the use of MQDT, can successfully account for the spectroscopy of high-L Rydberg states. It is reasonable to expect that the same a priori treatment should give a good description of the structure of the great majority of H_2 Rydberg states, most of which are as yet unobserved. Considering the potential complexity of this problem, and the simplicity of the model, this is quite remarkable, and it suggests that anyone who is concerned with the general behavior of H_2 Rydberg states should be aware of the model.

Of course, it represents only part of the complete description of these states, but it appears to contain a great deal of the most important physics.

The long-range interactions which dominate the spectroscopy of "almost all" H_2 Rydberg states may also be expected to play an important role in the dynamic behavior of these states. The process of vibrational auto-ionization and its inverse dielectronic recombination (electron capture stabilized by vibrational or rotational excitation of the H_2^+ core) makes an illustrative case study. Early theoretical treatments of vibrational auto-ionization in H_2 Rydberg states were motivated by observations of the process in nP states.[4] Although one of the earliest treatments of the subject did in fact invoke a long-range mechanism,[15] it was soon established that the dominant mechanism for nP vibrational auto-ionization was a short-range one,[16] which could be interpreted as inelastic scattering of quasi-free electrons from the core. The auto-ionization rate, from this mechanism, was found to be proportional to the probability of the Rydberg electron to be within the core ($r \leq \rho/2$). Encouraged by the success of this model, it was soon applied to predict the auto-ionization rate of higher L states.[17] As could be expected, the auto-ionization rate was predicted to decrease dramatically for the relatively non-penetrating higher L states. This prediction, which is still widely cited,[18] is actually spectacularly incorrect, not because of any error in the calculation, but because it neglects the contribution of long-range interactions, which rapidly become dominant for higher L states. for instance, the short-range model predicts an auto-ionization lifetime for the $v = 1$, 10G state of 100 seconds. This is eleven orders of magnitude slower than the 10^{-9} seconds lifetime calculated with the long-range model.[19),10] The rate, in the long-range model, may be calculated with a simple "golden rule" argument, involking the matrix element of Eq. (2) between the $(v = 1)$10G state and $v = 0$ continuum states, which is precisely the mechanism proposed earlier by Russek.[15] This process, and the closely related process of rotational auto-ionization via long-range interactions have been discussed in detail by Eyler.[19] A possible physical interpretation is "photo-ionization" of the Rydberg electron through the time-varying quadrupole electric field of the vibrating and rotating core. As this case study indicates, the neglect of long-range interactions in describing the dynamics of H_2 Rydberg states may lead to serious errors.

Studies of the pre-dissociation of high-L Rydberg states are in a much less satisfactory state, partly because there is very little clear experimental data as yet. From our studies of n = 10 Rydberg states, for instance, we can conclude only that for $v = 0$, high L n = 10 states, predissociation is very slow ($\tau \geq 10^{-6}$ sec.). As far as I know, at the present time all existing theoretical treatments of this process invoke a short-range mechanism involving coupling between attractive and repulsive potential curves of H_2. These models have been reasonably successful in describing predissociation of low-L Rydberg states, although some intriguing puzzles remain.[2d)] For higher-L states, the rate of predissociation due to this mechanism would be expected to drop rapidly, since the non-penetrating Rydberg electron has very little probability of being at the small radial distances required for the curve crossing. There is, however, no reason to assume that the same mechanism will still be dominant for high-L states. Indeed, it seems quite conceivable that a completely different mechanism, involving long-range interactions could govern predissociation of these states. One possible such mechanism is the "electric field induced dissociation of H_2^+ ", due to the electric field of the distant Rydberg electron. An electron in the 6F state, for instance, exerts a considerable electric field on the H_2^+ ion within its orbit ($E \cong 10^6$ V/cm). Electric fields of this size may be suffient to dissociate H_2^+ with appreciable probability via the tunneling mechanism described some time ago by Hiskes.[20)]

Dissociative recombination of H_2 via an intermediate Rydberg state may be viewed as a compound process involving

1) capture of a free electron into a Rydberg state, with the capture stabilized energetically by vibrational and/or rotational excitation of the H_2^+ core.
2) predissociation of the Rydberg state.

The role of high-L Rydberg states in process 1) may be assessed using the long-range model of vibrational/rotational auto-ionization.[19] However, since the net rate for dissociative recombination involves the competition between these two processes,[21)] the second of which is not well understood for high-L Rydberg states, it is not yet possible to reach definitive conclusions about the possible role of high-L Rydberg states in this compound process. It is certain, however, that thereotical treatments which exclusively invoke short-range interaction mechanisms cannot be relied upon to accurately assess the answer to this question. In addition, if external electric fields are present in the region of recombination, so that Rydberg states of different L are strongly mixed, improved understanding of the behavior of the high-L states will be required to assess the result of the more complex recombination process.

It is also interesting to speculate as to what will occur if process 1) is temporarily stabilized by rotational excitation of the H_2^+ core. This process proceeds very efficiently for low energy electrons via long range interactions.[22)] One would think, however, that the inverse process would also proceed rapidly, much more rapidly than radiative decay, for instance, and the result would be very little recombination. It may be important to note, however, that the process of rotational auto-ionization must obey certain strict angular momentum selection rules ($\Delta N = 0$, $\Delta L = 0, \pm 2$), which imply that different members of the EFS multiplet should have very different rates of auto-ionization. In one example, for instance, recombination into the (0, 2)34F_N states can occur only into states of odd N and will be mostly into the (0, 2)34F_3 state. The other (0, 2)34F states, even though they are very close in energy to these states which are populated by recombination, are strictly forbidden to auto-ionize. It would appear that even extremely gentle collisions might lead to equalization of the populations of the EFS states, thus thus increasing the fraction of Rydberg states formed in recombination which decay optically and stabilize the recombination.

ACKNOWLEDGEMENTS

The experimental work at the University of Notre Dame has been carried out in collaboration with W. G. Sturrus, E. A. Hessels, and P. W. Arcuni, and supported by the National Science Foundation under Grant PHY87-09707. The author also wishes to acknowledge many helpful conversations with E. E. Eyler.

REFERENCES

1. H. Hus, F. Yousif, C. Noren, A. Sen, and J. B. A. Mitchell, Phys. Rev. Lett. 60, 1006 (1988).
2. a) W. A. Chupka and J. Berkowitz, J. Chem. Phys. 51, 4244 (1969)
   b) G. Herzberg and Ch. Jungen, J. Mol. Spectrosc. 41, 425 (1972)
   c) P. M. Dehmer and W. A. Chpuka, J. Chem. Phys. 65, 2243 (1976)
   d) Wallace L. Glab and Jan P. Hessler, Phys. Rev. A35, 2102 (1987).
3. U. Fano, Phys. Rev. A2, 353 (1970).
4. R. S. Berry and S. E.Nielson, Phys. Rev. A1, 395 (1970).
5. E. E. Eyler and F. M.Pipkin, Phys. Rev. A27, 2462 (1983).
6. W. G. Sturrus, E. A. Hessles, P. W. Arcuni, and S. R. Lundeen, Phys. Rev. A 38, 135 (1988).
7. David M. Bishop and Brenda Lam, Chem. Phys. Lett. 134, 283 (1987).
8. R. D. Knight and Liang-guo Wang, Phys. Rev. Lett. 55, 1571 (1985).
9. E. E. Eyler, R. C. Short, and F. M. Pipkin, Phys. Rev. Lett. 56, 2602 (1986).
10. G. Herzberg and Ch. Jungen, J. Chem. Phys. 77, 5876 (1982).
11. Edward S. Chang, Stanley Pulchtopek, and E. E. Eyler, J. Chem. Phys. 80, 601 (1984).
12. W. G. Sturrus, E. A. Hessels, and S. R. Lundeen, Phys. Rev. Lett. 54, 792, (1985).
13. W. G. Sturrus, E. A. Hessels, and S. R. Lundeen, Phys. Rev. Lett. 57, 1863 (1986).
14. W. G. Sturrus, E. A. Hessels, P. W. Arcuni, and S. R. Lundeen, Bull. Am. Phys. Soc. 33, 924 (1988).
15. A. Russek., M. R. Patterson, and R. L. Becker, Phys. Rev. 167, 17 (1968).
16. F. H. M. Faisal, Phys. Rev. A4, 1396 (1971).
17. Y. H. Band, J. Phys. B7, 2072 (1974).
18. Susan Tarr, James A. Schiavone, and Robert S. Freund, Phys. Rev. Lett. 44, 1660 (1980).
19. E. E. Eyler, Phys. Rev. A34, 2881 (1986).
20. John R. Hiskes, Phys. Rev. 122, 1207 (1961).
21. J. N. Bardsley, J. Phys. B1, 349 (1968).
22. Robert C. Stabler, Phys. Rev. 131, 1578 (1963).

DISSOCIATIVE RECOMBINATION IN MEDIUM DENSITY PLASMAS

W.G. Graham*

SRI International

Menlo Park

Ca.94025 U.S.A.

ABSTRACT

The role of dissociative recombination (DR) in medium density plasmas is discussed with particular reference to hydrogen discharges, as used in intense ion sources, and more complex molecular plasmas used in many present day technological applications. Experimental measurements of parameters which influence the significance of DR are reported. It is found that under particular operating conditions dissociative recombination can be the dominant ion-electron pair loss mechanism and that in general dissociative recombination can play an important role in the chemistry of these plasmas.

INTRODUCTION

Plasma discharges (the terms are increasing used interchangeably) result when sufficient energy is provided to a gas, liquid or solid to create ion-electron pairs from the constituent atoms or molecules. This ionization process is balanced by the subsequent recombination of the ion-electron pair. The discharge reaches a self-sustained steady state when the generation and loss processes balance one another. The energy input to create the plasmas can be in many forms, however generally the same basic atomic processes are involved in the initiation and sustaining of the plasma.

The plasma environment has many useful technological applications. In this paper the role of dissociative recombination

*Permanent address: University of Ulster, Coleraine, Northern Ireland.

in the chemistry of some typical technologically important plasmas will be considered and measurements which reflect on the significance of DR in the plasma reported. Firstly in a hydrogen discharge initiated and sustained by electrons emitted from a biased hot filament. This most simple molecular system will be used to explore in detail the role of dissociation recombination (DR). Secondly a more complex molecular system of the type used in plasma etching and initiated and sustained by applied radio frequency fields will be briefly discussed.

These plasmas have the common features of low electron temperatures and dominant molecular ion species, suggesting an environment where dissociative recombination may be important and even dominant as an electron-ion pair loss mechanism. Electron-ion pairs can recombine in several ways; electrons and ions can be lost through diffusion to the walls and by recombination in the volume of the discharge. Electrons may additionally be lost by dissociative attachment to form negative ions while positive ions can be lost through mutual neutralization and dissociative attachment with negative ions.

A simple rate equation can be used to describe the ion density in a discharge with a single molecular positive ion species and no negative ions.

$$dn_+/dt = nf(e)\, n(0)\,\overline{\sigma v}_i - n_+ v_+/L - n_+ n_e \overline{\sigma v}(DR) \qquad (1)$$

where n_+ is the ion density, n_e is the thermal electron density, nf(e) is the fast electron density, n(0) is the neutral gas density, v_+ is the ion velocity, L is the typical dimension of the system, $\overline{\sigma v}_i$ is the ionization rate and $\overline{\sigma v}(DR)$ is the dissociative recombination rate. The concept of there being two electron groups will be discussed later.

For dissociative recombination to dominate the electron-ion loss processes then

$$n_+ n_e \overline{\sigma v}(DR) > n_+ v_+/L \quad (2)$$

$$ne > v_+/L\overline{\sigma v}(DR) \quad (3)$$

This will be dependent on the ion energy or velocity, the electron energy, the dimensions of the system and the species in the discharge. These in turn depend on the nature and geometry of the discharge.

HYDROGEN DISCHARGE

Medium density hydrogen discharges are used as sources of intense beams of both positive[1)] and negative hydrogen ions[2)] for use in neutral beam heating of nuclear fusion devices[1)]. Detailed models of the atomic processes in these sources have been produced[3,4)], however, in considering the role of dissociative recombination various experimental observations must be considered.

The main production mechanism for negative ions is thought to be through dissociative attachment of low energy electrons to vibrationally excited molecules[4)]. Figure 1 shows a schematic of an ion source designed to optimize negative hydrogen ion production and in which measurements reported here have been obtained. This device has been described in detail elsewhere[5)]. It consists of a stainless steel box with dimensions 19 x 19 x 24 cm with permanent magnets of pole strengths of 3.6 kG, spaced 4.5 cm apart on the outside walls to form a line cusp symmetry. The symmetry is broken on opposing walls forming a one-dimensional transverse "filter field", of approximately 50 G, accross the source.

The discharge is produced by electrons emitted from two tantalum filaments heated by a D.C. current. The filaments are biased negatively with respect to the vessel wall which forms the anode.

This configuration produces two distinct plasma regions, separated by the transverse magnetic filter field. In the driver region, where the filaments are located, the plasma is produced and the vibrational levels of the hydrogen molecules are populated by fast electron impact. The plasma and vibrationally excited molecules can diffuse accross the filter field but fast electrons are excluded. The lower electron temperature in this extracter region allows a high density of negative ions to be established through dissociative attachment.

(a) Positive Ion Species.

The distribution of positive ion species is strongly dependent on the plasma parameters. It has been found that in this, and other[6)] medium density discharges H_3^+ is the dominant ion species, due to the high reaction rate for

$$H_2^+ + H_2 \rightarrow H_3^+ + H \tag{4}$$

however under certain operating conditions large percentages of protons have been found[3)].

(b) Positive Ion Energy.

Positive ions acquire energy during their formation, from the plasma electric fields and from energy exchange in collisions. It has been estimated[7)] that at low pressures the H_3^+ energy could be up to 1 eV. Experimentally the ion energy can be inferred from measurements of the ion confinement time and spatial distribution[8)]. These measurements indicate that the ion energy lies between 0.06 eV and 0.2 eV, corresponding to H_3^+ velocities of between 2 and 4 x 10^5 cm s^{-1}.

(c) Electron Energies.

The electron energy distribution in many discharges is not Maxwellian. Figure 2 shows the electron energy distribution function measured[9)] along the line AB in the discharge shown in Fig. 1. The influence of the filter in removing "fast" electrons is apparent. In figure 3 EEDFs typical of the driver and extractor regions are shown. In the extractor region the EEDF can be represented by a single temperature kT_e of about 1 eV, and a corresponding electron density n_e. In the driver region however the EEDF is better represented by two Maxwellians, one characterized by a temperature kT_e of 2.5 eV and the other kT_{fe} of 15 eV. Corresponding thermal n_e and fast n_{fe} electron densities can also be used to characterize the EEDF.

Modelling of these discharges is further complicated by a complex spatial variation of the plasma parameters[10)].

d) Dissociative Recombination.

In the typical discharge described above the dimension L is of the order of 10 cm and the ion velocity is 3×10^5 cm s^{-1}. Assuming an electron temperature of 1 eV and that H_3^+ is the dominant ion then $\overline{\sigma v}(DR) \sim 7 \times 10^8$ cm^3 s^{-1}[11)]. This implies from equation 3 that dissociative recombination will be the dominant loss mechanism if the electron density is greater than about 5×10^{11} cm^{-3}. Such densities may not be compatible with such a low electron temperature. At higher electron temperatures, $\overline{\sigma \vec{v}}(DR)$ will be lower requiring a higher electron density for dissociative recombination to dominate, however, under these conditions there will be less molecular ions.

Nevertheless there is a range of densities and locations in the hydrogen discharges were DR can be dominant loss mechanism.

Aside from that DR can have other significant roles in hydrogen discharges in

- the depletion of low energy electrons, which are required for the production of negative ions through dissociative attachment.

$$e + H_2(v^*) \rightarrow H^- + H \quad (5)$$

- the production vibrationally excited $H_2(v^*)$.

$$e + H_3^+ \rightarrow H_2(v^*) + H \quad (6)$$

- the direct production of H^-

$$e + H_2^+ \rightarrow H^- + H^+ \quad (7)$$

$$e + H_3^+ \rightarrow H^- + H^+ + H \quad (8)$$

$$H^- + H_2^+ \quad (9)$$

- the production of hydrogen atoms.

The latter is generally a beneficial process in proton sources but deleterious for negative ion production since H atoms can strip the negative ions

$$H^- + H \rightarrow 2H + e \qquad (10)$$

and vibrationally de-excite molecules.

Dissociative recombination may also have potential application as a spectroscopic diagnostic due to the sensitivity of the H atom (nl) values to the electron energy and internal energy of the molecular ion.

Clearly for the role of DR to be fully understood in hydrogen discharges much more information on branching ratios is required.

PLASMA ETCHING DISCHARGES.

These find applications[12)] in lasers, microelectronics fabrication, (e.g. thin film deposition, etching and oxidation), diamond and high temperature superconducting film production.

In plasma etching the goal is to create submicron structures on a substrate by removal of material from unmasked sites. The rate of removal of material, the selectivity of removal for different materials and the final shape of the structure are important parameters in this technology.

The basic concept is to feed an inert gas into a discharge and there generate reactive radicals or atoms which will form volatile compounds with the substrate material.

The gases are usually halocarbons mixed with varying percentages of other gases. The discharges used are generally radiofrequency or more recently microwave driven.

The general range of operating conditions are with gas pressures of between 10^{-4} to 10 Torr, electron densities from 10^9 to 10^{12} cm^{-3} and with measured electron temperatures ranging from 1 to 10 eV. The high electron temperatures may however be an artifact of the measuring technique used in these time varying discharges[13)]. Manufacturers of etching equipment use a wide variety of gas mixtures and machine geometries usually determined by trial and error. There is still considerble progress to be made before an the complex chemistry in these discharges is understood.

a) Ion Species.

Since the gas mixtures are complex, there will be a wide variety of molecular ion species in any discharge.

b) Ion Energies.

Radio frequency discharges are generated by creating large electric fields in the sheaths of the discharge. Ions can acquire energy in these sheaths of up to about 80% of the applied rf voltage[14], and indeed secondary electron emission from bombardment of the electrodes by energetic ions, may be the major source of ionizing electrons in the discharge. The ions therefore can have a wide range of energies from thermal to that acquired in sheaths i.e. up to hundreds of eV. The ion energy will be spatially and temporally varying.

c) Electron Energies.

Reliable measurements of electron energy distribution functions in rf driven discharges is very difficult due to the effects of rf fluctuations on the local plasma-probe potential. The time-resolved technique used at the University of Ulster allows the effect of rf fluctuations on the plasma-probe potential to be eliminated[13,15].

It is found that, at least at frequencies of less than 1 MHz, the EEDF is generally non-Maxwellian and can be dominated by atomic collision processes. In Figure 4 the EEDF measured in a parallel plate 100 kHz nitrogen discharge is shown. The EEDF was measured at two times, 1.5 μs and 2.5 μs after the positive to negative zero crossing point of the potential applied to the driven electrode. There are two particularly interesting features.

The EEDF is clearly non-Maxwellian and shows a distinct minimum around electron energies of 4 eV. This has been predicted theoretically[16], being due to the large cross section for inelastic collisions involving vibrational excitation of the N_2 molecule in this energy range. It was thought that superelastic collisions would repopulate this energy region but under these operating conditions this is clearly not the case.

The electron temperature of the bulk of the electrons is found to be quite low ~ 0.4 eV. Similar results have been found in other gas discharges[13]. Most previous, time-averaged, measurements of electron temperatures in rf discharges had reported higher electron temperatures (~10-20 eV)[17,18,19].

d) Dissociative Recombination.

The wide range of molecular ion species and ion energies coupled with complex EEDF's makes the plasma chemistry of these important discharges hard to model. However the relatively low electron temperatures recently measured[13,15] would indicate that dissociative recombination will play a significant role in such discharges. The comments on the role of DR in H_2 discharges would be expected to apply to these complex molecular discharges i.e. atom and radical production, vibrational excitation leading to negative ion production and as a diagnostic.

CONCLUSION

Dissociative recombination can be the dominant electron-ion pair loss mechanism in medium density plasmas. The significance of DR in this role will depend on the geometry of the system, the gas or gas mixture used and the plasma parameters. Whether or not DR plays a dominant role in electron or ion loss, it will certainly influence the plasma chemistry and may have potential applications in diagnostics of the plasma.

Much more detailed information is required before dissociative recombination can be accurately incorporated into plasma chemistry models.

ACKNOWLEDGEMENTS

The experimental measurements reported here were made in collaboration with Dr. M.B. Hopkins and C.A. Anderson at the University of Ulster, Northern Ireland. Preparation of this paper was partially supported by NSF(PHY-8713309).

REFERENCES

1. See the review by Kunkel, W.B. "Neutral-Beam Injection" in Fusion, Vol. 1, B. edited by Teller, E. (Academic Press, New York 1981).
2. See, for example, "Production and Neutralization of Negative Ions and Beams". AIP Conf. Proc. No. 158 (AIP, New York, 1987).
3. Chan, C.F., Burrell, C.F. and Cooper, W.S., J. Appl. Phys. 54, 6119 (1983).
4. Hiskes, J.R. and Karo, A.M., J. Appl. Phys. 56, 1927 (1984).
5. Hopkins, M.B. and Graham, W.G., Rev. Sci. Instrum. 57, 2210 (1986).
6. For example, Leung, K.N. et al, Rev. Sci. Instrum. 49, 321 (1978).
7. Bruneteau, A.M. and Bacal, M., J. Appl. Phys. 57, 4342 (1985).
8. Hopkins, M.B., Ph.D. Thesis, University of Ulster (1987).
9. Hopkins, M.B. and Graham, W.G., J. Phys. D20, 838 (1987).
10. Graham, W.G. and Hopkins, M.B., In Preparation.
11. Peart, B and Dolder, K.T., J. Phys. B7, 1948 (1974).
12. See for example, Chapman B. Glow Discharge Processes (Wiley, New York, 1980) and Von Engel A. Electric Plamas: their nature and uses. (Taylor and Francis, London, 1983).
13. Anderson C.A., Graham, W.G. and Hopkins, M.B., Appl. Phys. Lett. 52, 783 (1988).
14. Kohler, K. Horne, D.E., Coburn, J.W., J. Appl. Phys. 58, 3350 (1985).
15. Hopkins, M.B., Anderson, C.A., Graham, W.G., "Time resolved electron energy distribution function measurements in a low frequency rf glow discharge" submitted for publication.
16. Capitelli, M., Geliberto, R., Gorse, Winkler, R. and Wilhelm, J. (To be published).
17. Kushner, M.J., J. Appl. Phys. 53, 2939 (1982).
18. Maudrill, D. et al., J. Phys. D 20, 815 (1987).
19. Cox, T.I. et al., J. Phys. D. 20, 820 (1987).

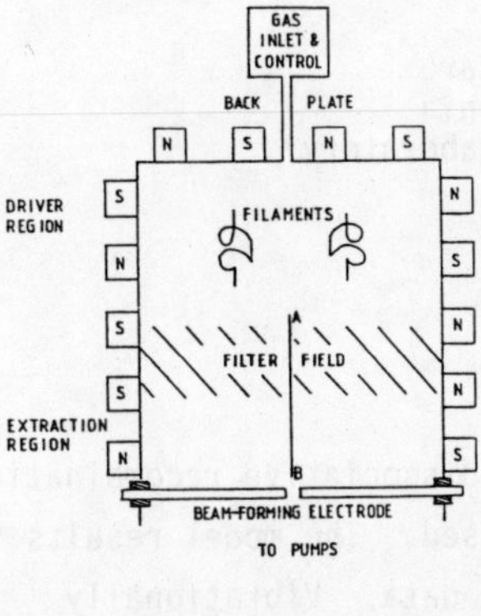

Fig 1. Schematic diagram of negative hydrogen ion source.

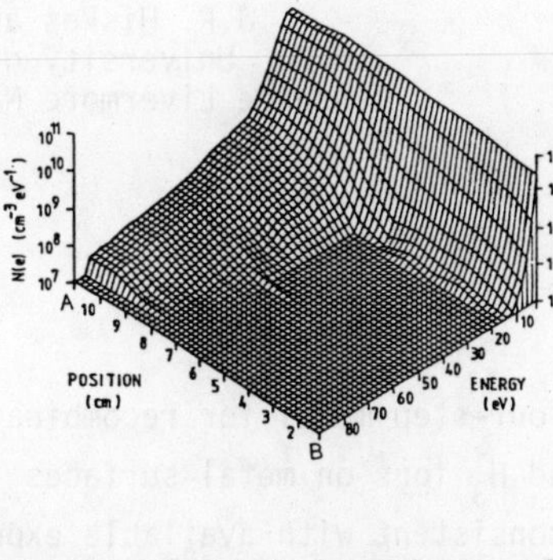

Fig 2. Spatial dependence of EEDF in negative hydrogen ion source. Discharge current = 10A, Discharge voltage = 60V and gas pressure = 2 m Torr.

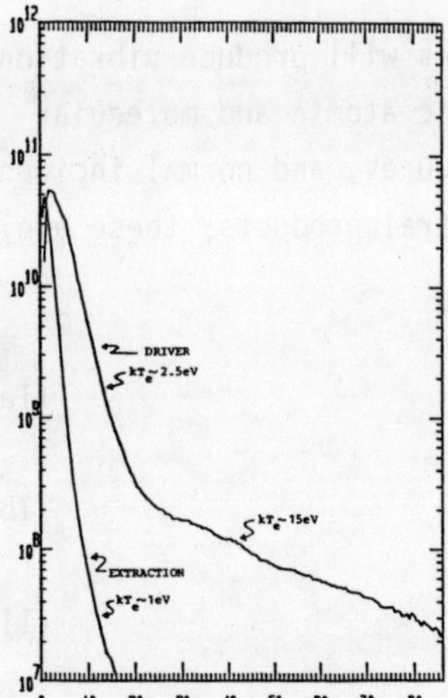

Fig 3.
EEDFs measured in the centre of the driver region and extraction region. Plasma conditions same as Fig 2.

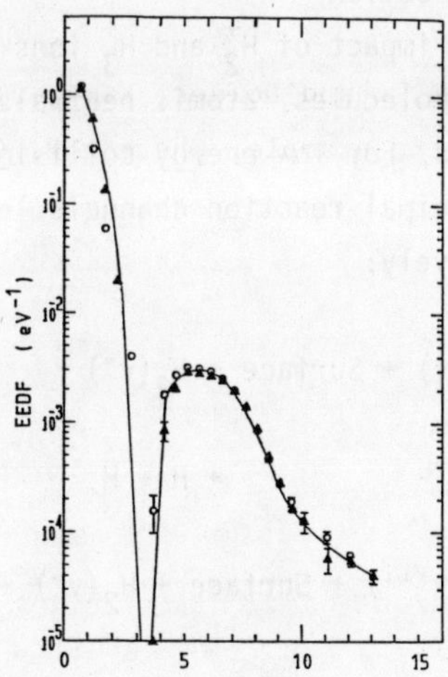

Fig 4.
Time resolved EEDF measured in 100 kHz N_2 discharge at 0,1.5µs and ▲, 2.5µs after zero crossing point of rf driving voltage. Rf power, 25 W; gas pressure 175 m Torr.

DISSOCIATIVE RECOMBINATION OF H_2^+ AND H_3^+ IONS ON SURFACES

J.R. Hiskes and A.M. Karo
University of California
Lawrence Livermore National Laboratory*

ABSTRACT

A four-step model for recombination and dissociative recombination of H_2^+ and H_3^+ ions on metal surfaces is discussed. The model results for H_2^+ are consistent with available experimental data. Vibrationally excited molecules from H_3^+ recombination are produced in a broad spectrum that enhances the excited level distribution. The application of this latter process to hydrogen negative ion discharges is discussed.

I. INTRODUCTION

The impact of H_2^+ and H_3^+ ions on surfaces will produce vibrationally excited molecules, atomic neutrals, and ionic atomic and molecular fragments. For low energy collisions, $E < 100$ eV, and normal incidence, the principal reaction channels lead to neutral products; these are, respectively:

$$H_2^+(v) + \text{Surface} \rightarrow H_2(v'') \qquad \text{(Ia)}$$

$$\rightarrow H + H, \qquad \text{(Ib)}$$

$$H_3^+(v''') + \text{Surface} \rightarrow H_2(v'') + H \qquad \text{(IIa)}$$

$$\rightarrow H + H + H. \qquad \text{(IIb)}$$

Here we have indicated that the molecular ions were initially formed with a distribution of vibrational levels. In this paper we shall assume these molecular ions retain their formation spectral distribution up to the point of interaction with the surfaces.

In a previous paper we have discussed a three-step process for H_2^+ recombination.[1] The incident ion is neutralized and electronically relaxed by a resonant electronic capture into the excited $b^3\Sigma_u$ state of H_2, followed by an Auger capture into the $X^1\Sigma_g$ ground state and ejection of the excited electron. The ground electronic state configuration, vibrationally excited, collides with the surface and experiences a "hard" scattering collision on the surface. For a low energy surface collision the diatomic system remains largely intact following the hard collision, the fraction of incident ions dissociating into atoms being dependent upon the vibrational level populated at the moment of impact. Of those incident ions surviving as bound molecules the residual population distribution may span the entire spectrum of bound states.

Molecular H_2 fragments have been observed by Willerding et al.[2] following glancing collisions of 200-600 eV H_2^+ ions off a nickel surface. These authors have rejected the three-step process mentioned above in favor of a direct Auger or resonance capture into the $X^1\Sigma_g$ ground state. In their model captures into the repulsive $b^3\Sigma_u$ state lead directly to dissociation into two free atoms.

Collisions of H_3^+ ions on Ni(111) surfaces have also been studied by Willerding et al.,[3] again at moderately high energies, 230-1065 eV, and for low-angle glancing collisions. These workers have identified reactions IIa, IIb and have presented evidence for possible vibrational excitation up to the v" = 5 level. This and other experiments[4] have shown that for energies up to several keV and for collisions at or near normal incidence the ionic component among the rebounding species will be insignificant as compared to the neutral yields via channels I and II.

In a recent paper Mitchell and Graham[5] have considered several dissociation processes leading to fragmentation of H_3^+ into H_2(v"), including low-energy (E $<$ few eV), normally incident surface collisions. They have concluded there is no appreciable excited vibrational population in the residual H_2 molecule following these surface collisions as compared with other dissociation processes.

In this paper we shall review process I and introduce some new features of the surface collision process. The discussion is extended to include low-energy surface collisions of H_3^+ ions, with particular emphasis on the residual population distribution of the fragment $H_2(v'')$ molecules, process IIa.

II. RECOMBINATION AND DISSOCIATIVE RECOMBINATION OF H_2^+ IONS

Here we extend the earlier three-step model to a four-step recombination model leading to formation of $H_2(v'')$ and atomic H fragments, with emphasis on low-energy, normally-incident collisions. This four-step model is shown schematically in Fig. 1.

As the H_2^+ ion approaches the surface it interacts with the surface first through the image field

$$\mathcal{E} = -\frac{1}{4}\,\frac{e}{z^2} = -\,\frac{1.29\times10^9}{(z/a_0)^2}\ \text{volts cm}^{-1}, \tag{1}$$

where z is the distance measured from the image plane. When the ion has approached to within 113 a_0 of the image plane it experiences an electric field of 10^5 volts cm^{-1}. This field is sufficiently intense to electrically dissociate the uppermost vibrational level of H_2^+ (v = 18) to yield a proton and a ground state atom[6], according to

$$H_2^+(v=18) + \mathcal{E}(10^5 V/cm^{-1}) \rightarrow H^+ + H(1s). \tag{2}$$

The H continues to drift toward the surface while the H^+ is accelerated until it becomes Auger neutralized at approximately 1-2 a_0 from the surface plane. Both atoms are eventually reflected from the surface and appear as recombination products.

As the remaining ions continue to move inward, successively lower vibrational levels are electrically dissociated until they reach a point about 8 a_0 in front of the metal surface where the electric field has increased to 7×10^7 volts cm^{-1}; here the v = 7 level is electrically dissociated. At closer approaches an electron is resonantly or Auger

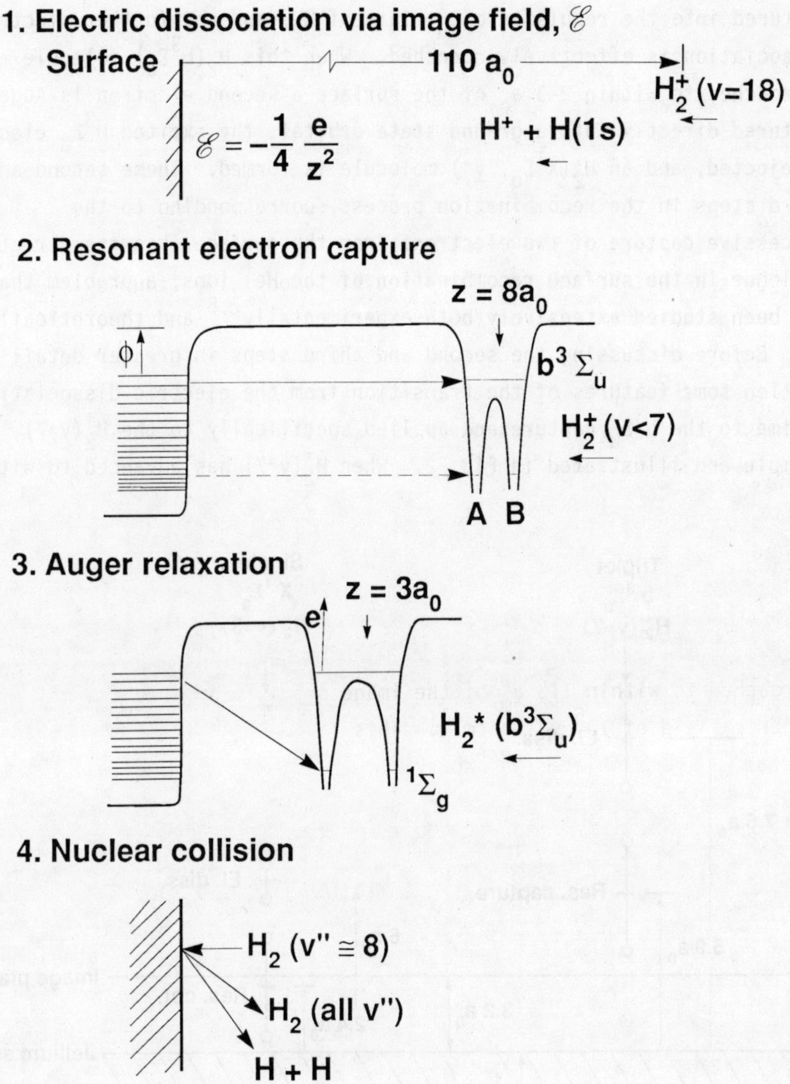

Figure 1. Stepwise recombination of H_2^+ ions on surfaces.

captured into the repulsive $b^3\Sigma_u$ state of H_2, and any further electrical dissociation is effectively quenched. When this $H_2(b^3\Sigma_u)$ molecule approaches to within 2-3 a_o of the surface a second electron is Auger captured directly into a ground state orbital, the excited $b^3\Sigma_u$ electron is ejected, and an $H_2(X^1\Sigma_g, \underline{v}")$ molecule is formed. These second and third steps in the recombination process, corresponding to the successive capture of two electrons from the surface, has its direct analogue in the surface recombination of the He^+ ions, a problem that has been studied extensively both experimentally[7,8] and theoretically.[9]

Before discussing the second and third steps in greater detail we mention some features of the transition from the electric dissociation regime to the $b^3\Sigma_u$ capture and applied specifically to the $H_2^+(v=7)$ example and illustrated in Fig. 2. When $H_2^+(v=7)$ has advanced to within

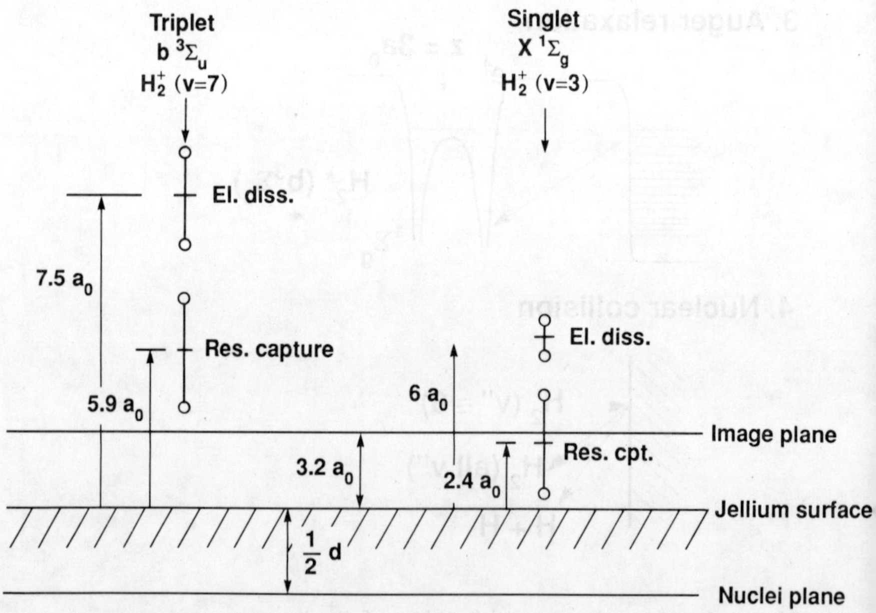

Figure 2. Electric dissociation vs resonant capture through triplet and singlet channels.

7.5 a_o of the surface the ion is unstable against electric dissociation and the two nuclei begin to separate under the influence of the electric field. In the c.m. system of the molecule the electric force is equal to $-1/2\xi e$. When the c.m. has advanced to approximately 5.9 a_o from the surface resonance capture to the $H_2(b^3\Sigma_u)$ state occurs.[11] From the point of onset of electric dissociation up to the resonant capture position the H_2^+ ion is "electrically stretched" by about one a_o. Inspection of the H_2^+ and H_2 potentials[12] indicates that internuclear separations larger than about 5 a_o will not lead to bound $H_2(v")$ molecules in the final electron capture. The internuclear spacing of v = 7, 4.2 a_o, plus the electric stretching leads eventually to an unbound diatomic system and the formation of two free atoms. Similar considerations for the $H_2^+(v=6)$ system lead to a loosely bound $H_2(v")$ configuration.

The fraction of the incident molecular ions that are electrically dissociated is obtained by summing the populations[13] of the $H_2^+(v)$ ions from v = 7 to v = 18,

$$f(1.,b^3\Sigma_u) = \sum_{v=7}^{18} p(v) = .139 \; ; \tag{3}$$

hence about 14% of the incident ions are lost to electric dissociation during this first stage, (1.), of the recombination process.

Resonant electron capture can also occur directly to the ground electronic $H_2(X^1\Sigma_g)$ state.[11] In this case resonant capture occurs much closer to the jellium surface, about 2-3 a_o, with the consequence that the H_2^+ in vibrational levels above about v = 3 will experience electric dissociation. For captures proceeding directly through the $X^1\Sigma_g$ state the fraction of incident ions lost to electric dissociation in step one is

$$f(1.,b^3\Sigma_u) = \sum_{v=3}^{18} p(v) = .56 \; , \tag{4}$$

a loss that is four times larger than for those captures proceeding via $b^3\Sigma_u$.

Calculations by Imke et al.[11] have shown that the orientation angle α can strongly influence the relative capture into the triplet $b^3\Sigma_u$ or the singlet $X^1\Sigma_g$ state, and these workers have calculated the relative final-configuration transition probabilities as a specific function of α. We can use these transition probabilities to obtain the total relative captures by integrating over all orientation angles α. For any given rotational state J, the spatial distribution obtained by summing over all orientations corresponds to an isotropic distribution. The ratio of captures into the respective states is then

$$\frac{P(b^3\Sigma_u)}{P(X^1\Sigma_g)} = \frac{\int_0^{\pi/2} P_T(\alpha)\sin\alpha d\alpha}{\int_0^{\pi/2} P_s(\alpha)\sin\alpha d\alpha} = 2.2, \tag{5}$$

The captures into the triplet state represent 69% of the total, and the singlet captures represent 31%. These triplet captures will in turn be relaxed by Auger capture into the ground state at 2-3 a_o from the jellium surface, with ejection of the excited $b^3\Sigma_u$ electron.

At this point there is now a mixture of $H_2(\underline{v}")$ and H atoms about to impinge upon the metal surface. These neutral species will in turn be reflected from the surface in the fourth and final stage of the recombination process as a new and final mixture of $H_2(v")$ and H.

We have developed a molecular trajectory model for the description of these $H_2(\underline{v}")$ collisions with surfaces.[14] For an arbitrary incident energy this model calculation identifies the two reaction channels

$$H_2(\underline{v}") + \text{Surface} \rightarrow H_2(v") \tag{6a}$$

$$\rightarrow H + H\ . \tag{6b}$$

We shall designate that fraction of incident $H_2(\underline{v}")$ molecules that survive as diatomic systems $H_2(v")$, by f_1. These fractions are listed in Table I as functions of the incident $H_2(\underline{v}")$ kinetic energy and the

Table I

$\underline{v''}$

E(normal)	2 f_1	2 f_2	6 f_1	6 f_2	8 f_1	8 f_2	10 f_1	10 f_2	12 f_1	12 f_2
1.0 eV	1.00	.00	1.00	.12	1.00	.42	.86	.56	.59	.45
4.0 eV	.86	.14	.86	.23	.44	.48	.35	.64	.64	.29

initial vibrational level $\underline{v}''$. Inspection of the table shows that for low incident energies and for not too large $\underline{v}''$ reaction (6a) is dominant. Also listed in the table is the fraction f_2, being the fraction of those in f_1 that lie in the portion of the vibrational spectrum, $6 \leq v'' \leq 14$. This fraction f_2 is of importance in the discussion of applications, Section IV. The distribution $H_2(v'')$ obtained for a $H_2(\underline{v}''=6)$ molecule and with kinetic energy equal to one and four electron-volts is shown in Fig. 3.

Willerding et al.[2] have identified the H_2 and H fractions resulting from low-angle, 260 eV H_2^+ collisions on Ni(111) surfaces. These 5° glancing-collisions correspond to a normal energy component, $E_\perp$ = 2 eV. Integration over the H and H_2 fragment energy distributions obtained in their experiment and taking into account that each dissociation process leads to two H atoms, we find that 40% of the recombination processes lead to H + H and 60% of the processes lead to H_2. This implies immediately that the single-step capture process leading to $H_2(b^3\Sigma_u)$ formation (found to be 69% of the total in Eq. (5)) followed by dissociation into H + H, as proposed by Willerding et al., cannot be the operative process in H atom formation. Nor does the direct 31% capture into $X^1\Sigma_g$ explain the observed large H_2 fraction.

On the other hand if we compare the 4-step and 3-step process operating in parallel, .86 x 69% = 59% result in H_2 formation up to and through the electronic processes in the 4-step model (Eq. (3) and (5)), and .44 x 31% = 14% (Eq. 4) lead to H_2 formation in the 3-step model,

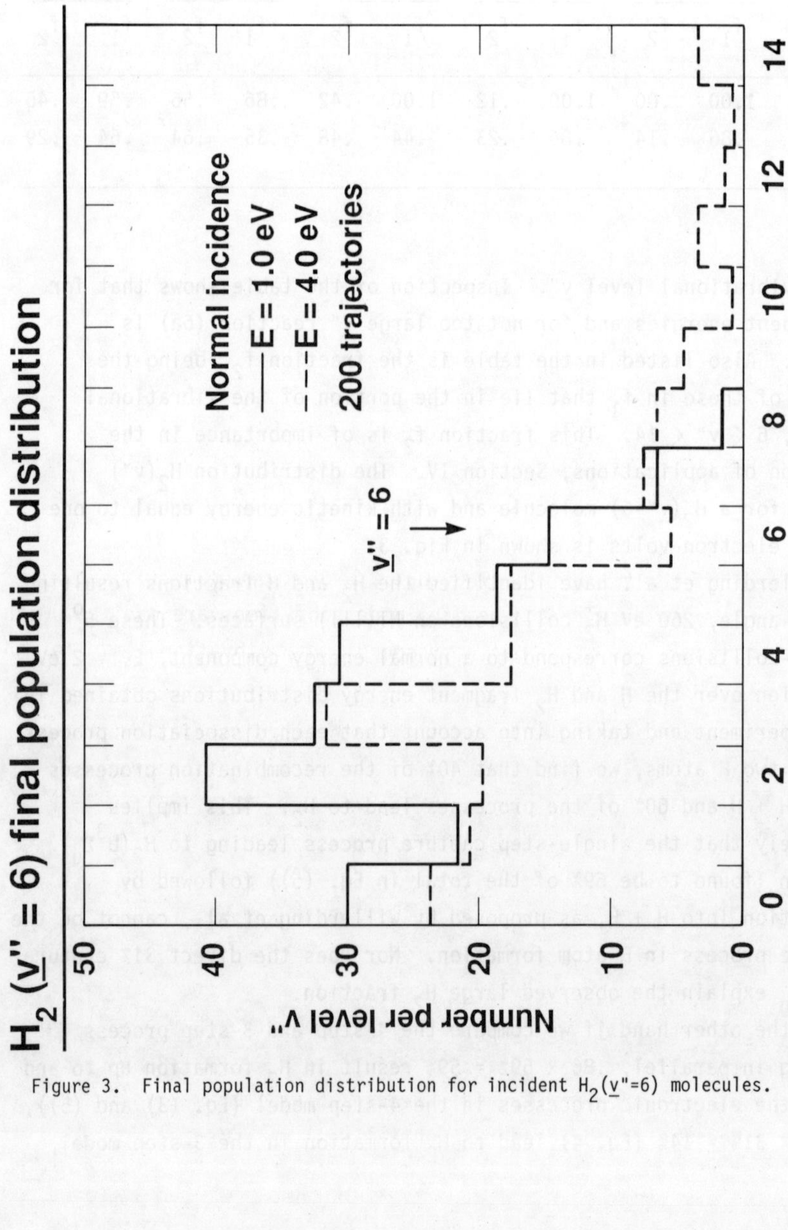

Figure 3. Final population distribution for incident H_2($\underline{v}$"=6) molecules.

for a total of 73% H_2 immediately prior to the surface scattering collision. Inspection of Table I shows that only those $H_2(\underline{v}'')$ in the highest $\underline{v}''$ are susceptible to dissociation in the final surface scattering, and if we attribute twenty percent to surface scattering dissociation into H + H, the large observed H_2 fraction is explained.

III. DISSOCIATIVE RECOMBINATION OF H_3^+ IONS

In a qualitative way the recombination of H_3^+ ions on surfaces is similar to that of H_2^+ recombination. But the complexity of the problem is compounded by the necessity to consider three relative vibrational motions rather than the single vibrational coordinate, R_N, in the H_2^+ problem. An essential point in the H_3^+ problem is that this ion is also formed with a broad vibrational population distribution.

The formation process leading to H_3^+ is

$$H_2^+(v) + H_2(v''=0) \rightarrow H_3^+(v''') + H \tag{7}$$

A semi-emperical model for this process has been developed by Anicich and Futrell.[15] These authors find that a broad spectrum of symmetric and degenerate asymmetric and bending mode vibrations are excited, with the maxima of the distributions corresponding to vibrational excitation energies in the range of one to two electron volts.

The recombination of H_3^+ is described by a four-step process culminating in the dissociation products $II_{a,b}$. These four steps, taken in sequence, are:

(i) Electric dissociation of $H_3^+(v''')$ by the image field proceeds according to

$$H_3^+(v''') + \mathcal{E} \rightarrow H_2(v'') + H^+. \tag{8}$$

The threshold is near 10^5 volts/cm corresponding to electric dissociation of the v'''=12 bending and asymmetric modes. Because of the deeper potential well of H_3^+ compared with that of H_2^+, something less than two percent of the H_3^+ ions are electrically dissociated prior to resonant electron capture.

(ii) Near 7-10 a_o from the surface resonant or Auger electron capture occurs to form the 2p $^2E'$ degenerate electronic ground state of H_3. Since we are concerned here with large amplitude vibrations causing large excursions away from the $2p^2$ $^2E'$ equilateral configuration, we shall discuss the electronics in terms of the two intersecting electronic surfaces of the Jahn-Teller system, $^2A_1$, $^2B_2$, developed from 2p $^2E' \rightarrow {}^2A_1$, $^2B_2$.

(iii) Following electron capture the unstable H_3 molecule while drifting toward the surface spontaneously divides according to $H_3(v''') \rightarrow H_2(\underline{v}'') + H$.

(iv) A "hard" nuclear collision scatters the dissociation fragments, $H_2(\underline{v}'')$, and H, from the surface to provide a new and final distribution, $H_2(v'')$ and H.

The electron capture process is briefly described by reference to the electronic energy surfaces shown in Fig. 4. In the upper figure is the H_3^+ ground state electronic surface appropriate to isosceles internuclear configurations.[16] In the lower figure is a parametric representation of the $^2A_1$, $^2B_2$ potential surfaces of H_3.[17] The vertical scale, Y, in atomic units, of the upper figure becomes the horizontal scale in the lower figure. An H_3^+ ion in its lowest vibrational level, near point O., transfer to the new point O. in the lower diagram following electron capture and a vertical Frank-Condon transition. Inspection of the lower figure shows that the point O. lies near the position of the v"=0,1 levels if the H_2 potential is taken as the projection of the surface onto the plane of the figure. We would expect this to lead to an H_2(v"=0,1) as a dissociation fragment. This is confirmed by an explicit calculation of the Frank-Condon factors linking the ground vibrational level of H_3^+ with the lower H_3 levels.[18]

Resonant electron capture

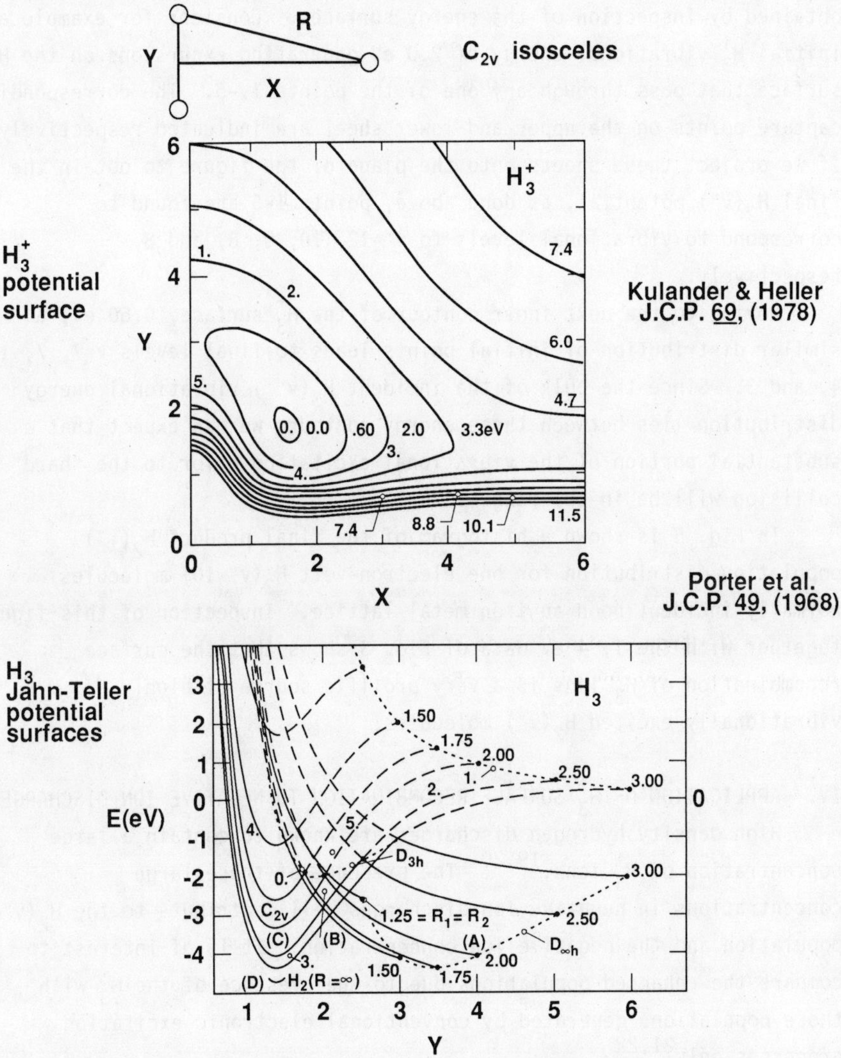

Figure 4. Potential energy contours for the H_3^+ and H_3 electronic states.

The Frank-Condon factors coupling the higher levels of H_3^+ to H_3 are not available but an estimate of the final v" distribution can be obtained by inspection of the energy surfaces. Consider for example an initial H_3^+ vibrational energy of 2.0 eV generating excursions on the H_3^+ surface that pass through any one of the points 1.-5. The corresponding capture points on the upper and lower sheet are indicated respectively. If we project these sheets onto the plane of the figure to obtain the final H_2(v") potential, as done above, points 1-5 are found to correspond to vibrational levels to v"=12, 10, 0, 8, and 8, respectively.

Moving to the next inner contour of the H_3^+ surface, 0.60 eV, a similar distribution of initial points leads to final levels v=7, 7, 1, 4, and 3. Since the bulk of the incident H_3^+(v"') vibrational energy distribution lies between these energy contours we can expect that a substantial portion of the vibrational excitation prior to the "hard" collision will be in the range v"=6-12.

In Fig. 5 is shown a histogram of the final product H_2(v") population distribution for one electron-volt H_2(v"=10) molecules normally incident upon an iron metal lattice. Inspection of this figure together with the 1, 4 eV data of Fig. 3 shows that the surface recombination of H_3^+ ions is a very prolific source of highly vibrationally excited H_2(v") molecules.

IV. APPLICATION OF H_3^+ SURFACE RECOMBINATION TO NEGATIVE ION DISCHARGES

High density hydrogen discharges are known to contain a large concentration of H_3^+ ions.[19,20] The presence of these large concentrations in negative ion discharges will contribute to the H_2(v") population and the negative ion concentration. It is of interest to compare the enhanced populations due to the presence of the H_3^+ with those populations generated by conventional electronic excitation processes only[21,22].

In a discharge with electron temperature kT, electron and H_3^+ density n, and chamber radius R_0, the H_3^+ ions flow to the chamber walls with a flux equal to the product of the H_3^+ density and the sound

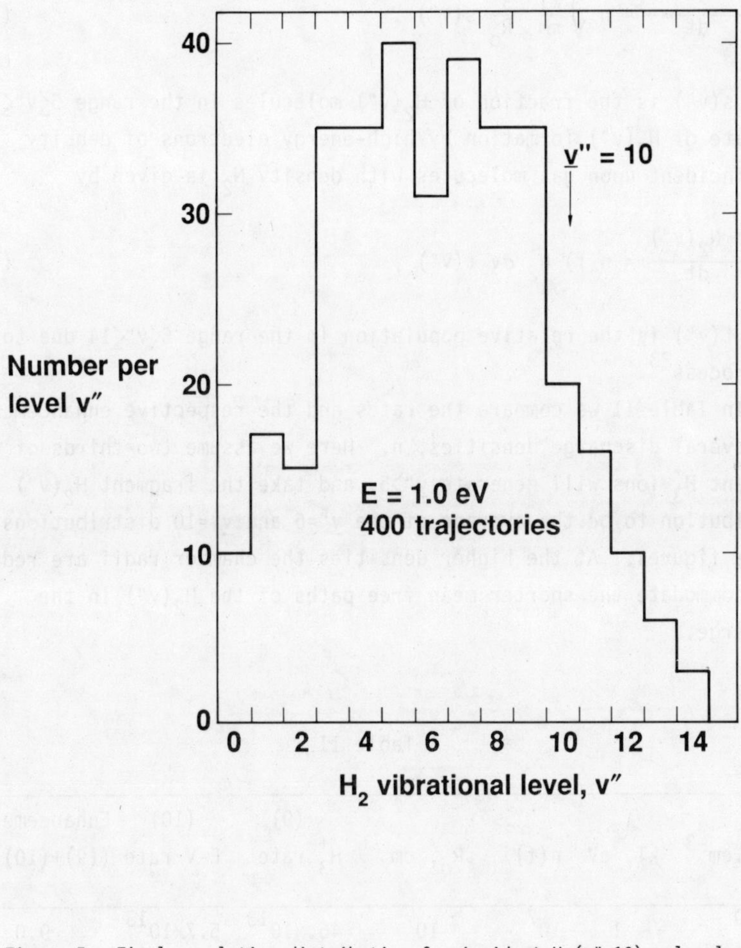

Figure 5. Final population distribution for incident $H_2(\underline{v}"=10)$ molecules.

velocity. The residual $H_2(v")$ fragments in turn fill the volume of the plasma chamber. The rate of $H_2(v")$ formation due to the H_3^+ surface process is then

$$\frac{d\,N_2(v")}{dt} = n\sqrt{\frac{kT}{\pi M}}\,\frac{3}{R_o}\,s(v")\ , \tag{9}$$

where $s(v")$ is the fraction of $H_2(v")$ molecules in the range $5 \lesssim v" \lesssim 14$. The rate of $H_2(v")$ formation by high-energy electrons of density $n(f)$ incident upon gas molecules with density N_2 is given by

$$\frac{d\,N_2(v")}{dt} = n(f)\,N_2\,\sigma v\,t(v")\ , \tag{10}$$

where $t(v")$ is the relative population in the range $5 \lesssim v" \lesssim 14$ due to the E-V process[23].

In Table II we compare the rates and the respective enhancements for several discharge densities, n. Here we assume two-thirds of the indicent H_3^+ ions will generate $\underline{v}">5$, and take the fragment $H_2(v")$ distribution to be the average of the $\underline{v}"=6$ and $\underline{v}"=10$ distributions shown in the figures. At the higher densities the chamber radii are reduced to accommodate the shorter mean free paths of the $H_2(v")$ in the discharge.

Table II

n, el.cm^{-3}	kT, eV	n(f)	R_o, cm	(9) H_3^+ rate	(10) E-V rate	Enhancement ((9)+(10))/(10)
10^{10}	1	10^7	10	$46.\mathrm{x}10^{13}$	$5.7\mathrm{x}10^{13}$	9.0
10^{11}	2	10^9	10	$7.3\mathrm{x}10^{15}$	$5.7\mathrm{x}10^{15}$	2.3
10^{12}	5	$5\mathrm{x}10^{10}$	3	$4.4\mathrm{x}10^{17}$	$3.2\mathrm{x}10^{17}$	2.4
10^{13}	5	10^{12}	1.5	$8.7\mathrm{x}10^{18}$	$7.0\mathrm{x}10^{18}$	2.2

Inspection of the table shows a large relative enhancement at lower densities diminishing toward the higher densities. The negative ion concentrations will be enhanced proportionately.

Earlier we have commented that the observed negative ion concentrations in the low-density discharges appeared to be at least an order-of-magnitude larger than could be explained by electron collision processes only.[24] The large H_3^+ enhancements calculated here for the low-density case may serve to remove the major portion of the discrepancy.

This work was performed under the auspices of the U.S. Department of Energy by the Lawrence Livermore National Laboratory under contract number W-7405-ENG-48 and USAFOSR contract No. AFOSR-ISSA 88-0002.

REFERENCES

1. J.R. Hiskes and A.M. Karo, AIP Conference Series Number 111, 125 (1984).
2. B. Willerding, W. Heiland, and K.J. Snowdon, Phys. Rev. Lett. 53, 2031(1984).
3. B. Willerding, K. Snowdon, and W. Heiland, Z. Phys. B 59, 435 (1985).
4. W. Eckstein, H. Verbeek, and S. Datz, Appl. Phys. Lett. 27, 527 (1975).
5. J.B.A. Mitchell and W.G. Graham, AIP Conf. Proc. No. 158, 39 (1987).
6. J.R. Hiskes, Phys. Rev. 122, 1207 (1961).
7. R.F. Stebbings, Proc. Roy. Soc. A241, 270 (1957).
8. H.D. Hagstrom, Phys. Rev. 89, 244 (1953).
9. A. Cobas and W.E. Lamb, Phys. Rev. 65, 327 (1944).
10. J.R. Smith, S.C. Ying, and W. Kohn, Phys. Rev. Lett. 30, 110 (1973).
11. U. Imke, K.J. Snowdon, and W. Heiland, Phys. Rev. B 34, 41 (1986); 34, 48 (1986).
12. T.E. Sharp, At. Data 2, 119 (1971).
13. S. Rothenberg and E.R. Davidson, J. Mol. Spectry. 22, 1 (1967).

14. A.M. Karo, J.R. Hiskes, and R.J. Hardy, J. Vac. Sci. Technol. A3(2), 1222 (1985).

15. V.G. Anicich and J.H. Futrell, Int. J. Mass Spectr. and Ion Processes, 55, 189 (1983/1984).

16. K.C. Kulander and E.J. Heller, J. Chem. Phys. 69, 2439 (1978).

17. R.N. Porter, R.M. Stevens, and M. Karplus, J. Chem. Phys. 49, 5163 (1968).

18. K.C. Kulander and J.C. Light, J. Chem. Phys. 85, 1938 (1986).

19. Y. Okumura, M. Oraki, M. Hanada, H. Horiike, T. Inoue, T. Kurishima, S. Matsuda, Y. Ohara, S. Tanaka, and K. Watanabe, Proc. IAEA Technical Committee Meeting on Negative Ion Beam Heating, Culham Laboratory, 15-17 July, 1987. M. Nightingale, Editor, and private communication.

20. J. Uramoto, AIP Conference Service No. 158, 319 New York (1987).

21. J.R. Hiskes, A.M. Karo, and P.A. Willmann, J. Appl. Phys. 58, 1759 (1985).

22. J.R. Hiskes, A.F. Lietzke, Proc. IAEA Technical Committee Meeting on Negative Ion Beam Heating, Culham Laboratory, 15-17 July, 1987. M. Nightingale, Editor.

23. J.R. Hiskes, J. Appl. Phys. 51(9), 4592 (1980).

24. J.R. Hiskes, Comments At. Mol. Phys. 19(2), 59 (1987).

POSSIBLE SYNTHESIS OF LARGE CLUSTER IONS OF ANTIMATTER AND THE ROLE OF DISSOCIATIVE RECOMBINATION, John T. Bahns, Pacific-Sierra Research Corp., 12340 Santa Monica Blvd., Los Angeles, CA 90025 and William C. Stwalley, Center for Laser Science and Engineering and Departments of Chemistry and Physics, University of Iowa, Iowa City, IA 52242-1294

ABSTRACT

The problem of producing the first "seed crystal" antihydrogen cluster ion ($\overline{H_N^+}$) is discussed. We first present an overview of large cluster ion synthetic techniques, starting from elementary constituents. The concept of a pseudowall is introduced and some examples discussed. Experimental details and projected capabilities of a pseudowall for large cluster ion studies are presented. Finally, we discuss dissociative recombination and how it relates to the synthesis of $\overline{H_N^+}$.

I. INTRODUCTION

The goal we have before us is an unprecedented one: to "synthesize" from atomic ("nanoscopic" as opposed to microscopic) constituents bulk macroscopic substances. Moreover, we are to do this for low density antimatter ($\bar{p}$ and e^+) in the <u>complete</u> absence of contact with ordinary matter, whether in the form of walls, background gas or even radioactive decay fragments. And moreover, we are to do this at technically challenging low temperatures and at unprecedentedly low pressures (not only under static conditions, but also as valves are opened, lasers sent through windows, etc.). Moreover, it is not sufficient merely that each step in the synthesis be scientifically feasible; we must examine questions of rate (throughput) and efficiency (losses) related to temporal financial feasibility as well.

We adopt only a single major strategy in addressing the problem of synthesis of bulk quantities of antimatter, the antimatter cluster ion concepts[1]. Other strategies include such concepts as storing antimatter in ordinary matter (explored by Larry Campbell[2]); mass production of ultralight ion traps ($m_{trap} << 10^9\ Nm_{\bar{p}}$), each of which contains a small number N of antiprotons; trapping of antihydrogen atomic gas using lasers and/or magnetic fields[3]; and magnetic levitation of an antihydrogen molecular solid.[3] The entire field is admirably surveyed in a recent proceedings.[4]

In what follows we first overview the synthesis for normal matter and then discuss the various steps in sequence. We then examine the pseudowall strategy to condensation. Next, we give a brief description of a pseudowall specifically designed for cluster ions that is currently in the initial stages of construction. Then we examine the impact of dissociative recombination. Finally we summarize our opinion of the primary issues which should be studied in the near future. Note also that normal matter ions are of considerable interest, not only scientifically, but also in connection with ionized or neutral beams for directed energy, fusion, solid state, medical and other applications.

II. OVERVIEW OF THE CLUSTER ION SYNTHESIS PROCESS

A. Background

Before examining individual steps in the cluster ion synthesis process[5],[6] it is worth giving a brief overview of the steps involved. Our initial assumptions were as follows:

1. A source of perhaps $10^7\ \bar{p}$/s is available at energies <1 eV.
2. A source of perhaps $10^7\ e^+$/s is available at energies <1 eV.
3. Intense sources of photons (microwave through vacuum ultraviolet) are available.

4. Intense sources of e^- of <1 eV energy are available.
5. Normal matter can be used to fully simulate antimatter in advance of antimatter experiments.
6. High densities of $\bar{p}$ and e^+ (say $\geq 10^{12}/cm^3$) are unavailable, so most steps must be second order or less in these antimatter constituents. If high density antimatter (say $10^{18}\ \bar{H}/cm^3$) can be achieved, third order processes become possible.

Note that existence of the proper formation steps for the various intermediate products is insufficient; one must also be able to trap and rapidly cool the intermediate product in the absence of collisions and also rapidly transfer the intermediate product to the next reaction chamber. While these latter steps are not too difficult for charged particles, they are quite difficult for neutral species (H and especially H_2). An overview of important reactions, and trapping, cooling and transfer techniques is given in Table I. Note the convention that the bar over an ion implies a change of all particles to antiparticles; thus $\overline{H^-}$ is positively charged while $\overline{H_2^+}$ is negatively charged. These processes have been extensively examined in the _Proceedings of the Cooling, Condensation and Storage of Hydrogen Cluster Ions Workshop_, J. T. Bahns, Editor[1], and our references are given there. In addition, a bibliography of several hundred manuscripts (published since 1966) dealing with the positive and negative hydrogen cluster ions is available[7].

B. The "Seed Crystal Concept

Before enumerating the synthetic steps, we note the catalyst ("seed crystal") strategy in our synthesis[5],[6]. Initially we must synthesize a single cold ($T \lesssim 1$ K) large cluster ion H_N^+ such that radiative stabilization via photon (γ) emission

$$H_N^+ + H \rightarrow H_{N+1}^+ + \gamma \qquad (1)$$

is highly efficient compared to loss processes such as

$$H_N^+ + H \rightarrow H_{N-1}^+ + H_2. \qquad (2)$$

We then grow the ion at a rapid rate to $H_{\sim 2N}^+$. The $H_{\sim 2N}^+$ ion is then fragmented e.g. using

Table I. Overview of Particle (and Antiparticle) Properties and Characteristics.

particle	e^-	$H^+(\equiv P)$	H	H^-	H_2^+	H_2	H_3^+	H_3	H_4^+	H_4	... $H_{N‡}^+$	$H_{N‡}$
antiparticle	e^+	$\overline{H^+}(\equiv \overline{P})$	$\overline{H}$	$\overline{H^-}$	$\overline{H_2^+}$	$\overline{H_2}$	$\overline{H_3^+}$	$\overline{H_3}$	$\overline{H_4^+}$	$\overline{H_4}$	... $\overline{H_{N‡}^+}$	$\overline{H_{N‡}}$
binding energy in eV	-	-	13.6	0.75	2.6	4.5	4.1	0	0.1	<0.01	... 0.008	0.008
products			H^+,e^-	H,e^-	H^+,H	H,H	H^+,H_2	H,H_2	H,H_3^+	H_2,H_2	H_{N-2}^+,H_2	H_{N-2},H_2
sensitivities†:												
λ < 0.2 μ VUV	-	-	x	x	x	x	x	x	x	x	x	x
0.2-0.7 μ UV-VIS	-	-	-	x	x	-	(x)	-	(x)	-	x	x
0.7-100 μ IR	-	-	-	x	-	-	(x)	-	(x)	-	x	x
λ > 100 μ microwave	-	-	-	-	-	-	-	-	-	-	x	x
radiative recomb. (e^-)	*	x	x	-	(x)	-	(x)	-	(x)	-	(x)	-
dissoc. recomb. (e^-)	*	-	-	-	x	-	x	-	x	-	x	-
assoc. detach. (H^-)	-	-	x	*	-	-	-	x	-	-	-	-
rxn. (H)	-	-	*	-	x	-	x	(x)	x	-	-	-
rxn. (H_2)	-	-	-	-	x	*	-		x	-	-	-
neutralization (H^-)	-	x	-	*	x	-	x	-	x	-	x	-
radiative assoc. (H, H_2)	-	-	-	-	-	-	-	-	-	-	x	x
trapping:												
Penning/Paul	x	x	-	x	x	-	x	-	x	-	x	-
laser	-	-	x	(x?)	x?	x?	-	-	-	-	-	-
inhomogeneous magnet	-	-	x	-	-	-	-	-	-	-	x	x
cooling:												
radiative	-	-	-	-	-	-	x	-	x	-	x	x
laser	-	-	x	-	x?	x	-	-	-	-	-	-
e^-	x	x	-	x	-	-	-	-	-	-	-	-
transfer:												
electrostatic	x	x	-	x	x	-	x	-	x	-	x	-
magnetic	-	-	x	-	-	-	-	-	-	-	-	x
laser	-	-	x	-	(x?)	(x?)	-	-	-	-	-	-

‡ N assumed large (> 10^3).

† All collision partners assumed to be cold (≲ 4K).

* Reactant.

$$H^+_{\sim 2N} + H^+ \rightarrow 2H^+_{\sim N} \quad (3)$$

$$(\text{or } H^+_{\sim 2N} + \gamma \rightarrow 2H^+_{\sim N} + e^-) \quad (4).$$

Thus our H^+_N "seed crystal" has served as a catalyst and rapidly converted N H atoms and 1 H^+ into a H^+_{N+1} ion (Figure 1a).

If good methods for synthesizing and manipulating H_2 can be found, equations analogous to (1)-(4), but with H_2 in place of H, could be used even for very small N (Figure 1b; N ~ 27 was estimated in[8]; recent measurements[9] suggest a smaller N ~ 19). This is because only ~0.008 eV in binding energy must be dissipated when H_2 is added to H^+_N, versus ~4.5 eV when H is added to H^+_N with N even. Hence significant efforts towards developing methods for H_2 production and manipulation are particularly needed and worth pursuing.

It is essential to stress the importance of this initial chunk of antimatter (like the caveman's "first flame", the alchemist's "philosopher's stone" or ice-IX in <u>Cat's Cradle</u>). Assuming the above growth and fragmentation equations ((1) and (3) or (4)) are correct, once that chunk has been produced (and backups produced and stored), the temporal and financial feasibility questions are reduced greatly. One is then <u>only</u> concerned with $\bar{p}$, e^+, H and the "seed crystal" (and if at all possible H_2 (figure 1b)). This is not true of other storage concepts (see Section I) except the magnetic levitation of antihydrogen "ice" (molecular solid). For this reason, the single most important issue in bulk storage via ions is to establish that equations (1) and (3) (or (4)) do in fact occur rapidly and efficiently for a reasonable N (10^3-10^6? for H or ~19 for H_2), with no significant side reactions (such as H_2 boil off). Of course, some other catalyst might be used if one can be found: some extraordinary form of ordinary matter such as H^--coated liquid helium; or "pseudowalls" of photons, positrons, antiprotons or other particles (see Section III).

C. Individual Reaction Steps

How then will we produce this first "seed crystal"? As previously discussed,[8] there are a number of conceivable options involving cluster ions: addition of H to positive ions (A^+), addition of H_2 to positive ions (M^+) and analogous processes for negative ions (A^- and

Figure 1a. Flow chart for bulk antimatter fuel production using $\overline{H}$ directly.

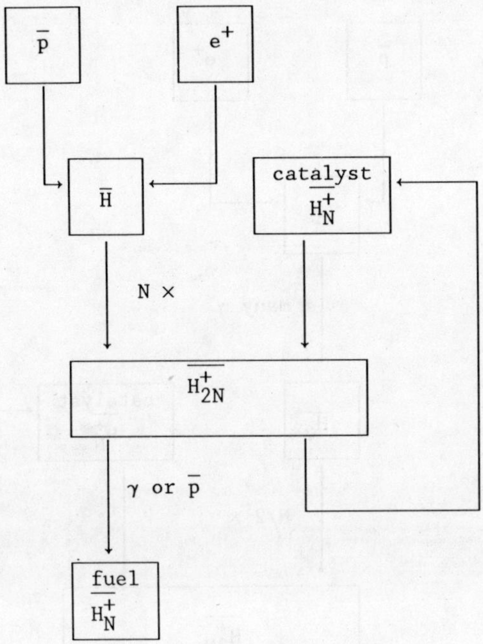

M^- respectively). However, although H^- is stable, H_2^- is not (yielding $H_2 + e^-$ spontaneously) and H_3^- is now thought to be unstable on both experimental[10] and theoretical[11] grounds. Although for some larger N, H_N^- must be stable (as $N \to \infty$, the processes $H_N^- \to H_N + e^-$ and $H_N \to H_N^+ + e^-$ require the same positive energy), the value of that N is not established. Thus only the positive ion processes will be considered here.

In what follows, we first list (Table II) options for production and storage of five potentially useful simple reagents: H, H^-, H_2^+, H_2 and H_3^+. We then list synthetic sequences involving H^+, e^-, photons and these reagents which lead potentially to the production of H_N^+. It

Figure 1b. Flow chart for bulk antimatter fuel production using $\overline{H}$ to make $\overline{H}_2$ and then using $\overline{H}_2$ for cluster growth.

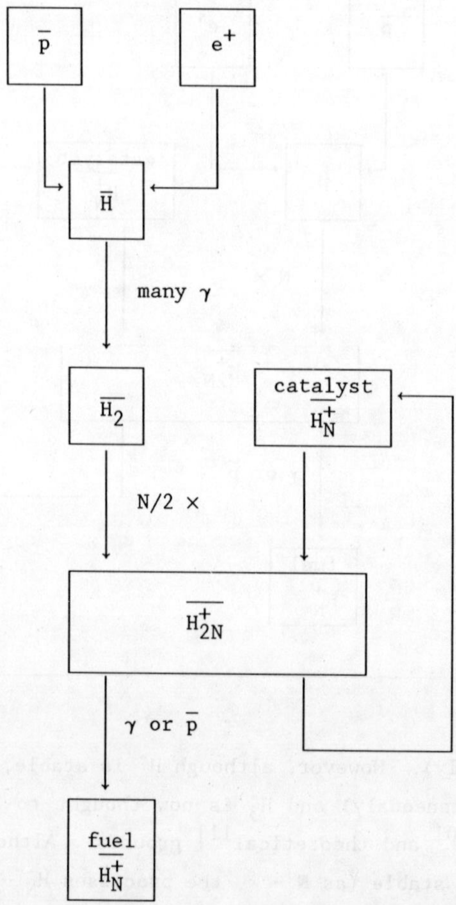

should be noted that there is a wide variety of possibilities, so we shall emphasize the simplest (e.g. Figure 2); determination of the optimal process deserves considerable further study.

Table II. Overview of possible steps in the "seed crystal" cluster ion synthesis process (* on letters indicate a currently favored process). For simplicity, substeps are given only in part 1.

---

1. Formation of H
   - a.* Stimulated Recombination
     - i. Cooling of H^+
     - ii. Cooling of e^-
     - iii. Stimulated recombination to H (2p or higher state)
     - iv. Radiative Decay of H (2p or higher state)
     - v. Cooling of H
     - vi. Trapping of H
     - vii. Transfer of H
   - b. Positronium Charge Transfer
     - i. Cooling of e^+
     - ii. Formation of positronium (e^+e^-)
     - iii. Charge Transfer of H^+ with (e^+e^-)
     - iv. Cooling of H
     - v. Removal of e^+
   - c.* High Density e^- Recombination
     - i. Formation of high density low temperature e^-
     - ii. Injection of H^+ into nested trap with e^-
     - iii. Formation of H^* (possibly H^- also)
     - iv. Decay of H^*
   - d. Spontaneous Recombination (analogous to but also inferior to 1.a.)
2. Formation of H^-
   - a. High e^- Density Recombination (see 1. c. above)
   - b. Photodissociation of H_2 to $H^+ + H^-$
   - c. Spontaneous Recombination of H + e^- (unlikely)
   - d. Collisional Ion Pair Formation ($H^* + H^* \rightarrow H^+ + H^-$)
3. Formation of H_2^+
   - a.* Associative Ionization ($H^* + H \rightarrow H_2^+ + e^-$)
   - b. Indirect Radiative Association

Table II (continued).

4. Formation of H_2
   a.* Reaction of $H_2^+ + H$
   b.* Indirect Radiative Association
   c. Associative Detachment ($H^- + H \rightarrow H_2 + e^-$)
   d. Direct Radiative Association (unlikely)
5. Formation of H_3^+
   a.* Reaction of H_2^+ with H_2
   b. Laser-Assisted Association of $H_2^+ + H$
   c. Laser-Assisted Association of $H^+ + H_2$
   d. Three-Body Recombination ($H_2^+ + H + H \rightarrow H_3^+ + H$)
6. Formation of H_N^+
   a. Laser-Assisted Association of H_{N-1}^+ and H
   b. Laser-Assisted Association of H_{N-2}^+ and H_2
   c. Sequential Three-Body Recombination

1. Formation of H. Formation of H is relatively straightforward by means of spontaneous or stimulated photorecombination; charge transfer from positronium to a proton; or three-body recombination of H^+ with a high e^- density. Mitchell[12] has reviewed the photorecombination and positronium charge transfer options in detail. He finds the stimulated photorecombination to be superior, and finds recombination to high n states possibly superior to the recombination to the 2p state. However, there is one important other option. This three-body recombination option was recently examined by Kells.[13] Based on the low temperature high e^- density device at La Jolla ($n_{e^-} \simeq 10^{10}/cm^3$, $T \simeq 10^{-3}$ eV) previously described,[14] he finds a production rate of $\geq 10^8$ $\bar{H}$/s possible using a LINAC source with e^+ accretion in a nested Penning trap.

2. Formation of H^-. Two-body recombination of an e^- with H to form H^- plus emission of a photon is presumably a very slow and unlikely process. However, three-body recombination ($2\,e^- + H \rightarrow H^- +$

Figure 2. Possible flow chart for antimatter catalyst ("seed crystal") production.

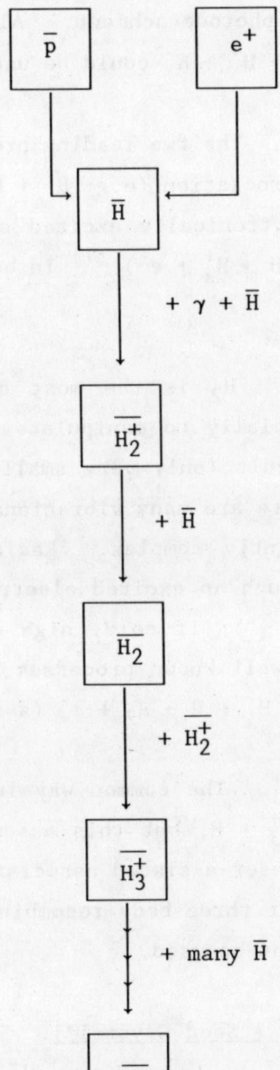

e^-) should occur fairly rapidly at high e^- density (although not so rapidly as ($2e^- + H^+ \rightarrow H + e^-$)). An alternative[15] is the ion-pair photodissociation process for H_2 ($H_2 + \gamma \rightarrow H^+ + H^-$). If H^- is an unwanted by-product to H formation by three-body recombination, it can be readily destroyed by photodetachment. Also the collisional ion pair process [16] $H^* + H^* \rightarrow H^+ + H^-$ could be used to produce H^-.

3. Formation of H_2^+. The two leading processes for H_2^+ formation are indirect radiative association (e.g. $H^+ + H + \gamma \rightarrow (H_2^+)^* \rightarrow H_2^+ + \gamma'$, where * indicates an electronically excited state[8]) and associative ionization (e.g. $H(3p) + H \rightarrow H_2^+ + e^-$).[17] In both cases, H_2^+ cooling is needed.

4. Formation of H_2. H_2 is the most difficult of the simple species to form and especially to manipulate (trap, cool, transfer). Magnetic traps are difficult (only very small nuclear and rotational magnetic moments) and there are many vibrational-rotational levels, so laser schemes are inherently complex. Radiative association[17] is possible, especially through an excited electronic state (e.g. $2H + \gamma \rightarrow H_2^*$ (e.g. $B^1\Sigma_u^+$) $\rightarrow H_2 + \gamma'$), if cold, high density H is available. Alternatives include the well known processes $H_2^+ + H \rightarrow H_2 + H^+$ (reaction/charge transfer) and $H^- + H \rightarrow H_2 + e^-$ (associative detachment).

5. Formation of H_3^+. The common way in which H_3^+ forms is via the reaction $H_2^+ + H_2 \rightarrow H_3^+ + H$, but this assumes the availability of H_2. Alternatively, the laser-assisted association reaction[8],[18] $H_2^+ + H + \gamma \rightarrow (H_3^+)^* \rightarrow H_3^+ + \gamma'$ or three-body recombination in high density H ($H_2^+ + H + H \rightarrow H_3^+ + H$) might be used.

6. Formation of H_N^+ ("Seed Crystal"). The same laser-assisted processes proposed for H_3^+ should work for H_N^+, but knowing the spectroscopy in detail for each N is a major undertaking. Conceptually simpler is an additional possibility which arises for the case of a high H density, namely sequential three-body recombination

$$H^+ + H + H \rightarrow H_2^+ + H$$

$$H_2^+ + H + H \rightarrow H_3^+ + H$$
$$\vdots$$
$$H_{N-1}^+ + H + H \rightarrow H_N^+ + H$$

for which there are many competing and complicating reactions such as

$$H_2^+ + H \rightarrow H^+ + H_2$$
$$H_2^+ + H_2 \rightarrow H_3^+ + H.$$

However, a net growth in ion size is expected regardless and collisional relaxation of vibrational energy can be significant. A particularly promising trapping technique for high H density (spin-polarized) is an inhomogeneous magnetic field[19],[20]; this was experimentally realized for Na.[21] Alternatives include laser traps[22] and hybrid laser-magnet traps[23], but for H they require a high power Lyman alpha source[24] and include a "recoil heating" mechanism for which cooling provisions must be made. The recently proposed microwave cooling technique[25] complements well known laser techniques (since the short wavelength laser radiation required can potentially destroy the molecular ions being produced).

III. THE PSEUDOWALL APPROACH

One of the primary difficulties in the cluster ion approach has been the general lack of rapid radiative pathways for the removal of recombination and condensation energy. The spontaneous radiative pathways are inefficient due to the large difference between collision times (~ 10^{-15}s) and the lifetimes for spontaneous radiation ($\gtrsim 10^{-9}$s). Stimulated radiative pathways have been looked at for just the smallest systems (H, H_2^+, H_2, H_3^+) and seem to work less well as the complexity of the system increases (only stimulated radiative association of H looks promising). The obvious way of overcoming these difficulties is to arrange the condensation in such a way that nonradiative collisional processes dominate (and radiative processes take care of themselves). To achieve this we propose the use of "pseudowalls".

A pseudowall is a cold, trapped, high density phase that may (or should) be thermally coupled to a heat sink (e.g. to electromagnetic fields if particles are charged; laser cooled if neutral). The trapped species could be positive, negative or neutral. For simplicity, in Table III we consider only pseudowalls consisting of a single species (and ignore spin statistics, e.g. for H_N^+). Cold here means kT << 0.008 eV (the desorption energy of H_2 from solid H_2) or T << 88 K. High density means that a particle hitting a pseudowall undergoes many collisions before it leaves. This idea is an extension of the liquid and solid pure electron plasmas that have been under development for some time by Driscoll and Malmberg at UCSD.[14]

To better define a pseudowall, we discuss a specific example; the important A^+ ion-neutral association reaction,

$$H + \{H_N^+\} \rightarrow \{H_{N+1}^+\}$$

where the braces denote a reactant is present at sufficient density to form a pseudowall. The reaction is characterized by an effective cross section Q. The pseudowall is characterized by a density n and length ℓ of longest dimension. The main requirement is that the mean free path L for an H atom propagating through the pseudowall be less than ℓ,

$$\frac{1}{nQ} = L < \ell .$$

If Q is 10^{-14} cm^{-2} and n is 10^{11} cm^{-3} (typical values for cluster ions smaller than 10^4), we find that $\ell > 10$ meters.

Three promising methods amenable to pseudowall condensation are outlined below:

1. Ion-neutral Association ("Seed Crystal" process)

   A. A^+ Method

   $$\{H_N^+\} + H \rightarrow \{H_{N+1}^+\}$$

   competing rxn.: $\{H_N^+\} + H \rightarrow \{H_{N-1}^+\} + H_2$

   B. M^+ Method

   $$\{H_N^+\} + H_2 \rightarrow \{H_{N+2}^+\}$$

   competing rxn.: $\{H_N^+\} + H_2 \rightarrow \{H_{N-2}^+\} + 2H_2$

2. "Seeded" H Atom (Growth of Initial "Seed Crystal")

   $$H^+ + \{H\} \rightarrow H_2^+$$
   $$H_2^+ + \{H\} \rightarrow H_3^+$$
   $$H_{N-1}^+ + \{H\} \rightarrow H_N^+$$

   competing rxn.: $H_2^+ + \{H\} \rightarrow H^+ + H_2$

   $$H_2^+ + H_2 \rightarrow H_3^+ + H$$

Table III. Low temperature high density pseudowall phases. Statistics associated with various spin states are not included.

Type of Particle	Statistics	Example(s)	Phases
Charged	Fermi-Dirac	H^+, e^-, H_N^+	Gas ($\Gamma \lesssim 1$) - Liquid ($2 \lesssim \Gamma \lesssim 155$) - Solid ($\Gamma \gtrsim 155$)
	Bose-Einstein	D^+	Gas/Condensed Phase?
Neutral	Fermi-Dirac	D	Gas - Fermi Liquid
	Bose-Einstein	H	Normal Gas - Bose-Condensed Gas ("superfluid")

3. Ion-Ion Recombination (using H^-)

$$2\{H^*\}\ (\text{or } H^{**} + H) \rightarrow H^+ + H^-$$
$$2\{H^*\}\ (\text{or } H^{**} + H) \rightarrow H_2^+ + e^-$$
$$H_2^+ + H^- \rightarrow H_3^+ + e^-$$
$$H_3^+ + H^- \rightarrow H_4^+ + e^-$$
$$H_{N-1}^+ + H^- \rightarrow H_N^+ + e^-$$
$$\text{competing rxn.:}\quad H_2^+ + \{H\} \rightarrow H^+ + H_2$$
$$H_2^+ + H^- \rightarrow H_2 + H$$
$$e^- + H_N^+ \rightarrow H_{N-2} + H_2$$

etc.

The main reactions which may take place in pseudowalls are illustrated in the reaction matrix of Fig. 3. Certain working assumptions were made in the construction of this matrix:

1. Many particles can be turned into pseudowalls.
2. Only two-reactant combinations are considered (reactant plus one or two pseudowall particles).
3. No nucleus larger than the proton.
4. Minimum dependence on radiative processes (photons are excluded).

5. No anions other than H^-.

6. No unstable or short-lived particles (for example, positronium, H_2^-, H_3^-, etc.).

Some reactions in Fig. 3 (with asterisk) could proceed spontaneously within the same pseudowall to grow large cluster ions. Recall that the cluster ions of sizes H_m^+ and H_a^+ are the threshold cluster ion sizes for ion-molecule (M^+) and ion atom (A^+) association, respectively. These have been postulated to exist at sizes around 19 to 27 and 2000, respectively.

The objective is to select the optimum pathway to these large cluster ions by connecting reactions and defining pseudowalls. One example (out of many) of a "containerless" condensation pathway is designated by the large arrows in Fig. 4.

The reactions indicated by the condensation pathway in Fig. 4 may be written specifically as:

1. $\{e^-\} + H^+ \rightarrow H$
2. $\{H\} + H^+ \rightarrow H_2^+$
3. $\{H\} + H_2^+ \rightarrow H_3$
4. $\{H\} + H_3^+ \rightarrow H_4^+$
5. $\{H\} + H_4^+ \rightarrow H_5^+$

Obviously, there are many pathways that may be identified in the reaction matrix and much work will be needed to sort through the possibilities.

IV. EXPERIMENTAL DESIGN FOR A CHARGED PSEUDOWALL

The system described here is similar to the one proposed recently by Kluge et al.[26] However, since it will contain an annular plasma, it has many of the design features developed by Wong and Rosenthal.[38]

The important components (from outside) are: Helmholz coils, superconducting magnet, outer vacuum wall, liquid helium baffle, trap wall, annular plasma (pseudowall) and central high voltage conducting rod (to charge neutralize).

Trap performance calculations have been performed for typical values of the trapping parameters:

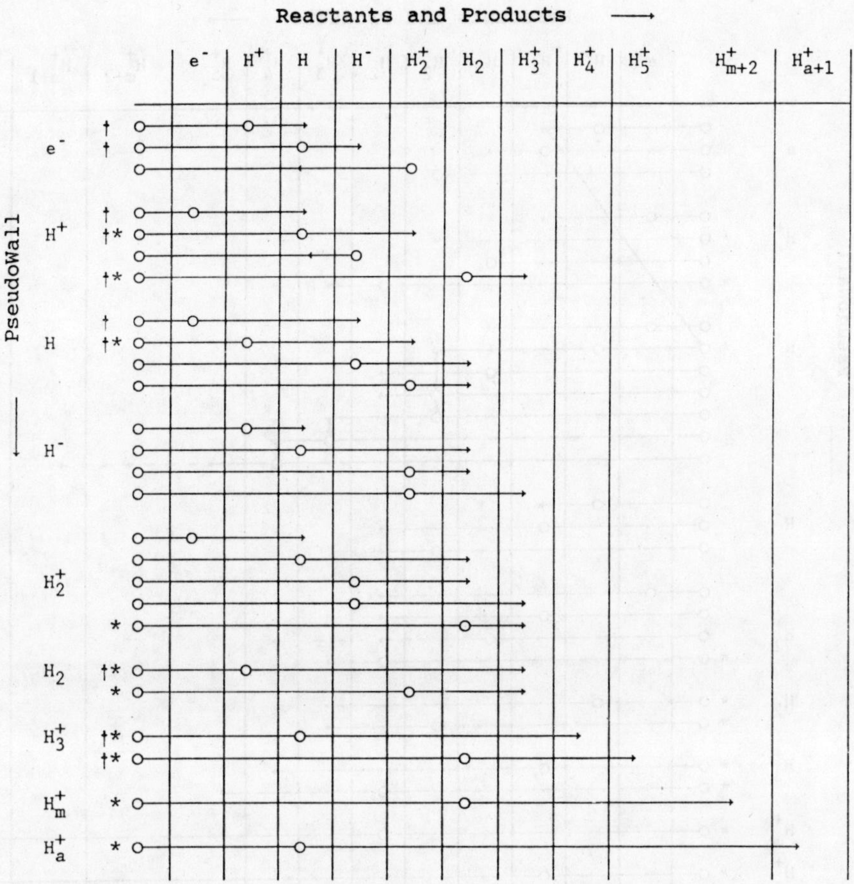

Figure 3. <u>Reaction</u> <u>Matrix</u> <u>Showing</u> <u>Containerless</u> <u>Condensation</u> <u>Routes</u>. In the first reaction, protons are added to a pseudowall composed of electrons, resulting in the production of cold hydrogen atoms. * Indicates reactions which could proceed spontaneously to grow large cluster ions. † Indicates reactions which nonradiatively could only proceed via three body processes.

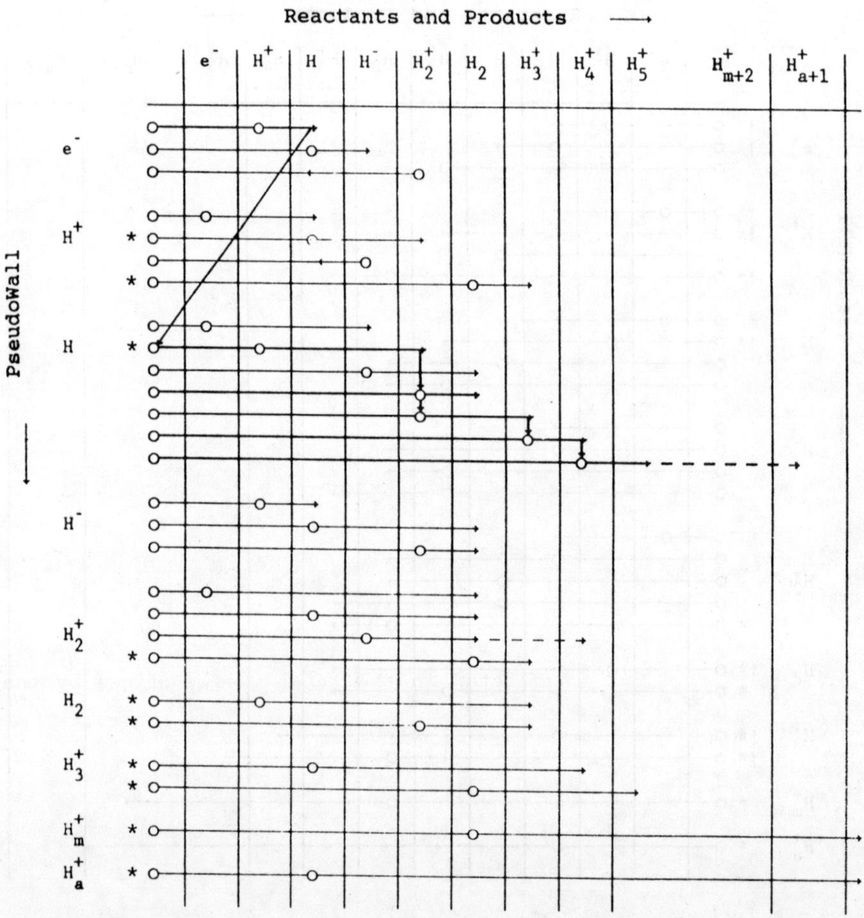

Figure 4. <u>Reaction Matrix Showing Containerless Condensation Routes</u>. In the first reaction, protons are added to a pseudowall composed of electrons, resulting in the production of cold hydrogen atoms. Then H_N^+ cluster ions of increasing size are added to a pseudowall consisting of cold hydrogen atoms.

B: The magnetic field (10 T).

Φ: The potential applied to the central conducting rod (1000 Kvolts).

a: The radius of the central conductor (0.3 cm).

d: The radius of the trap wall (5 cm).

r: The radius of the plasma (2.5 cm).

Δ: The thickness of the (annular) plasma (0.2 cm).

L: The length of the magnet (100 cm).

For a discussion of the theory the reader is referred to the work of Wong and Rosenthal.[27]

For this device, the plasma number densities and mass densities achievable have been calculated as a function of cluster number (number of proton masses). There is a sharp increase in mass density for the clusters larger than 10^6, with $>10^{-10}$ g/cm^3 for $N > 10^8$. This region indicates the dominance of electrostatic confinement for large cluster numbers (and the balance between the number of charges on the central conductor rod and the number of charges in the plasma).

The effect is also reflected in the total mass confined by the trap for cluster numbers beyond 10^6. Note that for a trap only 5 times larger than the device now being constructed by the University of Dayton for the Air Force Astronautics Lab (AFAL) a significant amount of energy is storable: if the mass were in the form of $\overline{H_N^+}$, the trapped mass of $\sim 1 \times 10^{-8}$ grams is equivalent to $\sim 10^9$ Joules!

Finally we note that this device will produce a cold plasma (due primarily to the inclusion of the liquid He baffle) of cluster ions that exhibit ordering (collective effects). The relevant parameter,

$$\Gamma = e^2/akT,$$

relates the coulomb repulsive energy to the thermal energy, where a is given by $4/3\ \pi a^3 n = 1$, e is the charge, T is temperature, n is density, and k is Boltzmann's constant. $\Gamma = 1$ defines a "gaseous" plasma; at $\Gamma = 2$ one has a "liquid"; and finally at $\Gamma \simeq 155$ one obtains a "solid".[28] One desires the largest density and therefore the largest Γ obtainable. For example, for the described device, the achievable density of clusters will be roughly 10^8 cm^{-3} for a cluster number of 10^4. Since the temperature of liquid He is ~4 K, one obtains, for this example,

$$\Gamma \sim 120$$

which falls well within the liquid regime.

V. DISSOCIATION RECOMBINATION CONSIDERATIONS

How dissociative recombination affects the cluster ion synthetic method has been considered in detail by Mitchell.[12] Neutralization of the cluster ion is assured in all cases; however, the fragmentation depends upon a number of factors: (1) the relative kinetic energy of the collision, (2) the internal energy (vibration and rotation) of the cluster ion, and (3) the relative positions of the ground electronic state of the ion and the nearest doubly excited state of the neutral cluster. These factors have been studied, but only for the smallest cluster ions: H_2^+ and H_3^+.

The cross sections for the dissociative recombination of H_2^+ have been measured and vary between 10^{-16} - 10^{-14} cm^2 for the center of mass collisions in the range 0.01 - 10 eV. Cross sections for H_3^+ are presumably larger[12] although currently somewhat controversial (as discussed elsewhere in these proceedings). How the cross section scales with cluster size has been examined for water cluster ions[12] and was found to increase linearly with cluster size. Also, the rate coefficient for H_5^+ recombination has been measured, and is large (cross section ~ 1.5×10^{-13} cm^2 at 300 K). Cross sections for the larger cluster ions are not known. Presumably the cross sections will increase linearly with cluster size until the cluster becomes so large that the recombination energy can be absorbed into normal modes of vibration to yield a neutralized but vibrationally hot cluster.

Neutralized antimatter cluster ions (fragmented or not) would immediately leave the trap and annihilate on inner trap surfaces, releasing more neutral background matter into the trap environment. Naturally, this could lead to a self-propagating failure mode and is certainly of serious concern.

The likelihood of this failure mode depends to a large extent on the likelihood of positrons entering the trap. Fortuitously, it turns out that this is very unlikely if the trap is of the Penning type. The negative anticluster ions H_N^+ are surrounded by a solenoidal magnetic field which means the positrons can only enter down the trap axis. However, the anticluster ions are confined axially by negatively charged end plates. Thus, only positrons with adequate kinetic energy may enter the trap, and the cross sections for recombination even in this event will be quite small. It seems, therefore, that annihilation due to neutral background gas within the trap may be a more serious problem.

VI. SUMMARY OF MAJOR ISSUES

We feel that the concept of a pseudowall as applied to the problem of condensing antimatter into cluster ions offers additional advantages which should be explored. The most important issues in the cluster ion approach from our perspective are summarized in Table IV. The appropriate response in terms of research effort is then proposed in Table V. We have not included critical "engineering" issues such as how to maintain low temperature and ultralow pressure while carrying out a variety of manipulations (trapping, cooling, transferring, diagnosing, etc.) and how to provide for safety and obtain minimal failure rates.

Table IV. Summary of Major Issues

1. Given a recyclable "seed crystal", how can bulk antimatter best (throughput, efficiency) be grown from added $\bar{H}$ or $\overline{H_2}$? (i.e. do equations (1)-(4) hold? for what N? what is the best (fastest) cooling method for H_N^+?)
2. How can a "seed crystal" best be grown? (e.g. can $\bar{H}$ or $\overline{H_2}$ be stored in a "pseudowall" at high density (10^{18} H/cm^3 have been stored at low temperature), e.g. in a magnetic trap?)
3. How best can $\bar{H}$ be produced from $\bar{p}$ and e^+? (i.e. is the Kells' approach[13] with a charged pseudowall (e.g. 10^{10} e^+/cm^3 at 4.2 K) feasible?)
4. How best can $\bar{H}_2$ be produced and manipulated?

Table V. Recommended <u>Initial</u> Research Efforts Corresponding to the Major Issues in Table IV.

	normal matter	antimatter
1. "Seed Crystal" Process (Figure 1)	a. H_N^+ growth demonstrations [equations 1, 2] with H, H_2	
	b. H_N^+ fragmentation demonstrations [equations 3, 4] with H^+, γ	
	c. H_N^+ radiative/nonradiative cooling rates determination	
2. "Seed Crystal" Growth (e.g. Figure 2)	a. H_N^+ stepwise growth (for small N) optimization using H only and using both H_2 and H	e. cold high density $\overline{H}$ pseudowall
		f. cold $\overline{H}_2$ production
	b. Cold high density H pseudowall	
	c. H^+ injection into cold high density H	
	d. H_2 indirect radiative association	
3. $\overline{H}$ formation	a. cold high density e^- pseudowall	d. cold high density e^+ production
	b. H^+ injection into cold high density e^-	e. $\overline{p}$ injection into cold high density e^+
	c. $H^+ + e^- + \gamma$ stimulated recombination experiment	

VII. ACKNOWLEDGMENTS

We wish to thank Drs. G. Rosenthal, A. Y. Wong, J. B. A. Mitchell, and C. F. Driscoll for helpful discussions. Partial support of this work by the Air Force Office of Scientific Research is gratefully acknowledged.

VIII. REFERENCES

1] Bahns, J. T., Editor, Proceedings of the Cooling, Condensation, and Storage of Hydrogen Cluster Ions Workshop (SRI International, Menlo Park, CA, 1/8-9/87) (University of Dayton Research Institute, Dayton, 1987).

2] Campbell, L. A., "Normal Matter Antiproton Storage Program" in Ref. 4] below.

3] Forward, R. L., "Antiproton Annihilation Propulsion", Air Force Rocket Propulsion Laboratory Report, AFRPL TR-85-034 (1985).

4] Augenstein, B. W., Mills, F. E., Bonner, B. E. and Nieto, M. M., Editors, Proceedings of the Antiproton Science and Technology Workshop (Rand Corporation, Santa Monica, CA, 10/6-9/87) (World Scientific Publishing, Singapore, 1988)

5] Stwalley, W. C., "Large Hydrogen Cluster Ions", in Ref. 1] above, p. 39.

6] Stwalley, W. C., "The Synthesis of Large Cluster Ions from Elementary Constituents: A Possible Route to Bulk Antimatter", in Ref. 4] above.

7] Stwalley, W. C., "Bibliography of Hydrogen Cluster Ions", in Ref. 4] above.

8] Bahns, J. T., Sando, K. M., Tardy, D. C., and Stwalley, W. C., "Proceedings of the Hydrogen Cluster Ion Study Group", in Ref. 1] above, p. A1.

9] Kirchner, N. J. and Bowers, M. T., "Fragmentation Dynamics of Metastable Hydrogen Ion Clusters H_5^+, H_7^+, and H_9^+: Experiment and Theory", J. Chem. Phys. 91, 2573 (1987)

10] Bae, Y. K., Coggiola, M. J., and Peterson, J. R., "Search for H_2^-, H_3^-, and other Metastable Negative Ions", Phys. Rev. A 29, 2888 (1984).

11] Michels , H. H. and Montgomery Jr., J. A., "The Electronic Structure and Stability of the H_3^- Anion", Chem. Phys. Lett. 139, 535 (1987).

12] Mitchell, J., "The Role of Electron-Ion Recombination in Bulk Antimatter Production", in Ref. 1] above, p. 1435.

13] Kells, W. P., "e^+ Accumulation for $\overline{H}$ Production" in Proceedings of the Intense Positron Beams Workshop (6/17-19/87), E. H. Ottewitte and W. P. Kells, Editors, in press.

14] Driscoll, C. F., "Containment of Single-Species Plasmas at Low Energies", in Low Energy Antimatter, D. B. Cline, Editor (World Scientific, Singapore, 1986), p. 184.

15] Chupka, W. A., Dehmer, P. M. and Jivery, W. T., "High Resolution Photoionization Study of Ion-Pair Formation in H_2, HD, and D_2", J. Chem. Phys., 63, 3929 (1975).

16] Mitchell, B. "The Molecular Antihydrogen Problem Solved?", Mirror Matter 3, No. 11, 4 (1988).

17] Weiner, J., "Associative Two-Body Condensation in Laser-Cooled Sodium and Hydrogen", in Ref. 1] above, p. 47.

18] Saxon, R., "Overview of Hydrogen Clusters", in Ref. 1] above, p. 27.

19] Pritchard, D. E., "Cooling Neutral Atoms in a Magnetic Trap for Precision Spectroscopy", Phys. Rev. Lett. 51, 1336 (1983).

20] Hess, H. F., "Evaporative Cooling of Magnetically Trapped and Compressed Spin-Polarized Hydrogen", Phys. Rev. B34, 3476 (1986).

21] Migdall, A. L., Prodan, J. V., Phillips, W. D., Bergeman, T. T. and Metcalf, H. J., "First Observation of Magnetically Trapped Neutral Atoms", Phys. Rev. Lett. 54, 2596 (1985).

22] Chu, S., Bjorkholm, J. E., Ashkin, A. and Cable, A., "Experimental Observation of Optically Trapped Atoms", Phys. Rev. Lett. 57, 314 (1986).

23] Stwalley, W. C., "A Hybrid Laser-Magnet Trap for Spin-Polarized Atoms", Prog. Quant. Electr. 8, 203 (1984).

24] McIlrath, T. J., "Long Duration Coherent Lyman-Alpha Sources", in Ref. 1] above, p. 195.

25] Silvera, I. F., "Microwave Cooling of a Gas of Atoms: Spin-Polarized Hydrogen", (preprint) (1987).

26] Kluge, H. J., Schnatz H. and Schweikhard, L., "Penning Trap for Studying Cluster Ions", Phys. D - Atoms Molecules and Clusters, 3, 189, 1986.

27] Rosenthal, G., Dimonte, G., and Wong, A. Y., "Stabilization of the Dichotron Instability in an Annular Plasma", Phys. Fluids, 30, 3257 (1987).

28] Malmberg, J. H. and O'Neil, T. M., "Pure Electron Plasma, Liquid and Crystal", Phys. Rev. Lett. 39, 1333 (1977).

Rydberg State Stark Spectroscopy of Plasmas

J. R. Shoemaker, B. N. Ganguly and A. Garscadden
Air Force Wright Aeronautical Laboratories
Wright-Patterson Air Force Base, Ohio 45433

The measurement of the magnitude and direction of electric fields in plasmas by the response of laser excited high principal quantum number states is discussed. The technique has a wide dynamic range, from a few volts cm^{-1} to tens of kilovolts cm^{-1}. Experimental results for various regions of low pressure helium discharges are reviewed. Revisions are suggested in the application of the Inglis-Teller method.

INTRODUCTION

The energy gained by an electron in the electric field in a plasma determines its influence on dissociation, excitation, ionization and recombination. It is possible to measure the local electric field by means of Langmuir probes provided the plasma is uniform, that there is no net space charge and there are no species that react with the probe material (e.g. causing secondary emission or chemical reactions). However, these conditions are rather constraining. This paper reports on the optical measurement of the local electric field and its derivative. Several groups have recently applied Stark spectroscopy (1-5) to measure with high sensitivity the electric field in plasmas in various gases. The electric field in the different regions of the glow discharge (6) have been measured in this laboratory using high principal quantum number Rydberg state Stark spectroscopy (RS^3) in helium. Advantage is taken of the fact that in many discharge plasmas long lived metastables are present at concentrations of 10^{10} cm^{-3} or greater. A tunable laser is used to excite the high n- levels from the helium metastable 2^3S, 2^1S levels. Optogalvanic detection is used to measure the spectral characteristics that reflect the plasma environment.

In this study, we have investigated the polarization-dependent RS^3 of the singlet and triplet states of atomic helium from n=14 up to n = 46 to measure the electric field in low current discharges. Calculations have been performed for hydrogen and helium Rydberg states to provide a library of reference spectra of line positions and line intensities for the non-equilibrium plasma conditions of interest to plasma processing and lasers. The triplet states of atomic helium have small positive quantum defects and the energy levels are nearly degenerate for $l > 2$. The Stark spectra of high n states, therefore approach hydrogen-like especially for

$\Delta m_j=\pm1$ polarization. Experiments with a linearly-polarized, narrow band (0.3 cm^{-1}) laser, permit comparison of the experiment and theory for both $\Delta m_j=0$ and $\Delta m_j=\pm1$ transitions.

BACKGROUND

The interaction of an atom with external fields or with other atoms is determined by its electrons. In an unexcited atom the electrons occupy a volume with characteristic length of the order 10^{-8}cm. When radiation is selectively absorbed by the atom, an electron is excited to a higher level and the electron is moved away from the nucleus. The scale length becomes larger and the influence of the Coulomb field becomes weaker, thereby increasing the atom's sensitivity to external fields. The mean radius scales as n^2 and the energy levels (wrt ionization) scale approximately as n^{-2}. At larger n all atomic spectra become approximately hydrogenic, however, small measurable differences remain because of the polarization of the core of the atom. The differences are related to the hydrogen levels by the quantum defect. The quantum defects for helium (Wing, et.al., 1973) are quite small (7). The radiative lifetime of a Rydberg state is given approximately by $t=n^3\ 10^{-10}$ secs. Therefore, for large n, the levels become metastable. On the other hand, the cross sections for many collisional processes tend to scale as n^4, and under many plasma conditions the collisional frequencies of neutrals and charged particles with a Rydberg atom are very large. Therefore, when a Rydberg state is created in a discharge the excitation is either rapidly redistributed or the atom can be ionized by the neutrals if it is within a few kT (gas), or by electrons if it is within a few kT (electrons) of the ionization level. Hence, laser excitation to a Rydberg level results in a modulation of the ionization which in turn causes a change in the discharge current. This signal is termed the optogalvanic response.

EXPERIMENTAL APPROACH

A schematic of the experimental arrangement is shown in Figure 1. A Q-switched, frequency-doubled or-tripled Nd:YAG laser pumps a dye laser using DCM or Coumarin 500 dye. The linearly polarized dye laser output ($0.15cm^{-1}$ FWHM) is upconverted using a potassium dihydrogen phosphate (KDP) crystal, which is continuously angle-tuned (Inrad autotracker) to produce the UV output which is required to excite the selected high n atomic helium np^1P or np^3P states from the $2s^1S$ or $2s^3S$ metastable states respectively. The laser polarization can be adjusted with respect to the axis of the discharge. The dye laser pump beam and the UV beam are spatially separated using a Pellin-Broca prism. The wavelength of the dye laser is calibrated using the optogalvanic response of either an argon or a neon hollow cathode discharge. The laser beam is mildly focused to 0.3mm x 2mm spot size. The laser energy was typically 100 microjoules/pulse and its pulsewidth was 5 nsecs. Translation stages permitted examination of the cathode sheath, the negative glow and the positive column of the discharge.

RESULTS

The atomic helium spectrum for Rydberg states n=26 and higher (from 38280 to 38400 cm^{-1}) is shown in Figure 2 for a measurement in the negative glow of a 2 mA discharge in helium. In the negative glow, the applied field is low as ionization is provided by the beam electrons from the cathode (6). The Rydberg states are well-resolved initially and as expected, are closer and closer at higher n and finally merge into a series limit for the plasma conditions, at approximately n=46. The optogalvanic response of higher levels falls off. This is primarily caused by the decrease in transition probability (proportional to n^{-3}). Figure 3 illustrates the spectrum obtained in the positive column of the discharge (diameter 6mm, but approx 8mm in the observed region due to window stems). The field in the positive column is expected to be higher than in the negative glow as the diffusion losses have to be balanced by local ionization. There are several Stark-induced features of this spectrum. First, there is a variation in the linewidth of the transitions, and the linewidths are significantly larger than those of the negative glow spectrum. Secondly, the series advance into the continuum now occurs at significantly lower n (approximately n=35). Thirdly, there is the appearance of satellite structure in the optogalvanic spectrum, which are the normally forbidden 2S-nS transitions. The linear Stark line broadening, under the quasi-static approximation is given by:

$$\left|\Delta w\right| = \frac{3\hbar}{2emZ}\left(n_i^{*2} - n_f^{*2}\right)F$$

where Δw is the HWHM of the transition, n_i^* and n_f^* are the initial and final state quantum numbers, and F is the total electric field in atomic units. Therefore, the net electric field can be calculated, with useful redundancy, from the width of the series of Rydberg lines that are measured. Also, following several text references we estimated the net electric field from the series termination, using the equation $F(\text{v/cm})=1.23\times10^9\ n_m^{-5}$ where n_m is the principal quantum number whose intensity is identical to the background continuum signal level, i.e., the highest-n resolved line (however, see below). This method gave F=6 volts/cm in the negative glow and F=23 volts/cm in the positive column for the examples shown. If measurements are made off-axis in the positive column region of the discharge, the amplitude of the satellite structure increased and the series termination occurred at lower n; also the linewidths were larger off-axis than on-axis even though the ion density is decreasing. These effects occur because the Stark spectroscopy measures the net electric field, which in a bounded plasma, includes the ambipolar radial field (6). Thus, it is possible also to obtain information about the magnitude of the ambipolar field for axially symmetric situations since on-axis one has only the applied axial field, and it can be subtracted from the net field measured off-axis.

The progression of the series termination with the increase in radial field off-axis is consistent with the calculated magnitude of the net static Stark effect as the ion density of the plasma was less than $10^{11} cm^{-3}$. The appearance of the satellite structure is a consequence of the increasing radial field contribution to the net field. For the on-axis spectrum the laser polarization was arranged to be perpendicular to the applied axial field. Consequently, for the off-axis spectra, the laser polarization is no longer perpendicular to the E-field direction, i.e., the net electric field of the discharge is no longer purely axial. Hence $\Delta m_j=0$ transitions are Stark allowed and the forbidden S-S transitions are also observed in the optogalvanic response.

It is possible to exploit this observation to determine the electric field vector in the discharge. The laser polarization can be oriented to maximize the forbidden signal intensity for an off-axis position and hence to obtain the direction of the net field Fig 4 (top). A check can be made by turning the laser polarization orthogonal to the net field direction and the $\Delta m_j=0$ transitions are no longer observed, and the response is indicative of a pure $\Delta m_j=\pm1$ spectrum Fig 4 (bottom). The Stark spectra of helium with $\Delta m_j=\pm1$ polarization are also the more interesting to compare with theory because all of the energy levels have small quantum defects.

In the cathode sheath region of the discharge the electric field is much higher than in other regions of the discharge and also the cathode sheath field is orthogonal to the electrode surface over most of its diameter. The sheath fields cause Stark splitting of the helium lines, typically at 600 vcm^{-1} for n=12. At lower n the splitting is less resolved and at high n the manifolds will overlap. The Rydberg state Stark spectra with laser polarization parallel to the electrode surface have been measured from n=15 to 20 in the sheath region. As expected, the spectra show $\Delta m_j=\pm 1$ characteristics, indicating that the electric field direction in the sheath is orthogonal to the laser polarization. The intensity distributions of the high n Stark manifolds more closely approximate the predicted hydrogenic intensity distribution with increasing field.

The experimental results in the cathode sheath for the intensity distribution of the Stark manifold of any given state show that the linewidth of each n_1 and n_2 state (where n_1 and n_2 are the parabolic quantum numbers) near the middle of the Stark manifold is almost Doppler limited, whereas the linewidth increases monotonically as $|n_1-n_2|$ is increased,[2,5] at the wings of the spectrum. These characteristics of the measured Stark spectra are consequences of the finite electric field gradient across the diameter of the laser beam and the dominance of the first order Stark effect. The measured spectrum represents a summation of Stark levels resulting from the range of electric field values probed by the laser beam. Calculations of this effect are illustrated in Figure 5 for the n=19 Stark manifold and $\Delta m_j=\pm 1$ polarization. The assumed average applied field was 745 volts cm^{-1} with 20 volts cm^{-1} ($2\Delta F$) variation across the beam. It is noted that, in contrast to hydrogen, the helium Stark induced manifold splitting is not symmetric because of the quantum defect. The line positions near the center of the manifold are not sensitive to the electric field gradient whereas the lines at the edges of the manifold are broadened significantly for typical sheath gradients. By choosing appropriate n-manifolds in conjunction with a library of calculated model intensity distributions for different average fields F and different electric field gradients, we have found that it is often possible to obtain the local values of both F and dF/dz from measurements at one spatial location of the Stark manifold intensity distribution.

The average electric field value is obtained from the Stark energy level shifts of each n_1 and n_2 substate and in order to measure both F and dF/dz values from the measured Stark spectra, it is necessary to fit the measured linewidths and the intensity distributions for each n_1 and n_2 state with the calculated spectra. The diagonalization method of Foster was used to calculate the energies and intensities of the n_1-n_2 components of the Stark manifolds. This method assumes no interaction between adjacent n manifolds, a condition satisfied by choosing lower principal quantum numbers as the electric field increased.

The relative intensity distribution of each Stark component, for a given n state, is proportional to the squares of the eigenvectors (up to tens of kV/cm Stark induced mixing of the 2^3S lower state is negligible). The energy levels and the eigenvectors are calculated, as functions of the electric field, by diagonalizing an nxn matrix using sparse matrix methods. In order to simulate the effect of electric field averaging, the intensity distributions and the energy levels are calculated for a range of electric field values of a given Stark n manifold and each manifold is multiplied by a weighting factor corresponding to the laser beam intensity profile; the individual intensity distributions are then summed to produce calculated profiles which best fit the experimental data. A set of calculated Stark spectra for n=19 with $\Delta m_j = \pm 1$ polarization is shown in Fig 6. The average electric field is 340 V/cm; ΔF ranges from 0 to 60 V/cm. The Stark spectrum corresponding to the zero field gradient shows a fully resolved manifold and with increase in field gradients, only the Stark components near the middle of the manifold remain distinct. The dashed line in Fig 5 indicates the zero field energy of the 2^3S to 19^3P transition.

Experimental data for the n=19 Rydberg state and $\Delta m_j = \pm 1$ polarization measured at 0.4, 1.25 and 2.5mm from the cathode are given in Fig 7. The electric field at these locations is estimated from both the Stark splitting and from the Stark shift. Consistent results were obtained and these were 745, 575 and 340 volts cm^{-1} respectively for the above three positions. Figure 8 shows the calculated intensity profiles for the three values of electric field, at a constant value ΔF=60 volts cm^{-1} and assuming a flat-top intensity profile of the laser beam. The result that a constant electric field gradient best fits the experimental results agrees with a simple analysis (2) for the cathode fall in a rare gas at low currents. The experimental technique, of course, does not require that the electric field gradient be linear. As the method measures dF/dz, from Poisson's equation this also means that the net space charge in the sheath was measured spectroscopically.

These measurements were made with the laser beam sampling a diameter or chord of the rotationally symmetric discharge. The amplitude of the optogalvanic signal depends on a convolution of the laser intensity and the number of metastables. When diffusion is the principal loss mechanism for the metastables, their concentration profile is expected in the positive column to follow a Bessel function form. These circumstances, therefore assist in defining a small interaction volume for the diagnostic. It is possible to achieve point resolution by using two intersecting tunable laser beams (Lawler, et.al.); one to excite to an intermediate state and one to excite further to the selected Rydberg state.

If the net electric field vector is defined from the experimental measurements or the symmetry of the experiment, then it is also possible to use the ratio of the allowed to forbidden intensities as a measure of the electric field. The theory of the intensity transfer is given in Bethe and Salpeter (8). The sensitivity of the method is illustrated in Fig 9 for the helium 2^1S transitions; the sensitivity is higher for higher n-states. The sensitivity of the method is about 20 volts cm^{-1} if n=25 is observable with good signal to noise. One cannot extrapolate arbitrarily to higher n to obtain higher sensitivity because adjacent manifold overlaps will occur.

DISCUSSION

In order to check the field calculations based on the Inglis-Teller equation, computer simulations of the Rydberg series for hydrogen and helium under defined Stark fields were performed. The results for hydrogen were compared with experimental data from Nayfeh (8). These experiments were performed under uniform field conditions in a Stark cell. A comparison of Nayfeh's data and our computer simulation for the field free hydrogen Rydberg series showed excellent agreement. In theory, one should resolve an infinite number of bound states, in practice, the resolution is limited by the laser linewidth. Our simulation incorporated a linewidth of $1.0cm^{-1}$, matching Nayfeh's experimental resolution. The simulation also weights the intensity of each successive Rydberg state by $1/n^3$. Nayfeh's experimental data deviates from a $1/n^3$ intensity drop-off, presumably caused by laser power fluctuations.

The original Inglis-Teller paper (10) derives a relationship $F \propto n_m^{-5}$ between the field required to cause adjacent lines to partially merge and the principal quantum number where the effect occurs. However, this is not a very precise definition, since the manifold intensity profiles are also functions of n_1 and n_2 (which depend on n). Also, because of the large inverse power involved and their interest in high fields the authors anticipated a rather abrupt termination of the series. This has lead in the literature to the identification of the last observable transition, or series termination, to be synonymous with the principal quantum number n_m. In the applications of RS^3 one has the capability to measure low electric fields and the transition to series termination is quite gradual, hence a more exact criterion is useful. It is convenient to define F as the electric field value where the Stark shift of the upper and lower extrema of adjacent manifolds equals one-half o the field free separation of the states. (Other onsets such as a Rayleigh criterion overlap could be used.) This criterion is observed in the Stark spectra as the <u>onset</u> of the continuum background, rather than the last resolvable state. In our model calculations this criterion is modified by incorporating the finite experimental linewidth of the laser. The n that is defined by this criterion is lower than that defined by the last resolvable state, and therefore gives higher but more accurate values of the net electric field for the same experimental Rydberg series. Figure 10 shows a fit of the Inglis-Teller relationship corrected for instrumental resolution to the results of computer simulated spectra of the Rydberg series of hydrogen at different field values. Since n can only assume integer values, the curve is not continuous. The error bars result from an uncertainty in n at the continuum onset of ±1.

The frequency of a transition from a level n+ Δn to the level n, where n is sufficiently large that the quantum defects are negligible, is given approximately by

$$f = 6.6\,\Delta n \left(\frac{100}{n}\right)^3 \text{ GHz}$$

The transitions for Δn=1 and n=40 have mm wavelength and for n=300 have meter wavelength. Such state transitions with n greater than 700 have been observed from astrophysical sources using radio astronomy. Smirnov, et.al., (11) investigated Stark broadening in radio recombination lines towards the Orion nebula. Collisional widths were found to be proportional to $n^{4.4}$ for quantum levels n=60 to 174, in agreement with the current theory of Stark broadening. Calculations of Stark broadening gave electron densities of $10^4 cm^{-3}$, when density fluctuation effects are included. Thus, the techniques outlined above are applicable to a very wide range of plasma conditions and can provide information on applied fields, self-consistent fields in bounded plasmas and microfields due to statistical (Holtsmark) broadening.

References

1. Doughty, D. K., Den Hartog, E. A., and Lawler, J. E., Phys. Rev. Lett. 58, 2668 (1987) and references therein.

2. Ganguly, B. N., Shoemaker, J. R., Preppernau, B. L. and Garscadden, A., J. Appl. Phys., 61, 2778 (1987).

3. Gottscho, R. A., Phys., Rev. A 36, 2233 (1987) and references therein.

4. Derouard, J. and Sadeghi, N., Opt. Commun., 57, 239 (1986).

5. Shoemaker, J. R., Ganguly, B. N. and Garscadden, A., Appl. Phys Lett., 52, 2019 (1988).

6. Von Engel, A., "Ionized Gases," Chapter 8, Oxford University Press, Oxford, (1965).

7. Wing, W. H., Lea, K. R. and Lamb, Jr. W. E., in "Atomic Physics 3," ed S. J. Smith and G. K. Walters, Plenum Press, New York (1973).

8. Bethe, H. A. and Salpeter, E. E., "Quantum Mechanics of One- and Two- Electron Atoms," Springer-Verlag, Berlin (1957).

9. Nayfeh, M. H., Ng, K. and Yao, D., Figure 5, P 197 in "Atomic Excitation and Recombination in External Fields" eds M. H. Nayfeh and C. W. Clark, Gordon and Breach, New York (1985).

10. Inglis, D. R., and Teller, E., Astrophys. J., 90, 439 (1939).

11. Smirnov, G. T., Sorochenko, R. L., and Panokonin, V., Astron. Astrophys., 135, 116 (1984).

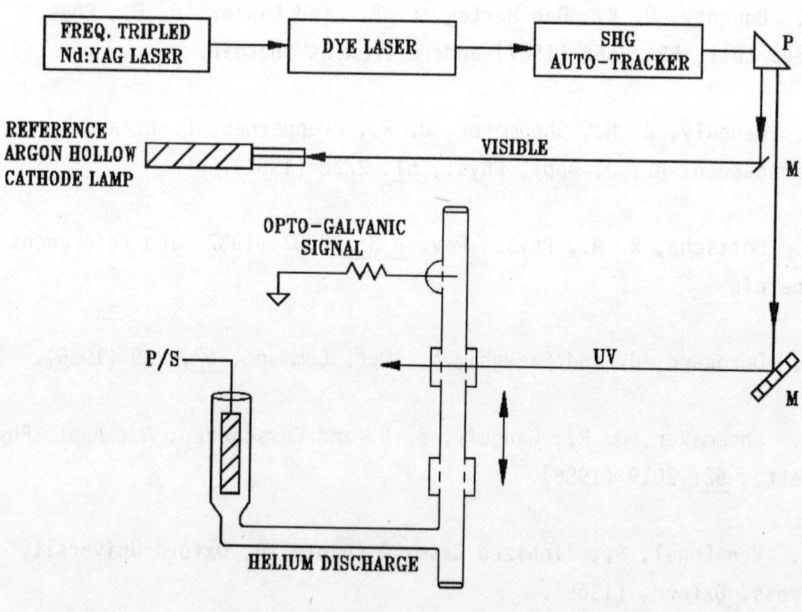

Fig. 1: Experimental Arrangement for RS ^{3}

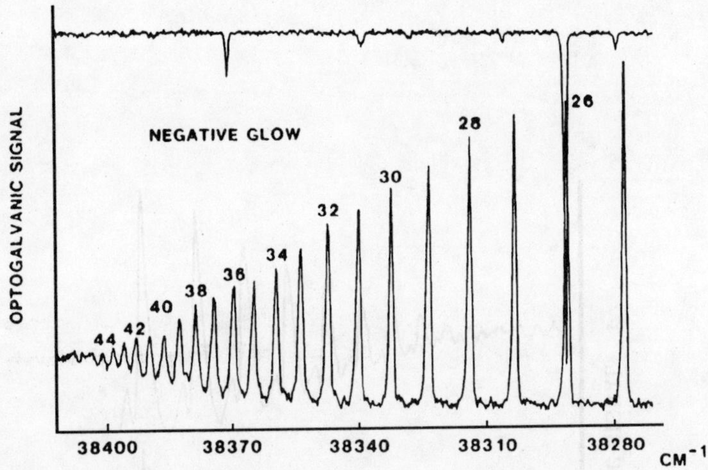

Fig.2: Optogalvanic Response in Negative Glow

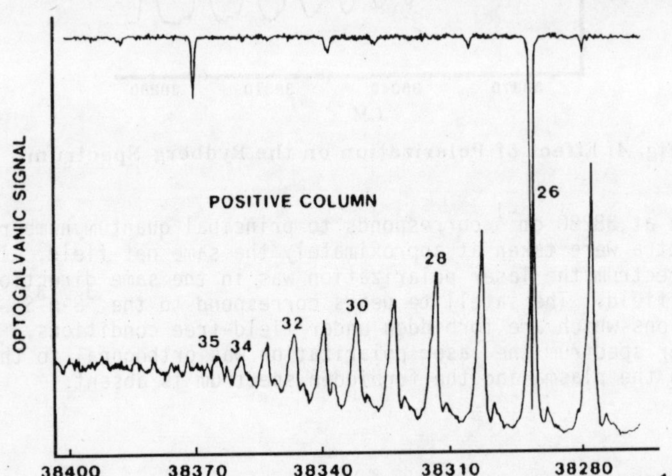

Fig.3: Optogalvanic Response in Positive Colomn

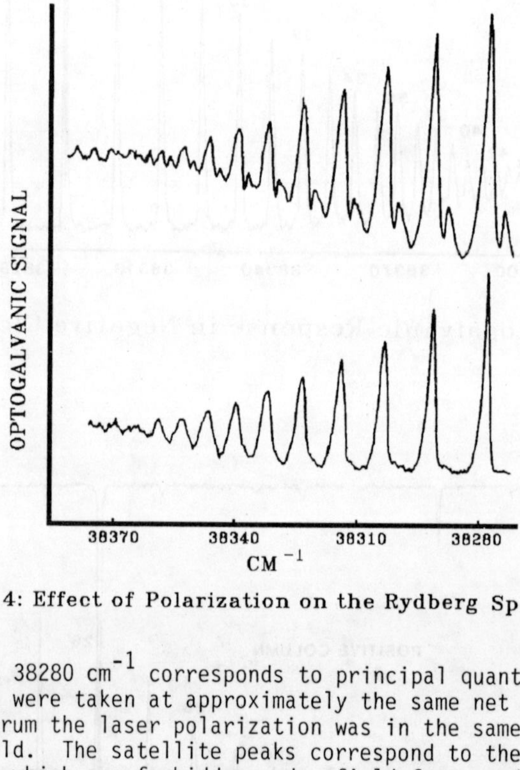

Fig. 4: Effect of Polarization on the Rydberg Spectrum

The line at 38280 cm^{-1} corresponds to principal quantum number n=26. The spectra were taken at approximately the same net field. In the upper spectrum the laser polarization was in the same direction as the net field. The satellite peaks correspond to the ^3S-n^3S transitions which are forbidden under field-free conditions. In the lower spectrum the laser polarization was orthogonal to the net field in the plasma and the forbidden spectrum is absent.

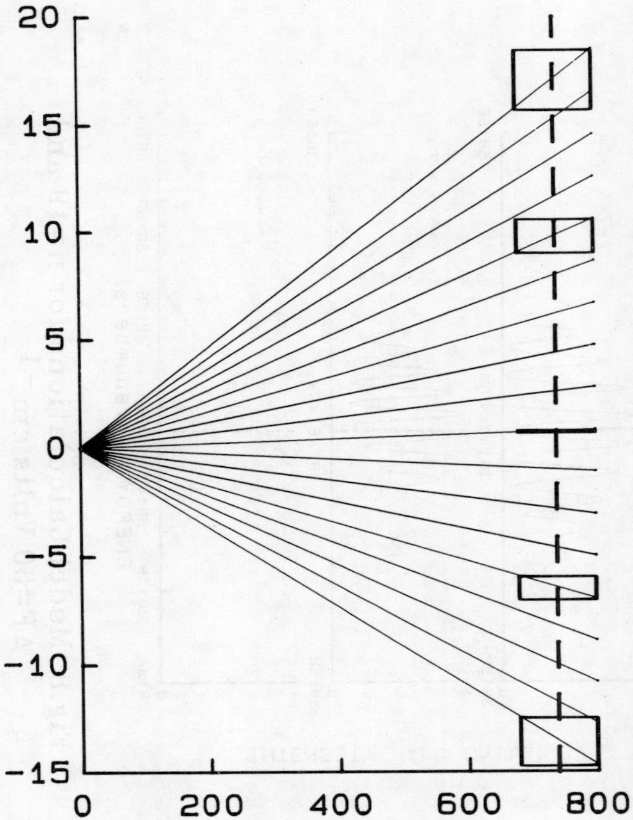

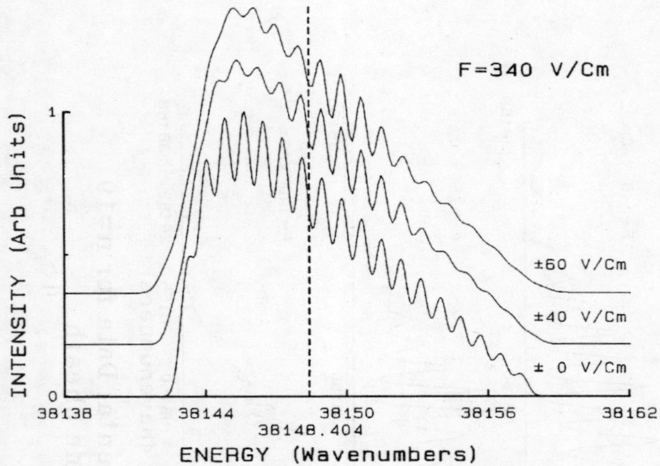

Fig.6: Effect of Different Gradient Values on Calculated Rydberg Spectra

Figure 5

Calculated Stark splitting of the n=19 manifold of helium vs electric field. Note that, in contrast to hydrogen, because of the quantum defect, the manifold splitting is not symmetric. The effect of a linearly varying electric field (average value 745 volts cm^{-1}) on the linewidth of different substrates is illustrated by the projections on the y ordinate.

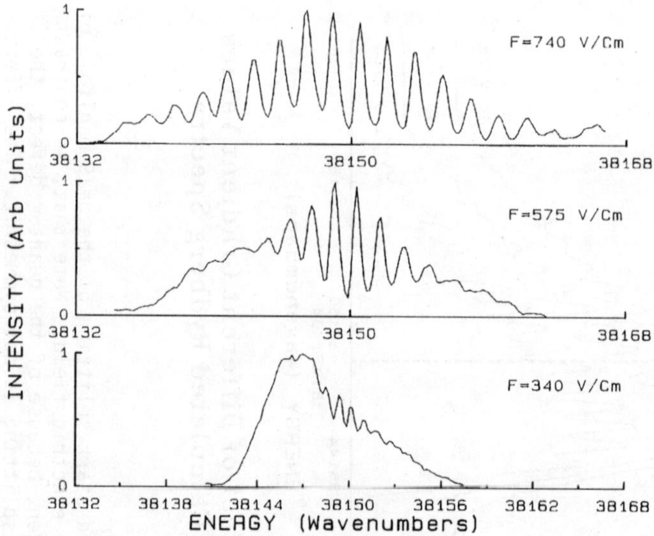

Fig.7: Experimental Data for n=19 in Cathode Sheath

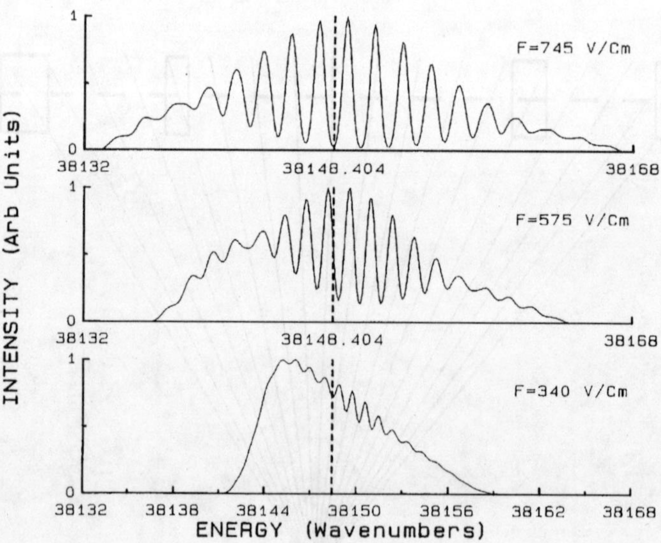

Fig.8: Model Calculations for n=19 and ΔF=60 Volts cm^{-1}

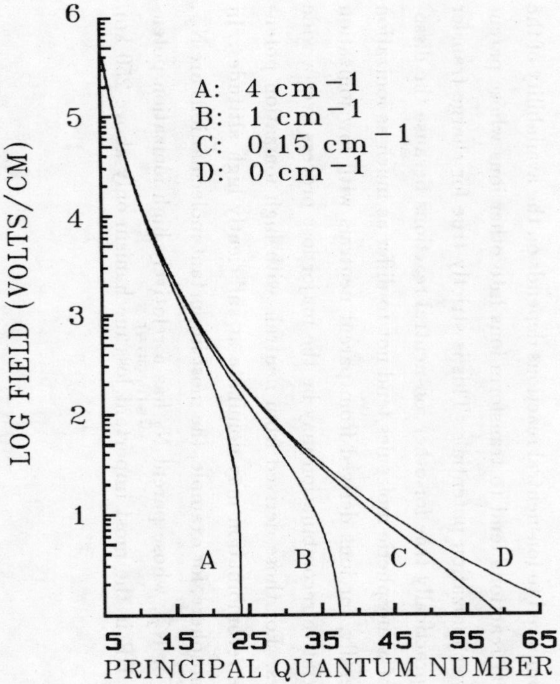

Fig 10: The corrected Inglis-Teller relationship defines the principal quantum number at a given applied electric field where the adjacent Stark manifolds begin to overlap. The different curves illustrate the effects of instrumental linewidth. The model calculations show that good spectroscopic resolution is required to give resolution in the electric field values. Curve C corresponds to a reasonable laser line halfwidth of 0.15 cm^{-1} and it provides an approximately constant exponent value (-5) over the range 1 to 100 volts cm^{-1}. The coefficient is 1.7×10^{9}. The curves at high n apply to hydrogen or helium ($\Delta mj = \pm 1$).

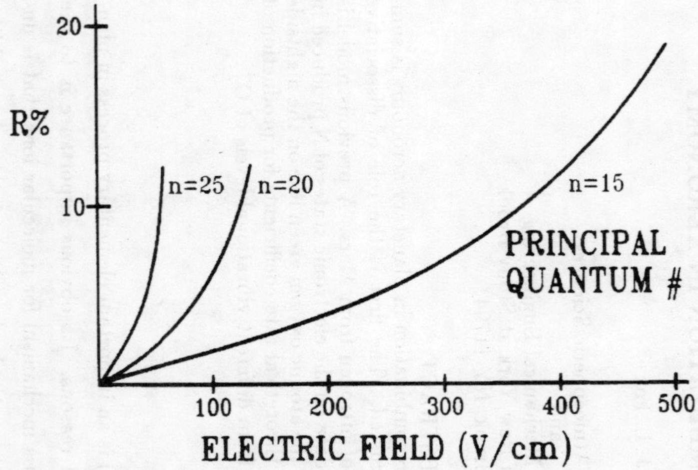

Fig.9: Ratio of Allowed to Forbidden Intensities: Helium S–P Singlets

DISSOCIATIVE RECOMBINATION IN AERONOMY

J. L. Fox

Institute for Atmospheric Sciences
and
Department of Mechanical Engineering
State University of New York at Stony Brook
Stony Brook, NY 11794

ABSTRACT

The importance of dissociative recombination in planetary aeronomy is summarized and two examples are discussed. The first is the role of dissociative recombination of N_2^+ in the escape of nitrogen from Mars. A previous model is updated to reflect new experimental data on the electronic states of N produced in this process. Second, the intensity of the atomic oxygen green line on the nightside of Venus is modeled. Use is made of theoretical rate coefficients for production of $O(^1S)$ in dissociative recombination from different vibrational levels of O_2^+.

1. INTRODUCTION

Dissociative recombination (DR) is an important elementary process in planetary upper atmospheres for several reasons. Its obvious importance is to the ionization structure: it provides a loss mechanism for molecular ions that is unavailable to atomic ions. Consequently molecular ions have shorter chemical lifetimes at high altitudes than atomic ions. The relative importance of dissociative recombination compared to ion-neutral reactions depends on the availability of the latter. Ion-neutral reactions tend to transform ions into other ions whose parent neutrals have lower ionization potentials. This is strictly true for charge transfer reactions and is practically true for other ion-neutral reactions because the dissociation energies of atmospheric molecules tend not to differ as much as ionization potentials. Generally, for ions derived from parent neutrals with low ionization potentials, dissociative recombination may be the major loss process over a wide range of altitudes. For those derived from neutrals with high ionization potentials, dissociative recombination may dominate at sufficiently high altitudes. In the terrestrial ionosphere, for example, the most important molecular ions are N_2^+, O_2^+ and NO^+. For N_2^+, whose parent N_2 has a relatively high ionization potential of 15.58 eV, DR is the most important loss mechanism only above 250 km.

Immediately below that altitude, the major loss process is reaction with O:

$$N_2^+ + O \rightarrow NO^+ + N. \qquad (R1)$$

O_2 has an intermediate ionization potential of 12.15 eV, and DR controls the loss of O_2^+ over a larger altitude range, beginning at about 190 km. The reaction

$$O_2^+ + N \rightarrow NO^+ + O \qquad (R2)$$

is the most important loss mechanism immediately below that altitude. NO has the lowest ionization potential of any of the major atmospheric species and consequently DR of NO^+ is the only loss mechanism available over the entire ionosphere.

Dissociative recombination may be an important source of airglow emissions, if the fragments are produced in excited states with short radiative lifetimes. In DR of CO_2^+,

$$CO_2^+ + e \rightarrow CO + O \qquad (R3)$$

55% of the CO molecules are produced in the excited $CO(a^3\Pi)$ state.[1] $CO(a^3\Pi)$ is a metastable state with a lifetime of about 8 ms[2,3]; it radiates to the ground $^1\Sigma$ state producing emission in the well-known Cameron Band System. The Cameron Bands are the brightest features in the ultra-violet dayglows of Venus and Mars. On Venus, approximately 5 kR of 20 kR total integrated intensity is due to DR of CO_2^+.[4] (One kiloRayleigh (kR) is an apparent column emission rate of 10^9 photons $cm^{-2}s^{-1}$.) In DR of O_2^+, the O atoms may be produced in various combinations of the electronic states ^3P, ^1D, and ^1S. $O(^1D)$ radiates to the ground state producing the 6300 Å "red line" and $O(^1S)$ to $O(^1D)$ producing the 5577 Å "green line". This source is important in the dayglows and nightglows of all the terrestrial planets and in terrestrial auroras. It is relatively more important in the nightglow, due to the absence of sources involving the direct interaction of solar radiation or photoelectrons with atmospheric gases.

The fragments produced in dissociative recombination may be important chemically. For example, ionization of N_2 followed by dissociative recombination of N_2^+ is a net source of odd nitrogen, an important family of species that are interconverted more rapidly than they are created or destroyed. The fragments may also be formed in metastable excited states whose chemistry is different from that of ground state species. For example, in DR of $N_2^+(v = 0)$, the yield of $N(^2D)$

has recently been measured by Queffelec et al. to be about 1.85.[5] $N(^2D)$ may radiate to the ground $N(^4S)$ state, producing emission at 5200 Å, but the transition probability is small, about $1.06 \times 10^{-5}s^{-1}$.[6] It is probable that the atom will be chemically destroyed before radiating. In the terrestrial atmosphere, the reaction

$$N(^2D) + O_2 \rightarrow NO + O \qquad (R4)$$

is an important source of NO,[7,8] as is the reaction

$$N(^2D) + CO_2 \rightarrow NO + CO \qquad (R5)$$

on Venus and Mars.[9,10] Reactions of ground state N atoms with O_2 and CO_2 are much slower than reactions (R4) and (R5).[11,12,13,14] Constantinides et al.[15] have shown that the reaction

$$N(^2D) + O^+ \rightarrow N^+ + O, \qquad (R6)$$

which is endothermic for ground state N atoms, is the most important source of N^+ above 250 km in the terrestrial ionosphere.

That the fragments produced in dissociative recombination may have large kinetic energies has several important consequences for planetary atmospheres. First, DR may be an important heat source for the neutral atmosphere. On Earth, most of the heating due to exothermic reactions is from dissociative recombination reactions.[16] On Venus, DR of O_2^+ is the dominant heat source above 135 km.[17] If the fragments are produced at a sufficiently high altitude and with sufficient kinetic energy, they may escape from the planet. (See the review by Hunten[18] for a description of thermal and non-thermal escape processes.) A common approximation, is that above the *exobase*, where the mean free path is equal to a scale height, the atmosphere is assumed to be collisionless . Then all atoms traveling upward with velocities greater than the escape velocity actually escape. For thermal escape, this approximation has been shown to produce essentially no error.[19] The cut-off velocity, v_{esc} is that for which the kinetic energy, $\frac{1}{2}mv^2$, is equal to the gravitational potential energy, mgr, where r is the distance of the exobase from the center of the planet. On Mars, the escape velocity is about 5 km/s, corresponding to an escape energy of 0.125 eV per amu. DR of O_2^+, CO^+ and N_2^+ are sufficiently exothermic in at least some channels to produce escaping O, C and N atoms. On Venus and Earth the escape energies exceed 0.5 eV per amu and none of the dissociative recombination processes are exothermic enough to produce escaping atoms. On Venus, dissociative recombination is implicated in an indirect way in the escape of H through the sequence[20)]

$$O_2^+ + e \rightarrow O^* + O^* \qquad (R7)$$

followed by the transfer of energy in a collision of the translationally hot O* with an H atom:

$$O^* + H \rightarrow O^* + H^*. \qquad (R8)$$

Cooper et al.[21)] have computed the fraction of H atoms that come away from (R8) with energy greater than the escape energy as 6.9 and 8.5% at 200 and 300 K, respectively.

Even if the atoms released above the exobase have insufficient energy to escape, they may travel to high altitudes producing a "corona" of hot atoms surrounding the planet. The hot O coronas of the terrestrial planets are produced mainly by the dissociative recombination of O_2^+ at the exobase. The hot O corona of the Earth has been discussed Yee et al.[22)] Such a corona was predicted for Venus and Mars by Wallis[23)] and distributions of hot O have been computed by Nagy et al.,[24)] Paxton,[25)] and Nagy and Cravens.[26)] The hot C corona around Venus has been studied by analysis of the limb profiles of the atomic carbon emission multiplets at 1561 and 1657 Å measured by the Pioneer Venus Orbiter (PVO) Ultraviolet Spectrometer.[25)] These features are produced at high altitudes by fluorescent scattering of sunlight by atomic carbon. The hot C is produced partly in dissociative recombination of CO^+, but mostly by (R7) followed by

$$O^* + C \rightarrow O + C^*, \qquad (R9)$$

a process that Paxton refers to as "sputtering", but which is more accurately termed "knock-on"[18)]

In the remainder of this article, two problems in planetary aeronomy in which dissociative recombination plays a major role will be described in detail. The first is the escape of nitrogen from Mars. In modeling this phenomenon, knowledge of the kinetic energies and therefore, indirectly, the electronic states of the N atoms produced in DR of N_2^+ is important. Second, the production of the atomic oxygen green line in the nightglow of Venus is discussed. Here the production rates of $O(^1S)$ from dissociative recombination of O_2^+ in different vibrational levels is required.

2. THE ESCAPE OF NITROGEN FROM MARS

The mass spectrometer on the Viking spacecraft measured an anomalous $^{15}N:^{14}N$ ratio of about 0.0060 in the Martian atmosphere; the value in the terrestrial atmosphere is 0.00368.[27,28] This isotope enhancement of a factor of 1.62 over the terrestrial value is presumably due to preferential escape of ^{14}N. The thermosphere of Mars is composed mostly of CO_2, with smaller amounts of N_2, O, CO, Ar, O_2 and NO. The model thermosphere used here is essentially the same as that used by Fox and Dalgarno[29] and the calculations described here are an updating of that model. At the *turbopause* (about 120 km), the mixing ratio of N_2 is about 2.5%.[30] Diffusion largely controls the distribution of species above the turbopause, so the mixing ratios of heavier species decrease with altitude and those of lighter species increase. Thus between the turbopause and the exobase at about 200 km, the ratio, f of $^{15}N^{14}N:^{14}N_2$ decreases. The exact value of the ratio at the exobase, f_c depends on the ratio at the turbopause, f_o, which is assumed equal to that in the bulk atmosphere, and on the value of the eddy diffusion coefficient, a measure of the strength of mixing in the thermosphere. McElroy et al.[31] computed the deficiency factor $R = f_c/f_o$ and reported a value of about 0.82. If we call $\phi(t)$ the $^{14}N_2$ flux at time t, $\mathcal{N}_{14}$ the column density of $^{14}N_2$ and $\mathcal{N}_{15}$ the column density of $^{15}N^{14}N$ then

$$\frac{d\mathcal{N}_{14}}{dt} = -\phi(t) \tag{1}$$

and

$$\frac{d\mathcal{N}_{15}}{dt} = -f_c\phi(t) = -Rf\phi. \tag{2}$$

Combining (1) and (2) we obtain an equation for the isotope ratio f:[27]

$$\frac{df}{dt} = \frac{\phi f(1-R)}{\mathcal{N}_{14}}. \tag{3}$$

The isotope enhancement produced by escape operating over a time t before the present can be obtained by integrating equation (3) backward in time. The effects on the escape rate of changes in the composition of the atmosphere and in the heights of the exobase and turbopause are taken into account as described by Fox and Dalgarno.[29]

The steady state ion densities are computed by solving the continuity equations including photochemistry and ambipolar diffusion and the results are shown in Fig. 1. Also shown are the O_2^+ and CO_2^+ densities derived from the Viking I Retarding Potential Analyzer and reported by Hanson et al.[32] The atomic oxygen density was not measured by the mass spectrometers and has been adjusted to fit the O_2^+ and CO_2^+ densities at the ion peak. A mixing ratio of 2% at about 130 km is derived. The N_2^+ density peaks above the exobase at approximately 215 km where it attains a value of about 180 cm^{-3}. The total rate of N_2^+ DR above the exobase is about 2×10^6 cm^{-2} s^{-1}, but not all channels are sufficiently exothermic for escape to occur. Dissociative recombination may produce N atoms in three channels:

$$N_2^+ + e \rightarrow N(^4S) + N(^2D) + 3.45eV \qquad (R10a)$$

$$N_2^+ + e \rightarrow N(^4S) + N(^2P) + 2.24eV \qquad (R10b)$$

$$N_2^+ + e \rightarrow N(^2D) + N(^2D) + 1.06eV \qquad (R10c)$$

Because the N escape energy is 1.74 eV, escape is possible only when the reaction proceeds via (R10a). If a statistical distribution of products is assumed, the fraction of dissociative recombinations proceeding by (R10a) is about 0.24. With this assumption, DR is the dominant N escape mechanism.[29,33] The measurements of Queffelec et al.[5] have shown, however, that the predominant channel is (R10c) and that the yield of $N(^4S)$ is less than 0.10. We therefore assume that an upper limit to the fraction of dissociative recombinations proceeding according to (R10a) is 0.10.

For most escape mechanisms, preferential escape of ^{14}N is assumed to occur only because the ratio $^{15}N^{14}N : ^{14}N_2$ is reduced at the exobase compared to that in the bulk atmosphere. The process of dissociative recombination via channel (R10a) itself, however, discriminates against ^{15}N. The escape energy of ^{14}N is about half the exothermicity of (R10a), but the escape energy of ^{15}N is 1.86 eV, much more than half the energy released. Wallis[23] computed the escape probabilities for both atoms and found that ^{14}N is about twice as likely to escape as ^{15}N. The exothermicities shown for (R10a–c) apply to DR of N_2^+ ions in the ground vibrational state. For vibrationally excited ions, the exothermicity is larger and the isotope discrimination due to the mechanism itself disappears! In order to model the evolution of the isotope ratio, it is necessary to determine the fraction of dissociative recombinations that occur from vibrationally excited states. Therefore we need to know the vibrational distribution of N_2^+ at the exobase and the rate

Table 1. The vibrational distribution of N_2^+ at the Martian exobase, 200 km.

v	Fraction
0	0.56
1	0.21
2	0.107
3	0.058
4	0.035
5	0.028

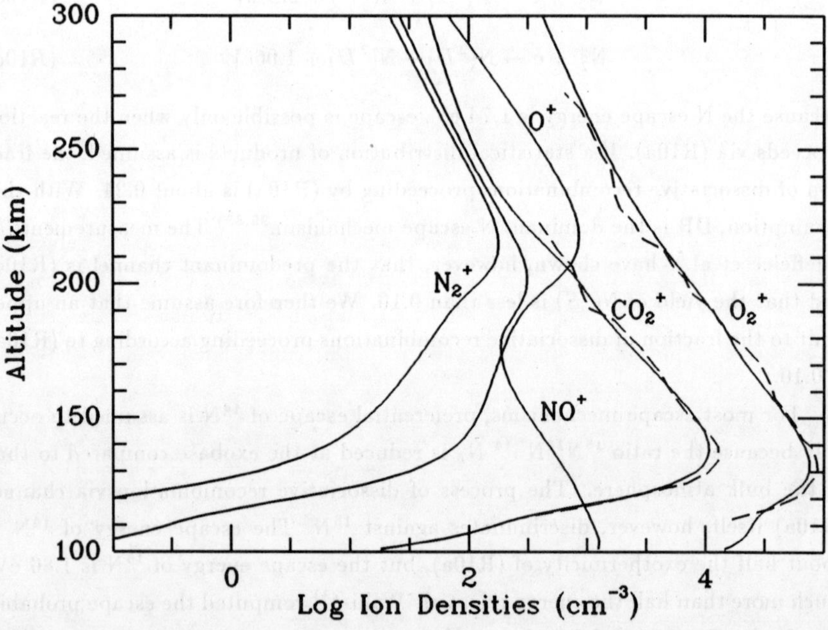

Fig. 1. Altitude profiles of the steady state ion densities for low solar activity. The solar zenith angle is 60°. The dashed lines are Viking 1 data from Hanson et al. (Ref. 32).

coefficients for dissociative recombination from individual vibrational levels. The latter are not available, although evidence suggests that the rate coefficient increases with increasing vibrational level.[34] We adopt the rate coefficient reported by Zipf[34] of $2.2 \times 10^{-7}(300/T_e)^{0.39}$ cm^3s^{-1} for dissociative recombination from all the vibrational levels ($v = 0 - 5$) included in the calculation. The model of Fox and Dalgarno[29] for the vibrational distribution of N_2^+ at the exobase on Mars has been updated (c.f. Fox and Dalgarno[35]) and the distribution presented in Table 1 was computed. The vibrational distribution at the exobase is essentially controlled by fluorescent scattering of sunlight in the Meinel and First Negative Band systems, which can be represented by

$$N_2^+(X^2\Sigma_g^+, v) + h\nu \rightarrow N_2^+(A^2\Pi_u, B^2\Sigma_u^+, v') \qquad (R11)$$

followed by

$$N_2^+(A, B, v') \rightarrow N_2^+(X, v'') \qquad (R12)$$

The fraction of N_2^+ ions in vibrationally excited levels is found to be about 44%, so the isotope discrimination operates only 56% of the time.

The ^{14}N escape fluxes from all sources are presented in Table 2. Values for both high and low solar activity were computed and averaged to obtain the escape fluxes used in the calculation. The values for photodissociation, photodissociative ionization, and photoelectron impact ionization are unchanged from those of Fox and Dalgarno.[29] The products of the reaction

$$O^{++} + N_2 \rightarrow N + N^+ + O^+ \qquad (R13)$$

are unknown, but exothermicity is large, about 10.9 eV, and it is assumed that an energetic N atom is produced in every reaction. The reaction

$$N_2^+ + O \rightarrow NO^+ + N \qquad (R1)$$

produces an N atom in the ^2D state at least 90% of the time.[36] In that case the exothermicity is not sufficient to allow escape of the N atom produced. We assume initially that 10% of the reactions (R1) produce a ground state N atom. With these assumptions, dissociative recombination is still the largest source of escaping N atoms, but photodissociation and reaction (R1) are comparable. The differential escape of ^{15}N and ^{14}N due to DR from the ground vibrational level,

Table 2. ^{14}N Escape Fluxes ($10^4 cm^{-2}s^{-1}$)

Process	Solar Activity		
	Low	High	Average
$N_2 + h\nu \rightarrow N + N$	5.9	14	10
$N_2 + h\nu \rightarrow N^+ + N + e$	1.7	6.0	3.9
$N_2 + e \rightarrow N^+ + N + 2e$	1.4	4.8	3.1
$N_2^+ + O \rightarrow NO^+ + N$	6.1	17	11
$O^{++} + N_2 \rightarrow N^+ + N + O^+$	0.6	1.7	1.1
$N_2^+ + e \rightarrow N + N$	4.0	22	13
TOTAL	20	66	42

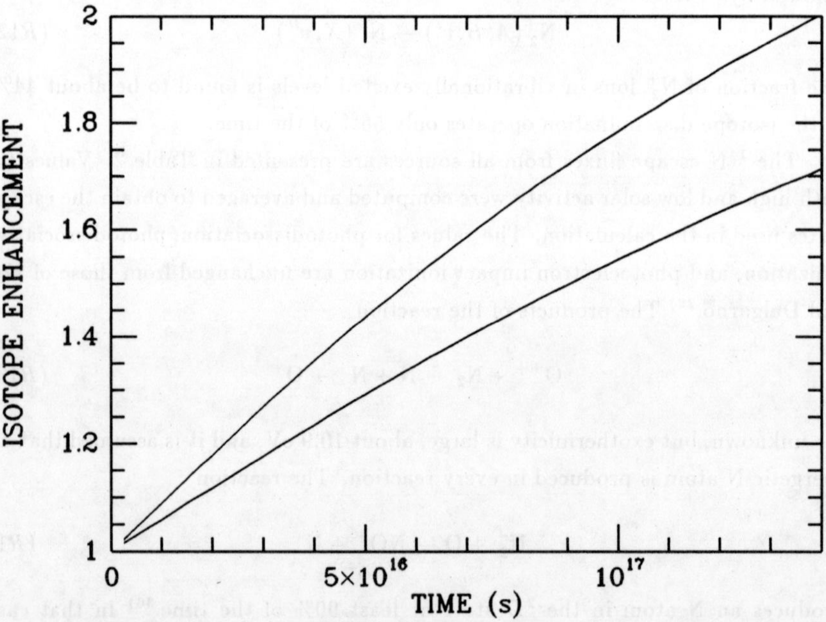

Fig. 2. Evolution of the isotope enhancement, the ratio ^{15}N:^{14}N relative to the terrestrial value. The upper curve was calculated assuming fluxes due to (R1) and (R10) equal to their upper limits, as explained in the text. The lower curve is for both fluxes reduced by half.

however, causes this mechanism to produce a much larger isotope enhancement than the other mechanisms.

Using the average escape fluxes shown in Table 2 and integrating equations (1) and (3) backward in time over the age of the solar system, 4.6×10^9 years (1.45×10^{17}s), we find an initial column density of 8.5×10^{22} N_2 molecules cm^{-2}, about 14 times that found in the present atmosphere. The resulting isotope enhancement is shown in Fig. 2. The isotope enhancement for the escape processes operating over 4.6×10^9 years is 2.0, somewhat larger than the measured value. An isotope enhancement of 1.62 is reached in 7.5×10^{16}s (2.4×10^9 years). The escape fluxes due to reaction (R1) and to dissociative recombination are actually upper limits, so we tested the sensitivity of the model to the assumed yields by reducing both fluxes by half. The resulting initial N_2 column density is reduced by one-third and the predicted isotope enhancement is 1.71. The difference between the measured and the model isotope enhancements is probably within the uncertainties in the model. The magnitude of the solar flux, for example, in the early solar system was undoubtedly different from the current values, but arguments have been presented for both larger[37,38] and smaller values.[39] The isotope enhancement predicted for a statistical distribution of product states in (R10) was much larger, about 2.51.[29] It was necessary then to invoke a mechanism to reduce the escape fluxes in the past, such as a denser CO_2 atmosphere, surviving over a period exceeding 2×10^9 years. Such an atmosphere has been predicted (e.g. Cess et al.[40], Pollack et al.[41]) to be consistent with an earlier presence of liquid water on the surface. While it is no longer necessary, a dense CO_2 atmosphere is not excluded by the current model, as long as it was lost early in the history of the solar system.

3. THE GREEN LINE OF ATOMIC OXYGEN ON THE NIGHTSIDE OF VENUS

Ultra-violet emission features of atomic oxygen at 1304 and 1356 Å have been seen on the nightside of Venus by the PVO Ultraviolet Spectrometer.[42] Fox and Stewart[43] have suggested that the emissions are due to the precipitation of soft electrons into the nightside atmosphere. Suprathermal electrons have been detected outside the atmosphere by the PVO Retarding Potential Analyzer[44] and by the plasma analyzers on the Soviet spacecraft Veneras 9 and 10.[45] The PVO measurements suggest that the electron energy spectrum approximates a Maxwellian distribution with a characteristic energy of about 14 eV. Impact of such electrons should also excite atomic oxygen to the ^1S and ^1D states, producing emission at 5577 and 6300 Å. The Venera 9 and 10 fly-by spacecraft carried visible spectrophotometers, but no 5577 Å emission was measured on the nightside of Venus and an upper limit of 10 R has been placed on its intensity[46] (Krasnopol'sky, private communication, 1985). This upper limit places another constraint on models for the auroral production mechanism. In order to evaluate the constraint, it is necessary to predict the rate of production of the green line from other sources. Dissociative recombination of O_2^+ (R7) is probably the most important source in the nightglow. The production of $O(^1S)$ in (R7) has been shown to depend on the vibrational quantum number of O_2^+. Abreu et al.[47] showed that the 5577 Å emission in the terrestrial nightglow depends on the ratio of the electron density to atomic oxygen density, a measure of the extent of vibrational excitation of O_2^+.[48] Guberman[49,50] has performed *ab initio* calculations of the rate coefficients for the production of $O(^1S)$ and $O(^1D)$ in DR from the lower vibrational levels of O_2^+. He has shown in particular that the rate coefficient for production of $O(^1S)$ is a strong function of vibrational level, increasing almost 2 orders of magnitude from $v = 0$ to $v = 2$. We have constructed a model of the nightside ionosphere of Venus, including a model of the vibrational distribution of O_2^+, and we have used Guberman's rate coefficients to predict the intensities of the red and green lines in the Venus nightglow.

The thermosphere of Venus is similar to that of Mars, and consists mostly of CO_2 at low altitudes, with small amounts of N_2, CO and O. The neutral densities we have used are taken from the model of Hedin et al.[51] for 165° solar zenith angle and $F_{10.7} = 200$. This model is based on measurements of the PVO Neutral Mass Spectrometer.[52] The nightside ionosphere is maintained largely by transport of atomic ions, primarily O^+, from the dayside ionosphere.[53,54] Cravens et al.[55] constructed a quasi-two-dimensional model that showed that the observed O^+ horizontal velocities from the dayside are sufficient to maintain the nightside ionosphere. Spenner et al.[56] showed that a flux of O^+ ions of $1 - 2 \times 10^8 \mathrm{cm^{-2}s^{-1}}$ over the nightside is sufficient to maintain the observed mean ion densities. The nightside ionosphere is, however, highly variable and almost always exhibits some degree of disturbance.[57,58] Sometimes the ionosphere disappears completely, leaving only a few scattered patches of plasma.[59] Our model ionosphere is designed to represent "full-up" relatively undisturbed conditions, when we expect the rate of DR to be substantial, although not necessarily a maximum. We have modeled this ionosphere by introducing a flux of O^+ of $1.0 \times 10^8 \, cm^{-2} s^{-1}$ at the top of our model thermosphere. Fluxes of other atomic ions, including C^+, N^+ and O^{++} were assumed proportional to their densities at high altitudes on the nightside as measured by the PVO Ion Mass Spectrometer.[60] Altitude profiles of the major ions in the resulting ionosphere are shown in Fig. 3. The effect of drag on the upward diffusion of molecular ions by the downward flux of O^+ is included in an approximate way by reducing the diffusion coefficients of the molecular ions by a factor of 10. This factor was chosen to reproduce the approximate shape of the measured O_2^+ profiles presented by Taylor et al.[60] The maximum O_2^+ density in our model is about $2.4 \times 10^4 \, cm^{-3}$.

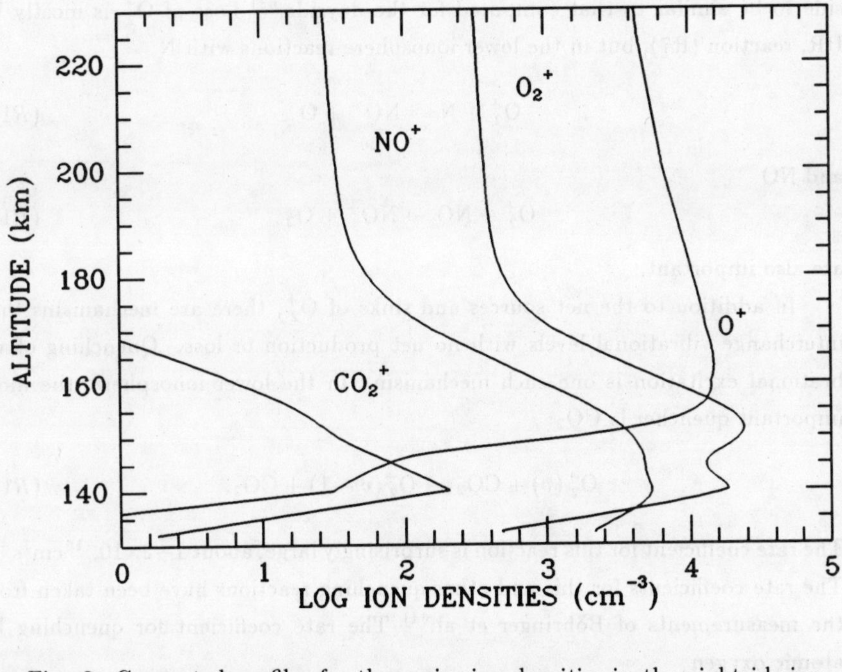

Fig. 3. Computed profiles for the major ion densities in the nightside ionosphere of Venus. The model was constructed by introducing a flux of O^+ of $1.0 \times 10^8 \, cm^{-2} s^{-1}$ at the top of the neutral atmosphere.

Vibrationally excited $O_2^+(v)$ is produced on the nightside mainly in the reaction

$$O^+ + CO_2 \rightarrow O_2^+(v \leq 5) + CO. \qquad (R14)$$

The rate coefficient for this reaction is large, about $9.4 \times 10^{-10}\text{cm}^3\text{s}^{-1}$.[61] In the dayside ionosphere, the major source of O_2^+ is

$$CO_2^+ + O \rightarrow O_2^+(v \leq 6) + CO, \qquad (R15)$$

which has a rate coefficient $k = 1.64 \times 10^{-10}\text{cm}^3\text{s}^{-1}$.[62] The nascent vibrational distribution produced in these reactions is unknown, so the energetically accessible vibrational levels are assumed to be populated equally. Since the distribution assumed for the production reactions is similar, and fluorescent scattering is not important for O_2^+, we expect the steady state vibrational distribution on the nightside to be similar to that computed for the dayside.[63] Loss of O_2^+ is mostly by DR, reaction (R7), but in the lower ionosphere reactions with N

$$O_2^+ + N \rightarrow NO^+ + O \qquad (R16)$$

and NO

$$O_2^+ + NO \rightarrow NO^+ + O_2 \qquad (R17)$$

are also important.

In addition to the net sources and sinks of O_2^+, there are mechanisms that interchange vibrational levels with no net production or loss. Quenching of vibrational excitation is one such mechanism. In the lower ionosphere, the most important quencher is CO_2:

$$O_2^+(v) + CO_2 \rightarrow O_2^+(v-1) + CO_2. \qquad (R18)$$

The rate coefficient for this reaction is surprisingly large, about $1-2\times10^{-10}\text{cm}^3\text{s}^{-1}$. The rate coefficients for this and other quenching reactions have been taken from the measurements of Böhringer et al.[64] The rate coefficient for quenching by atomic oxygen

$$O_2^+(v) + O \rightarrow O_2^+(v-1) + O \qquad (R19)$$

is unknown, but is probably large. We assume first a rate coefficient of $1 \times 10^{-10}\text{cm}^3\text{s}^{-1}$, but we test the sensitivity of the model to this rate coefficient by repeating the calculations with a nearly gas kinetic value of $6 \times 10^{-10} cm^3 s^{-1}$. Collisional excitation

$$O_2{}^+(v) + O \rightarrow O_2{}^+(v+1) + O \qquad (R20)$$

is the reverse of quenching and the rate coefficient is related to that for quenching by detailed balance, that is, $k_{20} = k_{19} \times e^{-\Delta E/kT}$, where ΔE is the energy difference between the vibrational levels.

We have solved the continuity equations for O_2^+, taking into account vibrational levels $v = 0-60$ of the ground electronic state, as was done for the dayside[63] and for the Earth.[65] Fig. 4 shows the fractions of O_2^+ in the first four vibrational levels as a function of altitude from 130 km to 200 km. The calculation of the

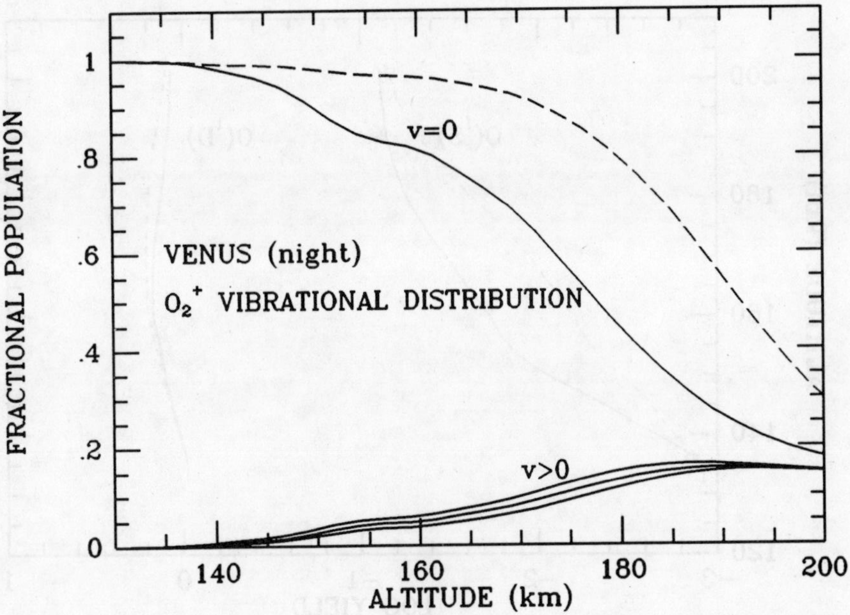

Fig. 4. Fractions of O_2^+ ions in vibrational states $v = 0 - 3$ as a function of altitude calculated with a photochemical steady state model, with an assumed value for k_{19} of $1 \times 10^{-10} cm^3 s^{-1}$. The curves labeled $v > 0$ are, from top to bottom, $v = 1$, $v = 2$ and $v = 3$. The dashed line is the fraction in $v = 0$ for $k_{19} = 6 \times 10^{-10} cm^3 s^{-1}$.

steady state ion densities includes transport by ambipolar diffusion, but the calculation of the vibrational distribution assumes photochemical equilibrium. Consequently, the distributions shown in Fig. 4 are probably valid only below about 170 km. The ions at higher altitudes are produced largely by transport from below. The dashed line shows the fraction in $v = 0$ for the larger rate coefficient for ($R19$), $k_{19} = 6 \times 10^{-10} \mathrm{cm}^3\mathrm{s}^{-1}$. It should be noted here that the number of vibrational quanta lost in (R19) is unknown and may be more than the assumed value of 1, because the reaction can occur by atom-interchange,[48] but the rate coefficient and the average number of vibrational quanta lost cannot be varied independently in a meaningful way.

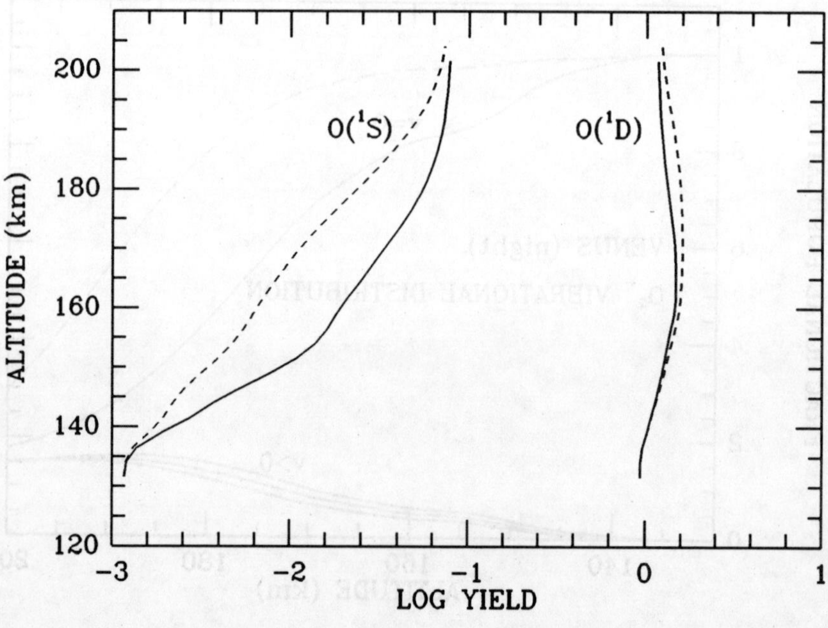

Fig. 5. Altitude profiles of the yields of $O(^1S)$ and $O(^1D)$ produced in dissociative recombination of $O_2^+(v)$. Solid lines are computed assuming $k_{19} = 1 \times 10^{-10} \mathrm{cm}^3\mathrm{s}^{-1}$. Dashed lines are for $k_{19} = 6 \times 10^{-10} \mathrm{cm}^3\mathrm{s}^{-1}$.

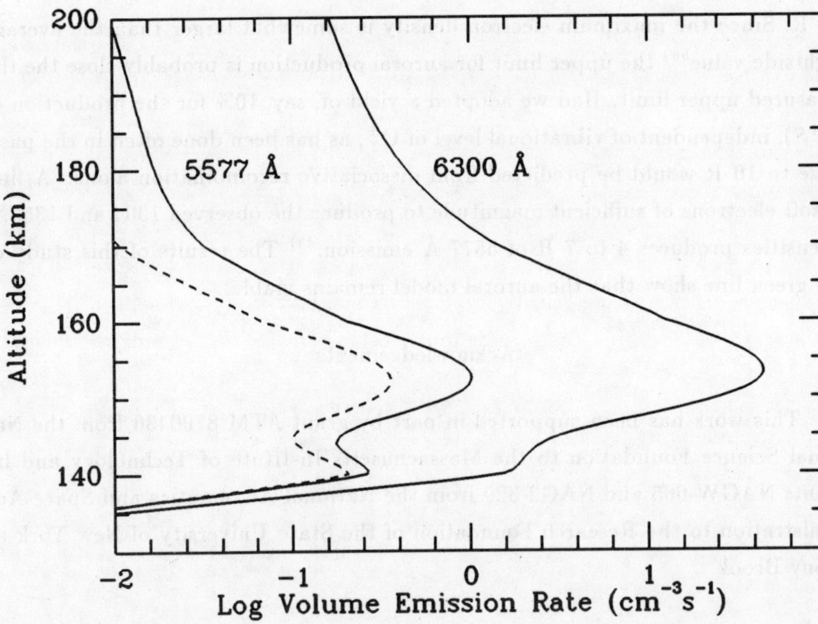

Fig. 6. Altitude profiles of the volume emission rates of the red and green lines of atomic O. Solid lines are computed assuming $k_{19} = 1 \times 10^{-10} \mathrm{cm}^3 \mathrm{s}^{-1}$. The dashed line is for $k_{19} = 6 \times 10^{-10} \mathrm{cm}^3 \mathrm{s}^{-1}$.

The vibrational distributions are combined with the rate coefficients from Guberman[49,50)] and the total DR rate coefficient from Alge et al.[66)] to give the yields for O(^1S) and O(^1D). For dissociative recombination from vibrational levels for which no information is available, the rate coefficient is assumed to be equal to that for the highest level for which data are available. Altitude profiles of the yields and the volume emission rates are shown in Figs. 5 and 6, respectively. The O(^1S) yield varies from much less than 1% to about 8% at high altitudes. Fig. 6 shows that most of the emission arises between 140 and 160 km where the ^1S yield is in the range 1 – 2%. The dashed lines show the yields and volume emission rates for the assumption of the larger value for k_{19}. The integrated overhead intensities are 1.2R for the green line and 46R for the red line. If O_2^+ is quenched by O at the larger rate, the integrated green line intensity is reduced to

0.5 R. Since the maximum electron density is somewhat larger than the average nightside value[58] the upper limit for auroral production is probably close the the measured upper limit. Had we adopted a yield of, say, 10% for the production of $O(^1S)$, independent of vibrational level of O_2^+, as has been done often in the past, close to 10 R would be predicted from dissociative recombination alone. A flux of soft electrons of sufficient magnitude to produce the observed 1304 and 1356 Å intensities produces 4 to 7 R of 5577 Å emission.[43] The results of this study of the green line show that the auroral model remains viable.

Acknowledgements

This work has been supported in part by grant ATM 8700436 from the National Science Foundation to the Massachusetts Institute of Technology and by grants NAGW-665 and NAG2-329 from the National Aeronautics and Space Administration to the Research Foundation of the State University of New York at Stony Brook.

REFERENCES

1. Wauchop, T. S. and H. P. Broida, Lifetime and quenching of $CO(a^3\Pi)$ produced by recombination of CO_2 ions in a helium afterglow. *J. Chem. Phys.,* 56, 330 (1972).

2. James, T. C., Transition moments, Franck-Condon factors, and lifetimes of forbidden transitions. Calculation of the intensity of the Cameron system of CO, *J. Chem. Phys.* 55, 4118 (1971).

3. Lawrence, G. M., Quenching and radiation rates of $CO(a^3\Pi)$, *Chem. Phys. Lett.* 9, 575 (1971).

4. Fox, J. L. and A Dalgarno, Ionization, luminosity and heating of the upper atmosphere of Venus, *J. Geophys. Res.* 86, 629 (1981).

5. Queffelec, J. L., B. R. Rowe, M. Morlais, J. C. Gomet and F. Vallee, The dissociative recombination of $N_2^+(v=0,1)$ as a source of metastable atoms in planetary atmospheres, *Planet. Space Sci.* 33, 263 (1985).

6. Garstang, R. H., Transition probabilities of auroral lines, in *The Airglow and Aurorae*, ed. by E. B. Armstrong and A. Dalgarno, p. 324, Pergamon, New York (1956).

7. Cravens, T. E., J. C. Gérard, A. I. Stewart and D. W. Rusch, The latitudinal gradient of nitric oxide in the thermosphere. *J. Geophys. Res.* 84, 2675 (1979).

8. McCoy, R. P., Thermospheric odd nitrogen 1. NO, $N(^4S)$ and $O(^3P)$ densities from rocket measurements of the NO δ and γ bands and the O_2 Herzberg I bands. *J. Geophys. Res.* 88, 3197 (1983).

9. Rusch, D. W. and T. E. Cravens, A model of the neutral and ion chemistry in the daytime thermosphere of Venus. *Geophys. Res. Lett.* 6, 791 (1979).

10. Fox, J. L., The chemistry of metastable species in the Venusian ionosphere, *Icarus* 61, 248, (1982).

11. Lin, C. L. and F. Kaufman, Reactions of metastable nitrogen atoms, *J. Chem. Phys.* 55, 3760 (1971).

12. Baulch, D. L., D. D. Drysdale, D. G. Horne and A. C. Lloyd, *Evaluated Kinetic Data for High Temperature Reactions. II. Homogeneous Gas Phase Reactions of the H_2–N_2–O_2 System*, Butterworths, London (1973).

13. Brown, R. and C. A. Winkler, The chemical behavior of active nitrogen. *Angew. Chem. intern. Edit.* 9, 181 (1970).

14. Piper, L. G., M. E. Donahue and W. T. Rawlins, Rate coefficients for $N(^2D)$

reactions *J. Phys. Chem.* 91, 3883 (1987).

15. Constantinides, E. R., J. H. Black, A. Dalgarno and J. H. Hoffman, The Photochemistry of N^+ ions. *Geophys. Res. Lett.* 6, 569 (1979).
16. Torr, M. R., P. G. Richards and D. G. Torr, A new determination of the ultraviolet heating efficiency of the thermosphere. *J. Geophys. Res.* 85, 6819 (1980).
17. Fox, J. L., Heating efficiencies in the thermosphere of Venus reconsidered, *Planet. Space Sci.* 36, 27 (1988).
18. Hunten, D. M., Thermal and non-thermal escape mechanisms for terrestrial bodies, *Planet. Space Sci.* 30, 773 (1982).
19. Chamberlain, J. W., Planetary coronae and atmospheric evaporation, *Planet. Space Sci.* 11, 901 (1963).
20. McElroy, M. B., M. J. Prather and J. M. Rodriguez, Escape of hydrogen from Venus, *Science,* 215, 614 (1982).
21. Cooper, D. L., J. H. Yee, and A. Dalgarno, Energy transfer in oxygen-hydrogen collisions, *Planet. Space Sci.* 32, 825 (1984).
22. Yee, J. H., J. W. Meriwether, P. B. Hays, Detection of a corona of fast oxygen atoms during solar maximum, *J. Geophys. Res.* 85, 3396 (1980).
23. Wallis, M. K., Exospheric density and escape fluxes of atomic isotopes on Venus and Mars, *Planet. Space Sci.* 26, 949 (1978).
24. Nagy, A. F., T. E. Cravens, J.-H. Yee and A. I. F. Stewart, Hot atoms in the upper atmosphere of Venus, *Geophys. Res. Lett.* 8, 629 (1981).
25. Paxton, L. J., Atomic carbon in the Venus thermosphere: observations and theory, *PhD thesis*, University of Colorado (1983).
26. Nagy, A. F. and T. E. Cravens, Hot oxygen atoms in the upper atmospheres of Venus and Mars, *Geophys. Res. Lett.* 15, 933 (1988).
27. Nier, A. O., M. B. McElroy and Y. L. Yung, Isotopic composition of the Martian atmosphere. *Science* 194, 68, (1976).
28. Nier, A. O. and M. B. McElroy, Composition and structure of Mars upper atmosphere: Results from the neutral mass spectrometers on Viking 1 and 2, *J. Geophys. Res.* 82, 4341 (1977).
29. Fox, J. L. and A. Dalgarno, The escape of nitrogen from Mars, *J. Geophys. Res.,* 88, 9027 (1983).
30. Nier, A. O., W. B. Hanson, A. Seiff and M. B. McElroy, Composition and struc-

ture of the Martian atmosphere: Preliminary results from Viking I. *Science* 193, 786 (1976).

31. McElroy, M. B., T. Y. Kong and Y. L. Yung, Photochemistry and evolution of Mars atmosphere: a Viking perspective, *J. Geophys. Res.* 82, 4379 (1977).
32. Hanson, W. B., S. Sanatani and D. R. Zuccaro, The Martian ionosphere as observed by the retarding potential analyzers, *J. Geophys. Res.* 82, 4351 (1977).
33. Fox, J. L. and A. Dalgarno, The production of nitrogen atoms on Mars and their escape, *Planet. Space Sci.*28, 41 (1980).
34. Zipf, E. C., The dissociative recombination of vibrationally excited N_2^+ ions. *Geophys. Res. Lett.* 7, 645 (1980).
35. Fox, J. L. and A. Dalgarno, The vibrational distribution of N_2^+ in the terrestrial ionosphere, *J. Geophys. Res.* 90, 7557 (1985).
36. Frederick, J. E., and D. W. Rusch, On the chemistry of metastable atomic nitrogen in the F region deduced from simultaneous satellite measurements of the 5200-Å airglow and atmospheric composition, *J. Geophys. Res.* 82, 3509 (1977).
37. Canuto, V. M., J. S. Levine, T. R. Augustson and C. L. Imhoff, UV radiation from the young sum and oxygen and ozone levels in the prebiological palaeoatmosphere, *Nature* 296, 816 (1982).
38. Zahnle, K. J. and J. C. G. Walker, The evolution of solar ultraviolet luminosity, *Rev. Geophys. Space Phys.* 20, 280 (1982).
39. Newman, M. J. and R. T. Rood, Implications of solar evolution for the Earth's early atmosphere, *Science* 198, 1035 (1977).
40. Cess, R. D., V. Ramanathan and T. Owen, The Martian paleoclimate and enhanced atmospheric carbon dioxide, *Icarus* 41, 159 (1980).
41. Pollack, J. B., F. J. Kasting, S. M. Richardson and K. Poliakoff, The case for a wet, warm climate on early Mars, *Icarus* 71, 201 (1987).
42. Phillips, J. L., A. I. F. Stewart and J. G. Luhmann, The Venus ultraviolet aurora: observations at 130.4 nm, *Geophys. Res. Lett.* 13, 1047 (1986).
43. Fox, J. L. and A. I. F. Stewart, The Venus ultraviolet aurora: a soft electron source, *in preparation* (1988).
44. Knudsen, W. C. and K. L. Miller, Pioneer Venus superthermal electron flux measurements in the Venus umbra, *J. Geophys. Res.* 90, 2695 (1985).

45. Gringauz, K. I., M. I. Verigin, T. K. Breus, and T. Gombosi, The interaction of electrons in the optical umbra of Venus with the planetary atmosphere–The origin of the nighttime ionosphere. *J. Geophys. Res.* 84, 2123 (1979).

46. Krasnopolsky, V. A., Excitation of oxygen emissions in the night airglow of the terrestrial planets, *Planet. Space Sci.* 29, 925 (1981).

47. Abreu, V. J., R. W. Eastes, J. H. Yee, S. C. Solomon and S. Chakrabarti, Ultraviolet nightglow production near the magnetic equator by neutral particle precipitation, *J. Geophys. Res.* 91, 11,365 (1986).

48. Bates, D. R. and E. C. Zipf, The O(^1S) quantum yield of O_2^+ dissociative recombination, *Planet. Space Sci.* 28, 1081 (1980).

49. Guberman, S. L., The production of O(^1D) from dissociative recombination of O_2^+, *Planet. Space. Sci.* 36, 47 (1988).

50. Guberman, S. L., The production of O(^1S) in dissociative recombination of O_2^+, *Nature,* 327, 408 (1987).

51. Hedin, A. E., H. B. Niemann, W. T. Kasprzak and A. Seiff, Global empirical model of the Venus thermosphere, *J. Geophys. Res.* 88, 73 (1983).

52. Niemann, H. B., W. T. Kasprzak, A. E. Hedin, D. M. Hunten and N. W. Spencer, Mass spectrometric measurements of the neutral gas composition of the thermosphere and exosphere of Venus, *J. Geophys. Res.* 85, 7817 (1980).

53. Knudsen, W. C., K. Spenner, K. L. Miller and V. Novak, Transport of ionospheric O^+ ions across the Venus terminator and implications, *J. Geophys. Res.* 85, 7803 (1980).

54. Knudsen, W. C., K. Spenner, and K. L. Miller, Antisolar acceleration of ionospheric plasma across the Venus terminator. *Geophys. Res. Lett.* 8, 241 (1981).

55. Cravens, T. E., S. L. Crawford, A. F. Nagy and T. I. Gombosi, A two-dimensional model of the ionosphere of Venus, *J. Geophys. Res.* 88, 5595 (1983).

56. Spenner, K., W. C. Knudsen, R. C. Whitten, P. F. Michelson, K. L. Miller and V. Novak, On the maintenance of the Venus nightside ionosphere: electron precipitation and plasma transport, *J. Geophys. Res.* 86, 9170 (1981).

57. Taylor, H. A., R. E. Hartle, H. B. Niemann, L. H. Brace, R. E. Daniell, S. J. Bauer and A. J. Kliore, Observed composition of the ionosphere of Venus: implications for the ionization peak and the maintenance of the nightside iono-

sphere, *Icarus* 51, 283 (1982).

58. Knudsen, W. C., K. L. Miller and K. Spenner, Median density profiles of the major ions in the central nightside Venus ionosphere. *J. Geophys. Res.* 91, 11,936 (1986).

59. Cravens, T. E., L. H. Brace, H. A. Taylor, C. T. Russell, W. L. Knudsen, K. L. Miller, A. Barnes, J. D. Mihalov, F. L. Scarf, S. J. Quenon and A. F. Nagy, Disappearing ionospheres on the nightside of Venus. *Icarus* 51, 271 (1982).

60. Taylor, H. A., H. C. Brinton, S. J. Bauer, R. E. Hartle, P. A. Cloutier and R. E. Daniell, Global observations of the composition and dynamics of the ionosphere of Venus: implications for the solar wind interaction. *J. Geophys. Res.* 85, 7765 (1980).

61. Smith, D., N. G. Adams and T. M. Miller, A laboratory study of the reactions of N^+, N_2^+, N_3^+, N_4^+, O^+, O_2^+ and NO^+ with several molecules at 300 K, *J. Chem. Phys.* 69, 308 (1978).

62. Fehsenfeld, F. C., D. B. Dunkin and E. E. Ferguson, Rate constants for the reaction of CO_2^+ with O, O_2 and NO; N_2^+ with O and NO; and O_2^+ with NO. *Planet. Space Sci.* 18, 1267 (1970).

63. Fox, J. L., The O_2^+ vibrational distribution in the Venusian ionosphere, *Adv. Space Res.* 5, 165 (1985).

64. Böhringer, H., M. Durup-Ferguson, E. E. Ferguson and D. W. Fahey, Collisional vibrational quenching of $O_2^+(v)$ and other molecular ions in planetary atmospheres, *Planet. Space Sci.*, 31, 483 (1983).

65. Fox, J. L., The O_2^+ vibrational distribution in the dayside ionosphere, *Planet. Space Sci.* 34, 1252 (1986).

66. Alge, E., N. G. Adams and D. Smith, Measurements of the dissociative recombination coefficients of O_2^+, NO^+ and NH_4^+ in the temperature range 200 – 600 K. *J. Phys. B.* 16, 1433 (1983).

AERONOMICAL DETERIMINATIONS OF THE QUANTUM YIELDS OF $O(^1S)$ AND $O(^1D)$ FROM DISSOCIATIVE RECOMBINATION OF O_2^+

Jeng-Hwa Yee, Vincent J. Abreu and William B. Colwell

Space Physics Research Laboratory
Department of Atmospheric, Oceanic and Space Sciences
The University of Michigan, Ann Arbor, MI 48109 USA

Abstract– Data from the Visible Airglow Experiment on the Atmosphere Explorer-E satellite have been used to determine the quantum yields of $O(^1S)$ and $O(^1D)$ from the dissociative recombination of O_2^+ based on a constant total recombination rate from each vibrational level. A range of values between 0.05 and 0.18 has been obtained for the quantum yield of $O(^1S)$ and shows a positive correlation with the extent of the vibrational excitation of O_2^+. The quantum yield of O(1D) has been measured to be 0.9±0.2, with no apparent dependence on the vibrational distribution of O_2^+.

1. INTRODUCTION

The dissociative recombination of molecular ions is a very important aeronomical process in the Earth's ionosphere. It is the only significant electron recombination or electron removal process in the ionosphere [1]. The reaction process is exothermic. The excess energy released provides a source of heat to the ambient atmosphere and is very important to the energy balance of the thermosphere, especially in the 200-250 km region [2]. Most importantly, airglow emissions which are produced from metastable atoms by dissociative recombination enable us to remotely monitor the thermospheric temperature and wind from the doppler width and shift of the emission line profile [3,4].

The dissociative recombination of O_2^+ with electrons leads to the production of atomic oxygen in the ground $O(^3P)$ state and in the $O(^1D)$ and $O(^1S)$ metastable states, i.e.

$$O_2^+ + e \rightarrow O(^3P,^1D) + O(^3P,^1D,^1S). \tag{1}$$

This process has been studied extensively through the 6300Å and the 5577Å emissions caused by the $O(^3P-^1D)$ and $O(^1D-^1S)$ transitions respectively. Although the nature of this process is readily understood and the rate coefficient has been obtained by many experimental measurements, there are aspects of this process that remain unexplained, specifically, the quantum yields of $O(^1S)$ per dissociative recombination of O_2^+.

In the nighttime ionosphere, the excited $O(^1S)$ and $O(^1D)$ atoms are believed to be produced mainly by the dissociative recombination of O_2^+ ions. The volume emission rates at 5577Å and 6300Å are given by

$$\eta\ (5577\ \text{Å}) = \frac{A_{5577}\ f_{1_S}\ k_d\ [O_2^+]\ [e]}{A_{1_S}}, \tag{2}$$

$$\eta\ (6300\ \text{Å}) = \frac{A_{6300}\ f_{1_D}\ k_d\ [O_2^+]\ [e]}{A_{1_D} + k_o\ [O] + k_{N_2}\ [N_2]}, \tag{3}$$

where A_{6300} and A_{5577} are the spontaneous emission rates for the production of 6300Å and 5577Å photons, A_{1_D} and A_{1_S} are the spontaneous emission rates for the ^1D and ^1S states, f_{1_D} and f_{1_S} are the quantum yields of $O(^1D)$ and $O(^1S)$, and k_d is the rate coefficient of the total dissociative recombination of O_2^+, assuming no variation with vibrational distribution. The sum in the denominator of Eq. (3) is over all quenching species X_i, and k_i is the quenching rate coefficient [5]. $[X_i]$ represents the number density of the species X_i. Note that the yield of $O(^1D)$ here includes both the direct production from the dissociative recombination process and the indirect process which considers cascading from the $O(^1S)$ state.

Regarding the yield of $O(^1D)$, Zipf [6] obtained a value of 1.0 in a laboratory plasma. Using the volume emission rates and the density profiles obtained during a rocket flight, Sharp et al. [7] deduced that f_{1_D} is only about 0.3 and obtained an N_2 quenching rate coefficient of $1.5\text{x}10^{-11}$ cm^3 s^{-1}. Values of 1.33 and $3\text{x}10^{-11}$ cm^3 s^{-1}, with estimated errors of 30%, were later obtained by Hays et al. [8] from an analysis of Atmosphere Explorer nightglow data. Since then a range of values between 0.9 and 1.3 for the quantum yield of $O(^1D)$ atoms has been obtained as various rate coefficients and transition probabilities are used [9]. As indicated in Table 1, the atmospheric measurements and the laboratory measurements are in agreement within the range of the probable uncertainties in the determinations.

On the other hand, the yield of $O(^1S)$ has been difficult to establish (Table 1). Laboratory measurements by Zipf [6] determined a value of 0.1 for the production efficiency. Similar values were inferred from the Atmosphere Explorer airglow measurements made during the day and evening by Frederick et al. [10] and Kopp et al. [11], and in the aurora by Sharp and Torr [12]. However, rather smaller values, approximately 0.03-0.04, were determined from the nightglow and twilight airglow measurements by Hernandez [13] and Hays and Sharp [14]. A smaller value of ~0.02 has also been obtained from the artificial auroral experiment Precede of O'Neil et al. [15]. After adopting the measured quenching rate coefficient of $O(^1S)$ by O_2 in his previous Precede 5577Å data, Zipf [16] deduced a value as low as 0.02 for the yield of $O(^1S)$ and suggested that f_{1S} is quite sensitive to the degree of vibrational excitation of O_2^+ ions. This new result

Table 1. $O(^1S)$ and $O(^1D)$ Quantum Yields from O_2^+ Recombination

Experiment	f_{1D}	Sources
Laboratory Plasma	1.0	Zipf [6]
Airglow (Rocket)	0.3	Sharp et al. [7]
Nightglow (AE)	1.33 ± 0.4	Hays et al. [8]
Nightglow (AE)	1.23 ± 0.42	Abreu et al. [23]
Nightglow (AE)	1.2	Abreu et al. [5]

Expermeint	f_{1S}	Sources
Laboratory Plasma	0.1	Zipf [6]
Post-sunset airglow	0.084 (+0.02 -0.033)	Frederick et al. [10]
Dayglow	0.094 ± 0.01	Kopp et al. [11]
Aurora	0.08	Sharp and Torr [12]
Equatorial nightglow	0.04 ± 0.005	Hernandez [13]
Midlatitude nightglow	0.03 ± 0.02	Hays and Sharp [14]
Twilight airglow	0.03 ± 0.02	Hays and Sharp [14]
Artificial Aurora	0.02	O'Neil et al. [15]
Precede Experiment	0.02	Zipf [16]
Nightglow (AE)	0.09 to 0.23	Abreu et al. [23]

confirmed his earlier claims [6] that the lower yields obtained in atmospheric studies could be interpreted as coming from atmospheric situations in which O_2^+ might be in low vibrational levels, while the higher values might be arised from high vibrational levels.

Theoretical calculations of the potential surfaces of O_2^+ by Guberman [17] show that only one repulsive curve resulting in an $O(^1S)$ atom is available for O_2^+ ions in the lowest vibrational levels. This repulsive $^1\Sigma_u^+$ curve crosses the ion potential curve just above the υ = 1 vibrational level of the O_2^+ ground state, making vibrational levels 1 and 2 the main contributors to the $O(^1S)$ production. Killeen and Hays [18] presented evidence from measurements of the line profile of 5577Å emission taken by the Fabry-Perot Interferometer on board the Dynamics Explorer (DE) satellite that the $^1\Sigma_u^+$, $O(^1S)+O(^1D)$, channel is preferred over the $5^3\Pi_g$, $O(^1S)+O(^3P)$ channel by a factor of four. For O_2^+ ions in the υ=1 and υ=2 states, the exothermicities are 1.06 eV and 1.29 eV respectively. Because the product of $O(^1S)$ atoms has higher translational energy, vibrational excitation of O_2^+ ions will significantly broaden the line shape of the high altitude 5577Å emission. Yee and Killeen [19] have re-analyzed the DE data used by Killeen and Hays [18] and treated the thermalization of $O(^1S)$ atoms more realistically by including elastic collisions as well as excitation exchange collisions. They found that if all the $O(^1S)$ atoms are produced in dissociative recombination of O_2^+ ions in υ=1 and υ=2, 60% of the $O(^1S)$ are produced from υ=2 and 40% from υ=1, substantiating the results of the potential curve calculation of Guberman [17].

The O_2^+ ions in the thermosphere are produced by charge exchange of O^+ with O_2. The excess energy of the reaction is capable of populating O_2^+ ions up to vibrational level 7. The efficiency of $O(^1S)$ production by the dissociative recombination of O_2^+ depends upon the lifetimes for the O_2^+ vibrational deactivation through collisions and loss through recombination. The extent of the vibrational excitation, r, beyond that appropriate to the ambient gas temperature would then be given by

$$r = \frac{\beta}{\alpha} \frac{[e]}{[x]}, \tag{4}$$

as suggested by Bates and Zipf [20], where α is the effective rate coefficient for the complete deactivation process by collisions with ambient species X, and β is the dissociative recombination rate coefficient. If one is to attribute the high values of f_{1S} obtained in the atmospheric measurements to the fact that the O_2^+ ions are vibrationally excited, Bates and Zipf [20] has argued that, the exothermic ion-molecular reaction

$$O_2^+ (v') + O \rightarrow O_2^+ (v < v') + O \tag{5}$$

has to be much slower than would be expected for such a process and the magnitude of r, in this case defined as $(\beta/\alpha)(n[e]/n[O])$, is large. By using the rate coefficient given by Torr and Torr [21] derived from the cross-section measurements of Walls and Dunn [22] for the dissociative recombination rate coefficient, one would expect a positive correlation between f_{1S} and $n[e]/n[O](300/Te)^{0.7}$ if the collisional deactivation rate α is a constant. This has been reported by Abreu et al. [23] from the data obtained from the visible airglow experiment (VAE) on board the Atmospheric Explorer satellite. However, a set of values of f_{1S} varying between 0.09 and 0.23, relatively larger than the ones previously determined, was obtained.

Guberman [24,25] reported the first theoretical calculations of the dissociative recombination rate coefficients for the production of $O(^1S)$ atom (v=0 to v=4) and $O(^1D)$ atom (v=0 to v=2). Table 2

Table 2. Calculated Production Efficiencies of $O(^1S)$ and $O(^1D)$ From Guberman [24,25] at Te = 800K assuming no dependence on v of the total recombination rate

Vibrational level	f_{1S}	f_{1D}
0	0.003(+0.0015,-0.001)	1.41(+0.13,-0.15)
1	0.056(+0.014,-0.012)	1.11(+0.12,-0.14)
2	0.16 (+0.02, -0.03)	0.79(+0.13,-0.16)
3	0.040(+0.026,-0.020)	Not Available
4	0.056(+0.011,-0.018)	Not Available

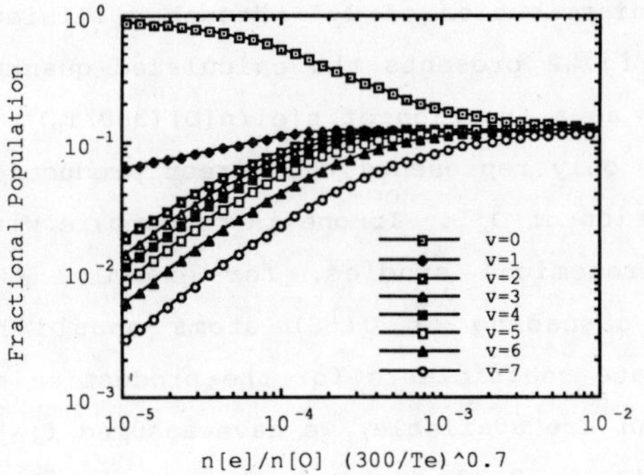

Figure 1. Calculated Vibrational Distribution of O_2^+

shows the production efficiencies of the $O(^1S)$ and $O(^1D)$ atoms at an electron temperature of 800K, derived from the total recombination coefficient of Walls and Dunn [22]. The quantum yield of $O(^1S)$ atoms increases from 0.003 for the O_2^+ ions at $\upsilon = 0$ to a maximum value of 0.16 at $\upsilon = 2$. The yield of $O(^1D)$ atoms, f_{1D}, however, has a maximum value of 1.41 for the O_2^+ ions at its ground vibrational level and decreases to 0.79 at $\upsilon = 2$. The effective quantum yields of $O(^1S)$ and $O(^1D)$ atoms determined from the airglow studies thus vary with the vibrational distribution of the atmospheric O_2^+ ions. Fig. 1 shows calculated vibrational distributions of O_2^+ ions in the nighttime thermosphere as a function of $n[e]/n[O](300/Te)^{0.7}$, using Model A of Fox [26,27] and an electron temperature of 800K. Five models of Fox [27], designated by letters A to E, which consider various $O_2^+(\upsilon)$ production and collisional redistribution mechanisms have been adopted (Table 3). Since molecular oxygen is much less abundant than the atomic oxygen in the region where the $O(^1S)$ and $O(^1D)$ atoms are mostly

Table 3. O_2^+ Vibrational Relaxation Models of Fox [27]

Model	$O_2^+(\upsilon)$ Production	O vibrational Quenching
A	Statistical	Medium ($1.0\times10^{-10}cm^{-3}sec^{-1}$)
B	Statistical	Slow ($1.0\times10^{-11}cm^{-3}sec^{-1}$)
C	Statistical	Fast ($6.0\times10^{-10}cm^{-3}sec^{-1}$)
D	$\upsilon = 7$	Medium ($1.0\times10^{-10}cm^{-3}sec^{-1}$)
E	Franck-Condon	Medium ($1.0\times10^{-10}cm^{-3}sec^{-1}$)

produced, vibrational level redistribution of O_2^+ through collisions with O_2 has been neglected. Fig. 2 presents the calculated quantum yields of $O(^1S)$ and $O(^1D)$ atoms as a function of $n[e]/n[O](300/T_e)^{0.7}$. Here the quantum yield of $O(^1D)$ only represents the direct production from the dissociative recombination of O_2^+. If one is to compare with the yield determined from aeronomical studies, for example, the indirect process due to the cascading of $O(^1S)$ atoms should be included. Since not all the rate coefficients for the production of $O(^1S)$ and $O(^1D)$ atoms from $O_2^+(\upsilon)$ are available, we have assumed f_{1S} = 0.056 for O_2^+ at $\upsilon > 4$ and f_{1D} = 0.79 for O_2^+ at $\upsilon > 2$. This assumption is apparently not correct, and the computed f_{1S} and f_{1D} only apply to situations when n[e]/n[O] is small or the O_2^+ ions are mostly populated in their lowest vibrational states. Qualitatively, however, the relationship between f_{1S} and f_{1D} with n[e]/n[O] is still valid. Although the magnitudes of f_{1S} and f_{1D} vary with different $O_2^+(\upsilon)$ distribution models, in general, f_{1S} shows a strong positive correlation with n[e]/n[O], while f_{1D} shows slightly the opposite.

The conclusion that f_{1S} and f_{1D} vary with the state of vibrational development in the recombining O_2^+ can be better examined

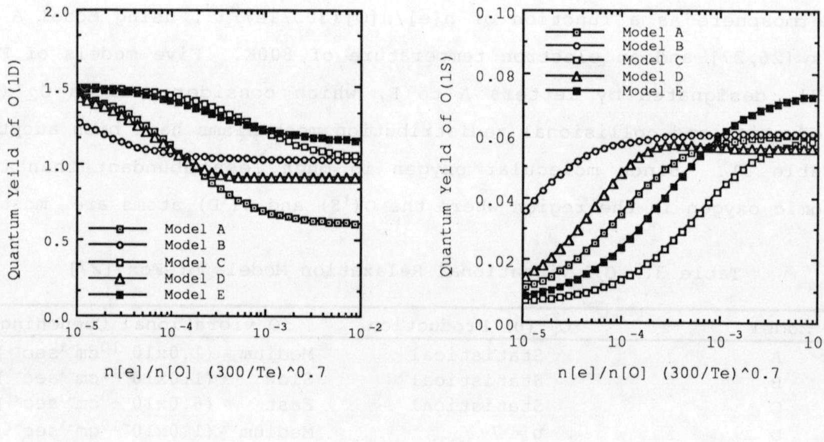

Figure 2. Calculated Quantum Yields of $O(^1D)$ and $O(^1S)$

by carefully analyzing a set of emission rates and density measurements obtained under various atmospheric conditions with an aid of the simple O_2^+ vibrational relaxation model described above. In this paper, we report the results of our analysis of the VAE nightglow data using the data reduction technique developed by Abreu et al. [23] and the O_2^+ vibrational relaxation model of Fox [28]. The analysis is similar to that of Abreu et al. [23], but the number of measurments have been extended and data rejection criteria improved.

2. DATA ANALYSIS

The quantum yields of $O(^1S)$ and $O(^1D)$ can be determined by using nightglow measurements of the volume emission rates at 5577Å and 6300Å from the relation,

$$f_{1_S} = \frac{A_{1_S}\ \eta\ (5577\ \text{Å})}{A_{5577}\ k_d\ [O_2^+]\ [e]}\ , \tag{6}$$

$$f_{1_D} = \frac{\left\{A_{1_D} + k_o\ [O] + k_{N_2}\ [N_2]\right\}\ \eta\ (6300\ \text{Å})}{A_{6300}\ k_d\ [O_2^+]\ [e]}\ . \tag{7}$$

The visible airglow experiment measures brightness, i.e., the integrated volume emission rate along the line of sight of the observation. In order to obtain the volume emission rate we used the technique of Abreu et al. [23] which involves the subtraction of measured brightness from consecutive satellite spins along the same direction in inertial space. Since the same volume in space is observed from both forward and backward directions, two measurements of the volume emission rate can be obtained and their average determines the volume emission rate used in this study. The Visible Airglow Experiment (VAE) consists of two photometers with different filters so that two emissions can be measured simultaneously. Of the various fixed filter combinations, one mode monitors both the 6300Å and 5577Å emissions. The O_2^+ and total ion densities used to calculate the quantum yields of $O(^1S)$ and $O(^1D)$ were simultaneously obtained from the Benett Ion Mass Spectrometer (BIMS) on board the AE satellite [28]. The measurements were normalized by using a constant normalization factor of 2.48, determined from the average ratio of the total ion density measured by BIMS to the total ion density measured by the Retarding Potential Analyzer (RPA) [29]. The neutral densities were obtained from the Open Source Spectrometer (OSS) [30]. The electron temperatures were given by the Cylindrical Electrostatic Probes (CEP) [31]. If the electron temperatures were not available, neutral temperatures obtained by the Neutral-Atmosphere Temperature Experiment (NATE) [32] were used with the assumption that $T_e \cong T_n$. This is a good approximation for the nighttime, low latitude F region.

In order to determine the magnitudes of f_{1S} and f_{1D} and clearly establish their correlation with r, a data set which covers a wide range of n[e]/n[O] is required. When the electron density is small, however, both the 5577Å and 6300Å emissions usually are weak, making the determination of f_{1S} and f_{1D} very difficult. In this analysis, most of data were obtained during the periods when the satellite was in the Appleton Anomaly and the electron density was large. We have also rejected the measurements while the computed volume emission rates of 5577Å (based on an f_{1S} of 0.08) and 6300Å (based on an f_{1D} of 1.2) are smaller than 0.5 and 2.0 photons/cm^3/sec respectively. The measurements of f_{1S} and f_{1D} with small r or n[e]/n[O] were therefore mainly obtained during the periods when there is significant amount of emissions and the atomic oxygen density is high.

Figure 3(a) shows the quantum yield of $O(^1D)$ as a function of $n[e]/n[O](300/T_e)^{0.7}$. A total of 2830 data points collected from 29 orbits have been used. It shows no observable dependence of f_{1D} on the extent of O_2^+ vibrational excitation and the average quantum yield is found to be 0.9±0.2. After correcting for the source due to cascading from the $O(^1S)$ atoms (obtained from the measured 5577Å volume emission rate), we derived the yield of $O(^1D)$ atoms produced directly from the dissociative recombination of O_2^+ ions and it is also shown in Figure 3(a). The cascading corrected yield derived here agrees fairly well with the theoretically calculated values based upon the specific rates of Guberman [24] and the total recombination rate of Walls and Dunn [22], assumed to be constant with $O_2^+(\upsilon)$.

Figure 3(b) shows the quantum yield of $O(^1S)$ as a function of $n[e]/n[O](300/T_e)^{0.7}$. A total of 1156 data points from fifteen orbits have been used. The quantum yield f_{1S} obtained range from 0.05 to 0.18 and increases with $n[e]/n[O](300/T_e)^{0.7}$. The range of temperatures present in the data varies between 660 K and 880 K. Unless there is a very strong temperature dependence in the rate coefficient of the O_2^+ vibrational deactivation process (reaction 5), the positive correlation between the $O(^1S)$ production efficiency and the extent of O_2^+ vibrational excitation, as the simple theoretical model has indicated, is clearly established. The magnitudes of f_{1S} determined from the observations, however, are larger by at least a factor of two, with large statistical deviations due to the uncertainties in the measured volume emission rates and the ion and electron densities.

As indicated in Eqs.(6) and (7), the accuracies of the determined f_{1S} and f_{1D} depend upon the accuracies of the volume emission rates and the neutral and plasma densities. Although the measurements

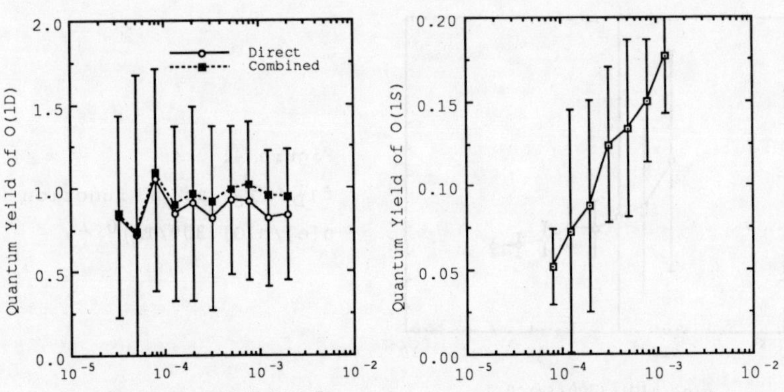

Figure 3. Observed Quantum Yields of $O(^1D)$ and $O(^1S)$

of O_2^+ and the total ion (or electron) densities of BIMS were used, they have been normalized using a constant of factor 2.48. The uncertainty of this normalization constant is found to be ±1.48 (or ~60%), and can be partly responsible for the large statistical standard deviation associated with the quantum yields determined here. The volume emission rates η(5577Å) and η(6300Å) used in this study were obtained simultaneously from a dual filter mode in which both the 5577Å and 6300Å emissions were monitored. Under this VAE operational mode, one can determine the ratio of $f1_D$ to $f1_S$ from

$$\frac{f_{1_D}}{f_{1_S}} = \frac{A_{5577}\left\{A_{1_D} + k_o\ [O] + k_{N_2}\ [N_2]\right\}\ \eta\ (6300\ Å)}{A_{6300}\ A_{1_S}\ \eta\ (5577\ Å)} . \qquad (8)$$

Note here that the plasma densities are not required in the determination of $f1_D/f1_S$ and in this fashion the uncertainty due to the lack of knowledge of the plasma densities is minimized. Fig. 4 presents the results of $f1_D/f1_S$ obtained as a function of $n[e]/n[O](300/T_e)^{0.7}$. The ratio of the quantum yields, $f1_D/f1_S$, decreases from ~23.0 to ~6.0 as $n[e]/n[O](300/T_e)^{0.7}$ increases from 8.0×10^{-5} to 2.0×10^{-3} and clearly shows an anti-correlation with the

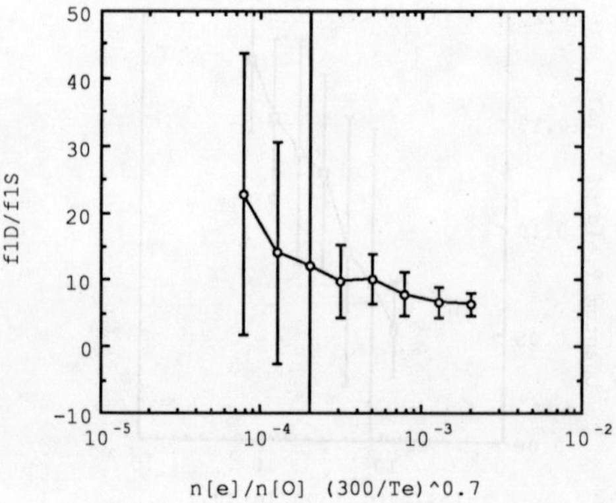

Figure 4:
$f1_D/f1_S$ as a function of $n[e]/n[O](300/T_e)^{0.7}$

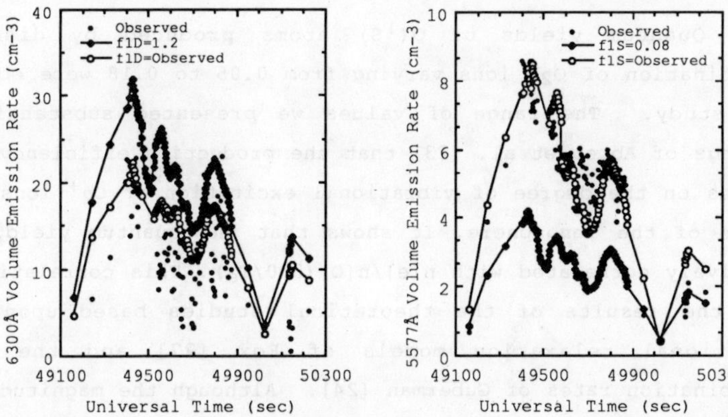

Figure 5. Comparison of Calculated and Observed 6300Å and 5577Å Volume Emission Rates During Orbit 11488.

extent of the vibrational excitation of O_2^+ ions. The ratio of ~5.0 obtained when the O_2^+ ions are highly vibrationally excited suggests that, if f_{1_D} is taken to be well-accepted value of 1.0 to 1.2, f_{1_S} should be still at least approximately 0.15, almost twice the values that Frederick et al. [10] and Kopp et al. [11] have derived.

Figure 5 shows the volume emission rates of 6300Å and 5577Å measured along the orbit 11488. The calculated values based upon constant quantum yields of 1.2 and 0.08 for the production of $O(^1D)$ and $O(^1S)$ atoms are indicated. Good agreement can be found between the calculated 6300Å volume emission rates and the observed ones. The calculated 5577Å emission rates based upon a constant quantum yield of 0.08, however, failed to account for the amount that was actually observed. If we consider the quantum yields f_{1_S} and f_{1_D} to vary with the degree of vibrational excitation of recombining O_2^+ ions and use the derived values presented in Figures 3(a) and 3(b), the agreement improves and the calculated emission rates become much closer to the observed values.

3. CONCLUSION

Quantum yields of $O(^1S)$ atoms produced by dissociative recombination of O_2^+ ions varying from 0.05 to 0.18 were obtained in this study. The range of values we presented substantiates the findings of Abreu et al. [23] that the production efficiency of $O(^1S)$ depends on the degree of vibrational excitation of O_2^+ ions in the F region of the ionosphere. It shows that the quantum yield, f_{1S}, is positively correlated with $n[e]/n[O](300/T_e)$. This correlation agrees with the results of the theoretical studies based upon the O_2^+ vibrational relaxation models of Fox [27] and the specific recombination rates of Guberman [24]. Although the magnitudes of f_{1S} obtained in this study are slightly smaller than the ones that Abreu et al. [23] have derived, both, however, are larger than the theory suggests if the total recombination rate varies very little with $O_2^+(\upsilon)$, and for large $n[e]/n[O](300/T_e)^{0.7}$ (high degree of O_2^+ vibrational excitation), they are larger than most of the values previously determined from the laboratory and airglow studies. The specific recombination rates calculated by Guberman [24] give a quantum yield of $O(^1S)$ of 0.16±0.02 at an electron temperature of 800K for the dissociative recombination of O_2^+ at $\upsilon = 2$ and much smaller values for $\upsilon \neq 2$. A quantum yield of 0.15 determined here suggests that the O_2^+ ions in the ionospheric altitudes be mostly populated at the vibrational level 2. This is unlikely, however, due to the frequent vibrational relaxation collisions with the ambient atomic oxygen. The 5577Å volume emission rates used in this study were obtained from the high sensitivity, wide angle VAE photometer channel. If the satellite was located above the peak of the 5577Å emission rate profile, due to its wide field of view (~3 degree), the 5577Å volume emission rates or the quantum yields would be over-determined by a maximum of 20 to 30 percent depending upon the slope of the $\eta(5577Å)$ profile. This correction would bring our measurements in better agreement with the results of Frederick et al. [10] and Kopp et al.[11], but they still would not be in complete agreement with the theory despite the uncertainties (~10%) in the specific rate coefficients of Guberman [24]. In this study, the determination of the $O(^1S)$ quantum yield depends critically on the assumption of a constant total recombination rate of $O_2^+(\upsilon)$. Unless the rate is smaller at higher υ, adopting the specific rate coefficients of Guberman [24], one would suggest from our measured f_{1S} that there exists an additional source of $O(^1S)$ atoms in the nighttime F region and this source should have similar characteristics to the dissociative recombination of O_2^+ ions.

In this study, the quantum yield of $O(^1D)$ atoms was measured to be 0.9±0.2 with no dependence on the $n[e]/n[O](300/T_e)^{0.7}$. This is in good agreement with the theory and the results previously determined from the laboratory and airglow studies. If both the $O(^1D)$ and $O(^1S)$ atoms at night are produced by the dissociative recombination process in the F region, our measurements showed that the ratio of their yields, f_{1D} to f_{1S}, varies between ~23.0 to ~6.0 (~30 to ~8 if field-of-view effect is corrected) and is anti-correlated with the extent of O_2^+ vibrational excitation.

Acknowledgments: The authors would like to thank Dr. Steve Guberman for his useful comments and providing his calculated results before they were published. This research was supported at the University of Michigan under NASA grant NAGW-496 and in part by the National Science Foundation under Grant No. ATM-8712631.

REFERENCES

1. Bates, D. R. and Massey, H. S. W., "The basic reactions in the upper atmosphere II. The theory of recombination in the ionized layers", Proc. R. Soc, 192, 1(1947).
2. Roble, R. G., Ridley, E. C. and Dickinson, R. E., "On the global mean structure of the thermosphere", J. Geophys. Res. 92, 8745 (1987).
3. Biondi, M. A. and Feibelman, W. A., "Twilight and nightglow specral line shapes of oxygen λ6300 and λ5577 radiation", Planet. Space. Sci., 16, 431(1968).
4. Hays, P. B. and Roble, R. G., "A technique for recovering Doppler line profiles from Fabry-Perot interferometer fringes of very low intensity", Appl. Optics, 10, 193(1971).
5. Abreu, V. J., Yee, J. H., Solomon, S. C. and Dalgarno, A., "The quenching rate of $O(^1D)$ by $O(^3P)$", Planet. Space Sci., 34, 1143(1986).
6. Zipf, E. C., "The dissociative recombination of O2_ ions into specifically identified final atomic states", Bull. Am. Phys. Soc., 15, 418 (1970).
7. Sharp, W. E., Rusch, D. W. and Hays, P. B.,"Dissociative recombination source for $OI(^1D)$ atoms", J. Geophys. Res. 80, 2876(1975).
8. Hays, P. B., Rusch, D. W., Roble, R. G. and Walker, J. C. G., "The OI(6300Å) airglow", Rev. Geophys. Space Phys., 16, 225(1978).
9. Links, R. and Cogger, L. L., "A Re-Examination of the OI 6300Å Nightglow", to appear in J. Geophys. Res. (1988).
10. Frederick, J. E., Rusch, D. W., Victor, G. A., Sharp, W. E., Hays, P. B. and Brinton, H. C., "The OI(5577Å) airglow: Observations and excitation mechanisms", J. Geophys. Res., 81, 22(1976).
11. Kopp, J. P., Frederick, J. E., Rusch, D. W. and Victor, G. A., "Morning and evening behavior of the F region green-line emission: Evidence concerning the sources of $O(^1S)$", J. Geophys, Res., 82, 4715(1977).

12. Sharp, W. E. and Torr, D. G., "Determination of the auroral O(^{1}S) production sources from coordinated rocket and satellite measurements", J. Geophys. Res., 84, 5345(1979).
13. Hernandez, G., "The signature profile of O(^{1}S) in the airglow", Planet Space Sci., 19, 468(1971).
14. Hays, P. B. and Sharp, W. E., "Twilight airglow, 1, Photoelectrons and [OI} 5577Å radiation", J. Geophys. Res., 78, 1163(1973).
15. O'Neil, R. R., Lee, E. T. P. and Huppi, E. R., "Auroral O(^{1}S) production and loss processes: Ground-based measurements of the artifial auroral experiment, Precede", J. Geophys, Res., 84, 323(1979).
16. Zipf, E. C., "The OI(^{1}S) state: Its quenching and formation by the dissociative recombinatio of vibrationally excited O_2^+ ions", Geophys. Res. Lett., 6, 881(1979).
17. Guberman, S. L., "Potential energy curves for the dissociative recombination", in *Physics of Ion-Ion and Electron-Ion collisions* (edited by Brouillard, F. and McGowan. J. W), 167(1983).
18. Killeen, T. L. and Hays, P. B., "O(^{1}S) from dissociative recombination of O_2^+: monthermal line profile measurements from Dynamic Explorer", J. Geophys, Res, 88, 163(1983).
19. Yee, J. H. and Killeen, T. L., "Thermospheric production of O(^{1}S) by dissociative recombination of vibrationally excited O_2^+", Planet. Space Sci., 34, 1101(1985).
20. Bates, D. R. and Zipf, E. C., "The O(^{1}S) quantum yield from O_2^+ dissociative recombination", Planet. Space Sci., 28, 1081(1980).
21. Torr, M. R. and Torr, D. G., "The dissociative recombination of O_2^+ in the ionosphere", Planet. Space Sci., 29, 999(1981).
22. Walls, F. L. and Dunn, G. H., "Measurement of total cross sections for electron recombination with NO^+ and O_2^+ using storage techniques", J. Geophys. Res., 8, 284(1974).
23. Abreu, J. A., Solomon, S. C., Sharp, W. E. and Hays, P. B., "The dissociative recombination of O_2^+: The quantum yield of O(^{1}S) and O(^{1}D)", J. Geophys. Res. 88, 4140(1983).

24. Guberman, S. L., "The production of O(^{1}S) from dissociative recombination of O_2^+", Nature, 327, 408(1987).

25. Guberman, S. L., "The production of O(^{1}D) from dissociative recombination of O_2^+", Planet. Space Sci., 36, 47(1988).

26. Fox, J. L., "The O_2^+ vibrational distribution in the Venusian ionosphere", Adv. Space Res., 5, 165(1985).

27. Fox, J. L., "The vibrational distribution of O_2^+ in the dayside ionosphere", Planet. Space Sci., 34, 1241(1986).

28. Brinton, H. C., Scott, L. R., Pharo, M. W. and Coulson, J. T. C., "The Benett ion-mass spectrometer on Atmosphere Explorer-C, -D, and -E", Radio Sci., 8, 323(1973).

29. Hanson, W. B., Zuccaro, D. R., Lippencott, C. R. and Sanatani, S., " The retarding-potential analyzer on Atmosphere Explorer", Radio Sci., 8, 333(1973).

30. Nier, A. O., Potter, W. E., Hichman, D. R. and Mauersberger, K. M., "The open-source neutral mass spectrometer on Atmosphere Explorer-C, -D, and -E", Radio Sci., 8, 271(1973).

31. Brace, L. H., Theis, R. G. and Dalgarno, A.,"The cylindrical electrostatic probes for Atmosphere Explorer-C,-D, and -E", Radio Sci., 8, 341(1973).

32. Spencer, N. W., Niemann, H. B. and Carignan, G. R.,"The neutral-atmosphere temperature instrument", Radio Sci., 8, 284(1973).

DISSOCIATIVE RECOMBINATION REACTIONS IN THE CHEMISTRY OF DENSE INTERSTELLAR CLOUDS

Eric Herbst
Department of Physics
Duke University
Durham, NC 27706, USA

ABSTRACT. Dissociative recombination reactions probably play a major role in the formation of polyatomic neutral molecules in dense interstellar clouds via gas phase reactions. Polyatomic ions in these clouds are synthesized by ion-molecule reactions. Unless electrons are not the major carrier of negative charge, neutral species are then produced via dissociative recombination reactions. Although chemical models of dense interstellar clouds have been constructed in which thousands of chemical reactions are incorporated, these models yield uncertain results due, among other reasons, to the current lack of available information on the neutral product branching ratios of dissociative recombination reactions. Scientists have formulated rather simple theories to predict to some degree the branching ratios of a large number of dissociative recombination reactions. In this paper, the simplified approaches used to make these predictions are discussed as are the sensitivity of gas phase chemical models to uncertain neutral product branching ratios.

1. INTRODUCTION

During the past twenty years, upwards of seventy different molecules have been discovered to exist in dense interstellar clouds. These molecules range in complexity from diatomic species to a thirteen atom nitrile $HC_{11}N$ and include many of the well-known simple organic reagents found in the terrestrial laboratory. Dense interstellar clouds are vast regions of concentrated matter located primarily in the planes of spiral-type galaxies. The matter is mostly in the form of gas, but typically 1 % is the form of small dust-like particles. Although gas densities are only in the range 10^4 - 10^6 cm^{-3}, the large size of dense clouds results in cloud masses that exceed thousands of solar masses. Cloud temperatures are customarily quite low (10 - 70 K) although warmer regions exist in the vicinity of gas collapsing to form stars. Molecules are customarily observed in the gas phase via rotational emission spectroscopy[1] in the millimeter region of the electromagnetic spectrum. Detailed observations show that the gas is primarily composed of molecular hydrogen with the next most abundant species, carbon monoxide, four orders of magnitude lower in abundance. In addition to high-resolution gas phase spectroscopy, there is some evidence based on broad infra-red spectral features for molecules adsorbed on the surfaces of the dust particles[2] and for a class of very large molecules or small dust grains called PAH's (polycyclic aromatic hydrocarbons) somewhat reminiscent of automobile exhaust soot[3]. Spectral characterization of the normal dust particles has proven difficult [4].

The chemical processes that lead to the formation of most interstellar molecules are thought to occur within the clouds since outside of these sources molecules would be photodestroyed in a relatively short time (≈ 30 years) by the harsh ultraviolet radiation field that pervades interstellar space. On the other hand, it is thought that dust particles and possibly PAH's as well are primarily stellar ejecta because these species are sufficiently large to withstand ultra-violet radiation. Molecules formed within dense interstellar clouds from precursor atoms, themselves the outflow of previous generations of stars, can be formed via gas phase processes or processes occurring on the surfaces of the dust particles. A recent review of the different chemical processes has been given by Winnewisser and Herbst[5]. Most astronomers are agreed that molecular hydrogen is formed solely by recombination of two hydrogen atoms on the surfaces of dust particles followed by thermal evaporation into the gas although it is unclear whether any other species observed to be in the gas were formed on dust particle surfaces. Rather, it has been shown that gas phase processes starting from molecular hydrogen can account for much if not all of the richness of interstellar chemistry[6]. The dominant gas phase reactions that occur at the low temperatures characteristic of interstellar clouds are ion-molecule reactions because these processes occur without activation energy in most instances. The major source of ionization is cosmic ray bombardment. A number of model calculations have shown that cosmic ray bombardment leads only to a small fractional ionization (10^{-7} - 10^{-8}) but that the ions produced are quite efficient in leading to the synthesis of polyatomic molecules[6)-8]. Some of the smaller polyatomic ions predicted to be abundant in dense interstellar clouds have subsequently been detected; for this and many other reasons the ion- molecule theory is on a firm footing.

Although sequences of ion-molecule reactions lead to polyatomic ions, the final step in the production of neutral species is dissociative recombination. Consider as an example the synthesis of water. The precursor ion H_3O^+ is produced via the reactions

$$H_2 + \text{Cosmic Ray} \longrightarrow H_2^+ + e + \text{Cosmic Ray} \quad (1)$$

$$H_2^+ + H_2 \longrightarrow H_3^+ + H \quad (2)$$

$$H_3^+ + O \longrightarrow OH^+ + H_2 \quad (3)$$

$$OH^+ + H_2 \longrightarrow H_2O^+ + H \quad (4)$$

$$H_2O^+ + H_2 \longrightarrow H_3O^+ + H \ . \quad (5)$$

The H_3O^+ ion does not react with molecular hydrogen but is depleted primarily via reaction with negatively-charged species. The still dominant view is that electrons are this dominant species, but there is a possibility that electrons are absorbed by PAH's leading to the dominance of PAH^- [9]. This highly conjectural view is not considered here. Although it is well known that the neutral products of rapid recombination reactions are dissociative in nature, the product branching ratios for this and all other important dissociative recombination reactions in interstellar chemistry are still unknown due to the extreme difficulties of both experimental and theoretical approaches. In the case of H_3O^+, it is uncertain how much water is produced and how much (if any) hydroxyl radical is formed. Since practitioners of interstellar chemistry need literally hundreds of these branching ratios to model dense interstellar clouds, scientists have used some simplified quasi-theoretical approaches to estimate these ratios. The treatments sometimes agree and sometimes disagree in their predictions. For want of better terminology, they will be referred to as the statistical, dissociative/predissociative, and valence bond approaches. In section 2 of this article, these methods are discussed, while in section 3, the sensitivity of models of dense interstellar cloud chemistry to uncertain dissociative recombination branching ratios is reviewed. A summary is contained in Section 4.

Before a discussion of theoretical approaches to branching ratios, it is worthwhile to comment on rate coefficients for dissociative recombination reactions.

With the exception of H_3^+ + e, the total rate coefficients for important dissociative recombination reactions would not appear to be in dispute. The case of H_3^+ is most interesting and will doubtless be discussed elsewhere in this volume. The rate of this reaction is not crucial for dense interstellar cloud chemistry because even if H_3^+ is not destroyed rapidly by reaction with electrons, it will be destroyed rapidly by reaction with CO.

2. TREATMENTS FOR DETERMINATION OF BRANCHING RATIOS

2.1 Statistical Model

Formulated by Herbst[10)], this approach has as its rationale the idea that excited states of polyatomic neutrals are rarely totally repulsive in all dimensions and that as the parent neutral molecule formed in the recombination fragments, the complexity of the potential surface/surfaces leads to a rather complex motion which can be approximated as ergodic. The products can then be derived using a variation of the well-known statistical phase space theory of Light and co-workers[11)]. In this approach, the probability of a given exit channel is proportional to the number of detailed quantum states that exist in the exit channel. Unfortunately, the theory is not parameter free, but depends on two parameters, one relating to the cross section of the inverse reaction and one relating to the usefulness of rotational energy in overcoming activation energy in the reverse reaction. Generally, the more exothermic the product channel, the more detailed states that exist in the channel that are consistent with the constraints of conservation of total energy and angular momentum. Therefore, the phase space theory generally leads to the prediction that more exothermic channels are the more probable. This implies that breaking of single bonds is preferred to the breaking of multiple bonds. Another general result is that there is rarely one product channel that is dominant. Phase space theory as formulated by Light can only account for channels in which there are two fragments; multi-fragment channels can be included as secondary processes in which an original fragment has sufficient internal energy to dissociate.

For the dissociative recombination reaction between H_3O^+ and electrons discussed above, the statistical theory predicts that the exit channels H_2O + H, OH + 2H, and OH + H_2 all occur with the OH/H_2O branching ratio ranging from 1 - 10 chiefly depending on the parameter dealing with rotational energy. Another interesting prediction concerns the important reaction between the linear ion $HCNH^+$ and electrons. This reaction[12)] has been postulated to produce the fragments HCN + H and HNC + H with unit branching ratio in order to account for the roughly equal concentrations of HCN and HNC in some of the colder interstellar clouds. The statistical theory predicts that HCN and HNC are produced at the same rate despite the greater exothermicity of the HCN channel because HNC has a greater density of vibrational states. It also predicts a sizeable amount of secondary decomposition to form CN.

Although the statistical approach is the simplest quantitative quantum mechanical theory that can be attempted, the demands of modellers for large numbers of dissociative recombination branching ratios are such that it is impractical to undertake statistical phase space calculations for all ions of interest.

2.2 Dissociative/Predissociative Approach

In this treatment, it is assumed that the lightness of hydrogen atoms results in preferential breaking of heavy atom-hydrogen bonds. Green and Herbst[13)] developed this idea following upon the classic work of Bates[14)] in which it was recognized that fragmentation along a repulsive surface must occur rapidly so that autoionization does not occur. The relatively rapid motion between hydrogen and heavy atoms as compared with motion between two heavy atoms for a given amount of kinetic energy leads to the conclusion that the heavy atom-hydrogen atom bonds are preferentially broken.

If dissociative recombination occurs via potential surfaces of transient neutral species in which some barrier penetration is required to produce fragmented products, hydrogen atom fragmentation will also be favored because of the efficiency of these light atoms in tunneling. Such behavior has been claimed by Gellene, Porter, and co-workers[15) - 17)] who performed a series of experiments on charge exchange collisions between beams of sodium or potassium atoms and beams of protonated ions such as H_3O^+, NH_4^+, and $CH_3NH_3^+$ in which the transient neutral species formed subsequently fragment. In these experiments H atom formation is dominant; for H_3O^+ and NH_4^+ no H_2 was detected while for $CH_3NH_3^+$ the branching ratio for CH_3NH_2 + H was 0.82 while that for CH_3 + NH_3 was only 0.18. Evidence for a predissociative mechanism comes from studies with deuterated and partially deuterated ions in which H atom fragmentation is shown to be much more rapid than D atom removal. Of course, charge exhange with Na is not the same process as recombination with electrons. A major difference is that the dissociating neutral parent has significantly less energy in the charge exchange process whether it dissociates directly or via a Rydberg state.

The viewpoint that H atom fragmentation dominates in dissociative recombination reactions has generally been utilized by modellers of interstellar chemistry because it is simple and can be applied to large numbers of reactions. However, it has generally *not* been assumed that only single hydrogen atom channels occur. There are two reasons for this. The first is that the statistical model leads to the prediction that more than one hydrogen atom can be removed from the transient neutral. The second is that if dissociative recombination reactions lead only to single H atom removal, it is very difficult to deplete interstellar neutral species.

The dominant interstellar ions that react with neutral species are H_3^+ and HCO^+; these ions react with neutral species A by protonating them:

$$A + H_3^+ (HCO^+) \text{ -----> } AH^+ + H_2 (CO) . \qquad (6)$$

If the AH^+ ion then reacts primarily with electrons to reform A with unit branching ratio, no net depletion of A takes place. To avoid this extreme assumption, most modellers have assumed that other channels exist for the $AH^+ + e$ reaction, in particular channels in which either two hydrogen atoms (or a hydrogen molecule) are produced if the ion possesses at least two hydrogen atoms. It is also customarily assumed that fragmentation leading to 2H or H_2 is as probable as fragmentation leading to a single hydrogen atom.

2.3 Valence Bond Approach

Bates[18)-19)] has recently extended his classic treatment of dissociative recombination of diatomic ions to polyatomic species. In his view, polyatomic ions are held together by localized valence bonds which are constant in strength from one ion to another[20)]. Dissociative recombination occurs via a radiationless transition in which an attractive valence bond is converted into a repulsive one by inverse-autoionization, followed by rapid bond breaking along a repulsive neutral surface. The probability of which bond is to be broken is proportional to the charge density surrounding the bond as well as other factors such as favorable or unfavorable

Franck-Condon factors. Only one strong valence bond is broken initially; however, weak polarization-type bonds can be simultaneously broken. In addition, subsequent secondary fragmentation can occur. Although Bates utilized simple valence bond ideas in estimating where the charge resides on polyatomic ions, Millar *et al.* [21], in their study of the sensitivity of dense cloud models to Bates' approach, utilized *ab initio* methods to determine charge densities in molecular ions.

For simple polyatomic ions such as H_3O^+ and NH_4^+, the approach of Bates leads to the prediction that the dominant channel is single hydrogen atom removal. Thus little or no OH can be formed from the recombination of $H_3O^+ + e$. For more complex ions in which charge is delocalized, the approach leads to the prediction that hydrogen atom-heavy atom and heavy atom-heavy atom bonds can both be broken, even if the heavy atom-heavy atom bond is a strong multiple one. This in turn leads to a wide assortment of products. Fragmentation via two hydrogen atoms or a hydrogen molecule is forbidden unless a polarization bond is broken simultaneously with a valence bond, as in the case of CH_5^+. This ion, considered by Bates to be well represented by the structure $CH_3^+{\cdot}H_2$, will fragment into $CH_2 + H + H_2$.

As an example of how the method can be used with larger ions, consider the case of $CH_3NH_3^+$. *Ab initio*[21] calculations show that the charge is distributed throughout the entire molecular ion with ≈1/3 on the methyl end and ≈2/3 on the ammonium end. Possible products in a dissociative recombination reaction include $CH_3NH_2 + H$, $CH_3 + NH_3$, and $CH_2NH_3 + H$. The first set of products derives from electron attack on the end of the molecular ion with the greater charge whereas the second set of products derives from electron attack upon the middle of the ion. The last set of products, which derives from electron attack upon the portion of the ion with the lesser charge and is only slightly exothermic, will undergo secondary decomposition, perhaps to $CH_2 + NH_3 + H$. The exact branching ratio cannot be determined although Millar *et al*[21] have generally assumed that the various neutral channels of this and analogous reactions are equally probable unless there is little positive charge density on the bond to be broken (as is the case for the N-H bond in $HCNH^+$).

It should be mentioned that although the approach of Bates is the only one to contain the "essential feature of (diatomic) dissociative recombination" [18], polyatomic molecules are complex entities and surprises can be expected when experimentalists finally study dissociative recombination branching ratios involving thermalized ions. Experimental results on H_3^+ and H_2O^+ [22)-23)] are not in agreement with Bates' approach but may not be directly comparable due to the non-thermal nature of the ions.

In Table 1 below, the estimated neutral product branching ratios for three important dissociative recombination reactions are compared for the various approaches discussed here. In addition, the values used in a typical model of dense interstellar clouds are included.

3. EFFECTS ON CHEMICAL MODELS OF DENSE INTERSTELLAR CLOUDS

Sternberg, Dalgarno and Lepp [25] investigated the effect on the static chemistry of dense interstellar clouds of varying the neutral product branching ratio OH/H_2O for the dissociative recombination of H_3O^+ + e. They found that if, as suggested by Bates, OH is not a major product, then the calculated OH concentration is significantly reduced to below observed levels. However, they were able to recover the observed concentration of hydroxyl radical via indirect photodissociation of water caused by ultra-violet photons produced by cosmic ray bombardment of clouds. This recovery argues strongly for such an indirect photodissociation if the OH is not produced by the dissociative recombination of H_3O^+.

Millar *et al.*[21] have published a gas phase model of dense interstellar cloud chemistry in which the effect of uncertain branching ratios for a wide variety of dissociative recombination reactions is investigated. Their model is a so-called pseudo time-dependent one, in which chemical abundances evolve from initial values under fixed physical conditions. These models show that for most organic species larger in size than four atoms, calculated concentrations reach a maximum roughly 10^5 yr from the onset of the gas phase chemistry before decreasing to much lower values when steady-state conditions are reached in ≈ 10^7 yr. Agreement with observation is superior when the calculated concentrations are at their peak ("early time"). In the work of Millar *et al.*, calculated concentrations at both early time and steady state are shown for three sets of branching ratios for dissociative recombination reactions. One set is the "typical" branching ratios used by many astrochemists, in which either one or two hydrogen atom fragmentation is favored. The second set is based on the view of Bates[18)-19)]. A detailed discussion of how Bates' theory was utilized is contained in a review by Herbst[26]. The third set is similar to the second except that a small amount of OH (10%) is allowed to be formed in the dissociative recombination of H_3O^+. Millar *et al.* reproduced the earlier work of Sternberg, Dalgarno, and Lepp[25]

TABLE 1 ESTIMATED PRODUCTS AND BRANCHING RATIOS OF SELECTED DISSOCIATIVE RECOMBINATION REACTIONS

	Approach			
Reactant	Statistical[a)]	Dissociative	Typical Model[b)]	Valence Bond[c)]
H_3O^+	$OH + H_2$ or 2H	$H_2O + H$	$OH + H_2$ or 2H	$H_2O + H$
	$H_2O + H$ (10-50%)		$H_2O + H$ (50%)	
$HCNH^+$	HCN + H (25%)	HCN + H[d]	HCN + H (33%)	HCN + H (33%)[e]
	HNC + H (25%)	HNC + H	HNC + H (33%)	HNC + H (67%)
	$CN + H_2$ or 2H		$CN + H_2$ or 2H	
$C_4H_2^+$	-----	$C_4H + H$	$C_4H + H$ (50%)	$C_4H + H$ (33%)[e]
			$C_4 + H_2$ or 2H	$C_3H + CH$ (33%)
				$2C_2H$ (33%)

Notes: Due to space limitations, some channels do not have branching ratios shown. These can be obtained by subtraction. The valence bond results for the last two ions are from reference 21); these authors have used *ab initio* calculations for selected ions. References: a) - see 10); b)- see 24); c) - see 18) - 19); d) - no branching ratio calculated; e) - see 21).

and showed the same effect on the OH abundance at early time. They expected that the major difference among the sets of branching ratios used would be that complex molecule concentrations would be reduced when the Bates viewpoint was used, since in general complex ions fragment into more channels, including channels in which heavy atom-heavy atom bonds are broken (see the case of $C_4H_2^+$ in Table 1). To some degree, the results show this expectation to be correct. In the following discussion, we ignore the third set of branching ratios utilized. In general, the results obtained with this set of branching ratios differ less from the first set than do the results obtained from the second set.

For hydrocarbons, Millar *et al.*[21] found an expected reduction in molecular abundances when Bates' (valence bond) branching ratios were utilized. However, in the main, the reduction is quite small (much less than one order of magnitude) at early time, which is the time at which the model best reproduces observation. At steady state, the reduction can be much larger. For the well-known interstellar radical species C_3H and C_3H_2, the reduction factors at early time were found to be 0.95 and 0.94 respectively, whereas the reduction factors at steady state were found to be 0.050 and 0.40 respectively. The reason why early time abundances are not affected significantly lies presumably in the fact that gas phase syntheses are sufficiently efficient at this time to withstand a loss in efficiency. Ion-molecule pathways to complex hydrocarbons are made more complicated and indirect when the branching ratios of Bates are utilized; this is because the breaking of heavy atom-heavy atoms bonds in polyatomic ions often necessitates the production of considerably larger ion precursors to form neutrals of any given size.

For organic molecules other than hydrocarbons, the story is a more complicated one. Use of the valence bond branching ratios leads to a large increase in the water abundance, especially at steady state, which actually enhances the abundances of some oxygen-containing species. For example, the abundance of methanol is enhanced by a factor of 2.3 at early time and by a factor of 22.1 at steady state. Finally, the important interstellar organo-nitrogen compound HCCCN is unaffected by the use of valence bond branching ratios at early time but increases at steady state by a factor of 8.3.

Based on these results, it would appear that the formation of complex molecules in dense interstellar clouds will not be understood until branching ratios for a variety of dissociative recombination reactions are available. In fact, even the abundances of simple molecules such as OH and H_2O are hostage to this lack of experimental information.

3.1 Effect On Deuteration Schemes

One of the most interesting aspects of the chemistry of interstellar clouds is the strong isotopic fractionation observed, especially as regards deuterium. Even though the overall deuterium to hydrogen ratio in dense clouds is typically at a value of $\approx 10^{-5}$, abundance ratios between singly deuterated and normal molecular species can approach 0.2 in the colder clouds. The reason for the strong fractionation is well understood and has to do with the low temperatures of interstellar sources[27)-29)]. The principal source of deuterium is the abundant molecule HD. This species is known to react with H_3^+ to produce H_2D^+:

$$H_3^+ + HD <====> H_2D^+ + H_2 ; \quad (7)$$

at low temperatures (< 20 K) the back reaction occurs so slowly due to a small endothermicity (≈ 200 K) that H_2D^+ is not depleted by reaction with H_2 but instead by reaction with species of lesser abundance such as carbon monoxide. The result is an abundance ratio between the deuterated and normal forms of H_3^+ that is quite large. The ion H_2D^+ can deuterate other neutral species AH via reaction sequences such as

$$AH + H_2D^+ \text{ -----> } AHD^+ + H_2 \quad (8)$$

$$AHD^+ + e \text{ -----> } AD + H, \; AH + D, \text{ etc.} \quad (9)$$

There are other isotopic exchange reactions similar to (7) that have greater endothermicities (≈ 500 K) in the backwards direction and lead to severe fractionation at temperatures as high as 70 K[29)]. If one examines processes (8) - (9), one can see immediately that the AD/AH abundance ratio will be affected by the branching ratio of (9). If, for example, H atom removal is favored due to either a barrier or just to the rapid motion of H atoms, then AD will be favored over AH and the abundance ratio AD/AH will be enhanced. If, on the other hand, channels other than AD + H are appreciable, the amount of AD produced will be diminished and the AD/AH abundance ratio will decrease. An argument in favor of the former (dissociative/predissociative) view has been made by Bell *et al.*[30)], who invoked the Gellene-Porter experiments[15)-17)] to explain the large abundance ratio (≤ 0.16) they observed for C_3HD/C_3H_2 in some well-known cold interstellar clouds at 10 K. However, as shown by Millar, Bennett and Herbst[29)], a large number of other abundance ratios between deuterated and normal species in these cold sources (normally about 0.05 or less) are calculated correctly in a pseudo time-dependent model with either the new valence bond branching ratios of Bates or the more typical branching ratios used by astrochemists in which single hydrogen atom removal competes with H_2 removal or double hydrogen atom removal. Both of these schemes for estimating branching ratios result in less fractionation than the dissociative/predissociative scheme. The large C_3HD/C_3H_2 abundance ratios observed by Bell *et al.*[30] cannot then be accounted for. For higher temperature sources such as the molecular cloud in Orion (≈ 70 K), Millar, Bennett, and Herbst[29] find that the typical branching ratios used by astrochemists tend to reproduce the data better than the valence bond values for species not located in the vicinity of star forming regions. Abundance ratios between deuterated and normal species in star forming regions such as the so-called "hot core" of the Orion nebula are not explained well by theory unless an earlier low temperature stage is assumed.

From the preceding argument, it can be seen that an exact treatment of isotopic fractionation in dense interstellar clouds requires knowledge of branching ratios for a variety of dissociative recombination reactions.

4. SUMMARY

Dissociative recombination reactions play a leading role in the chemistry of dense interstellar clouds. Polyatomic ions are synthesized through ion-molecule reactions; these ions are then neutralized via dissociative recombination reactions with electrons. In order to model the chemistry correctly, it is clearly necessary to know what the neutral product branching ratios of these reactions are. Given the absence of experimental or detailed theoretical studies of the neutral products of dissociative recombination reactions, modellers have been forced to use simplified means of estimating branching ratios. It remains to be seen which of these treatments is closest to the truth although the valence bond approach of Bates is the only one that at least contains the essence of diatomic ion dissociative recombination reactions.

The sensitivity of models of dense interstellar clouds to uncertain branching ratios of dissociative recombination reactions has been discussed. Both the concentrations of individual species and the ratio of concentrations between singly deuterated and normal species are affected by the choices of branching ratios. There is little indication from the observational data as to which branching ratios should be used since too many other uncertainties afflict chemical models. Progress in interstellar chemistry will be retarded until definitive results of the products of dissociative recombination reactions are available.

ACKNOWLEDGMENTS

E. H. acknowledges the support of his theoretical program by the National Science Foundation (U.S.) through grant AST-8715446. The ideas expressed on dissociative recombination branching ratios have been clarified by discussions with Sir David Bates. The author also expresses his thanks to Chung Leung and Tom Millar for their expert collaboration in model calculations of dense interstellar clouds.

REFERENCES

1) Irvine, W. M., Goldsmith, P. F., and Hjalmarson, Å, in *Interstellar Processes*, eds. D. J. Hollenbach and H. A. Thronson, Jr. (Reidel, 1987), p. 561.

2) Tielens, A. G. G. M., in *Laboratory and Observational Infrared Spectra of Interstellar Dust*, eds. R. D. Wolstencroft and J. M. Greenberg (Occasional Reports of the Royal Observatory, Edinburgh, No. 12, 1984), p.41.

3) Allamandola, L. J., Tielens, A. G. G. M., and Barker, J. R., *Astrophys. J. (Letters)* **290**, L25 (1985).

4) Savage, B. D. and Mathis, J. S., *Ann. Rev. Astron. Astrophys.* **17**, 73 (1979).

5) Winnewisser, G. and Herbst, E., *Topics in Current Chemistry* **139**, 119 (1987).

6) Leung, C. M., Herbst, E., and Huebner, W. F., *Astrophys. J. Suppl. Ser.* **56**, 231 (1984).

7) Prasad, S. S. and Huntress, Jr., W. T., *Astrophys. J. Suppl. Ser.* **43**, 1 (1980).

8) Millar, T. J., Leung, C. M., and Herbst, E., *Astronomy & Astrophys.* **183**, 109 (1987).

9) Lepp, S. and Dalgarno, A., *Astrophys. J.* **324**, 553 (1988).

10) Herbst, E., *Astrophys. J.* **222**, 508 (1978).

11) Light, J. C., *Discussions Faraday Soc.* **44**, 14 (1967).

12) Watson, W. D., *Rev. Mod. Phys.* **48**, 513 (1976).

13) Green, S. and Herbst, E., *Astrophys. J.* **229**, 121 (1979).

14) Bates, D. R., *Phys. Rev.* **78**, 492 (1950).

15) Gellene, G. I., Cleary, D. A., and Porter, R. F., *J. Chem. Phys.* **77**, 3471 (1982).

16) Gellene, G. I. and Porter, R. F., *J. Chem. Phys.* **81**, 5570 (1984).

17) Gellene, G. I. and Porter, R. F., *Accounts Chem. Res.* **16**, 200 (1983).

18) Bates, D. R., *Astrophys. J. (Letters)* **306**, L45 (1986).

19) Bates, D. R., in *Recent Studies in Atomic and Molecular Processes*, ed. A. E. Kingston (Plenum, 1987), p. 1.

20) Bates, D. R., *Int. J. Mass Spec. Ion Proc.* **80**, 1 (1987).

21) Millar, T. J., DeFrees, D. J., McLean, A. D., and Herbst, E., *Astronomy & Astrophys.* **194**, 250 (1988).

22) Mitchell, J. B. A., Forand, J. L., Ng, C. T., Levac, D. P., Mitchell, R. E., Mul, P. M., Claeys, W., Sen, A., and McGowan, J. Wm., *Phys. Rev. Letters* **51**, 885 (1983).

23) Vallée, F., Gomet, J. C., Rowe, B. R., Quéffelec, J. L., and Morlai, M., in *Astrochemistry*, eds. M. S. Vardya and S. P. Tarafdar (Reidel, 1987), p. 29.

24) Herbst, E. and Leung, C. M., *Mon. Not. Roy. Astron. Soc.* **222**, 689 (1986).

25) Sternberg, A., Dalgarno, A., and Lepp, S., *Astrophys. J.* **320**, 676 (1987).

26) Herbst, E., in *Rate Coefficients in Astrochemistry*, eds. T. J. Millar and D. A. Williams (Reidel, 1988).

27) Watson, W. D., in *CNO Processes in Astrophysics*, ed. J. Audouze (Reidel, 1977), p. 105.

28) Guélin, M., Langer, W. D., Snell, R. L., and Wootten, A., *Astrophys. J. (Letters)* **217**, L165 (1977).

29) Millar, T. J., Bennett, A., and Herbst, E., submitted to *Astrophys. J.*

30) Bell, M. B., Avery, L. W., Matthew, H. E., Feldman, P. A., Watson, J. K.G., Madden, S. C., and Irvine, W. M., *Astrophys. J.* **326**, 924 (1988).

DISSOCIATIVE RECOMBINATION IN INTERSTELLAR CLOUDS

John H. Black
Steward Observatory, University of Arizona
Tucson, AZ 85721, USA

and

Ewine F. van Dishoeck
Center for Cosmochemistry and Geochemistry
California Institute of Technology
Pasadena, CA 91125, USA

ABSTRACT. Molecular ions play a significant role in the chemistry and evolution of interstellar molecular clouds even though these regions are overwhelmingly neutral. The dissociative recombination process governs the abundances of many of these ions and of related neutral species. The gas-phase ion–molecule chemistry of the simplest species is summarized, with emphasis on those problems which are most sensitive to uncertain rates or product branching ratios of dissociative recombination processes. Examples of the kinds of information needed about dissociative recombination processes are presented. The importance of the H_3^+ ion and prospects for its direct observation are discussed.

1. INTRODUCTION

From the astronomer's point of view, the first dissociative recombination processes probably occurred more than 10^{10} years ago, when the Universe consisted of a uniform plasma bathed in thermal radiation at a temperature of the order of 10^4 K. The first chemical bonds were probably those of the molecular ions He_2^+ and then HeH^+, which formed by radiative association and were destroyed by dissociative recombination. Since that time, structure has arisen in the Universe and the births and deaths of more than one generation of stars have enriched it with a variety of elements heavier than the primordial hydrogen and helium: these are the raw materials for the grand and exotic chemistry that we witness in the interstellar medium at the present time.

With respect to dissociative recombination, the essential features of interstellar clouds are: (1) low density, typically a hydrogen nucleon number density

TABLE 1. Molecular Ions Identified in the Interstellar Medium

CH^+	HCS^+
H_2D^+	$HOCO^+$
HCO^+	$HCNH^+$
N_2H^+	SO^+
HOC^+ (?)	H_3O^+ (?)

$n_H \approx 10 - 10^5$ cm^{-3}; (2) low temperature, $T \approx 10 - 100$ K; (3) conditions that are far from thermodynamic equilibrium; (4) low fractional ionization, $x_e = n(e)/n_H \approx 10^{-8} - 10^{-4}$, maintained by external starlight and cosmic rays; (5) a composition of H, He, O, C, and N in ratios by number of nuclei $1 : 0.08 : 8 \times 10^{-4}\delta_O : 5 \times 10^{-4}\delta_C : 10^{-4}\delta_N$, plus smaller amounts of the other elements; and (6) the presence of dust particles ($\approx 0.1\mu$m diameter and accounting for 1% of the interstellar mass) and possibly of large molecules like polycyclic aromatic hydrocarbons (PAHs).

That dissociative recombination must play a role in interstellar chemistry is demonstrated by the existence of a number of molecular ions (see Table 1), some of which are otherwise relatively unreactive. Dissociative recombination certainly affects the abundances of many neutral molecules that are its products. Dissociative recombination controls the main contribution to chemical heating of the gas. Dissociative recombination may affect the overall ionization balance, although its importance is circumscribed in most regions where slowly recombining atomic ions may account for most of the free charge. In some species, notably H_2^+ and H_3^+, the recombination process may partly control the rotational and vibrational excitation of the ion, which may in turn affect its reactivity and observability. This review will concentrate on the ion chemistry of the diffuse and translucent interstellar clouds and thus on the simplest molecular species. The more opaque, denser molecular clouds and the chemistry of more complex molecules are discussed by Herbst and by Turner elsewhere in this volume.

2. INTERSTELLAR ION CHEMISTRY

It is conventional to distinguish between diffuse interstellar clouds and dark (opaque) molecular clouds. Empirically, the distinction is based primarily upon the total column density of hydrogen nuclei, N_H, or the directly related optical depth for visible (5500 Å) photons. In terms of models of cloud structure and theories of interstellar chemistry, the diffuse clouds have a chemistry that is limited mostly to diatomic and triatomic molecules and that is driven largely by external ultraviolet starlight. The photoprocesses cause the atomic species with first ionization potentials less than that of H to exist primarily in singly ionized form and limit the sizes of molecules in these regions.[1] The interiors of the opaque clouds are shielded from external starlight by the dust particles and by atomic and molecular absorptions in their outer layers: ions are produced by penetrating cosmic rays at a low rate (probably less than 10^{-16} s^{-1} per H nucleus), but chemical lifetimes are very long and the variety of observable molecules is very great (see recent reviews[2],[3]). In recent years, it has become useful to characterize an intermediate class, the translucent molecular clouds, which have thicknesses large enough to support a large variety and abundance of molecules but small enough so that ultraviolet starlight still dominates much of the chemistry.[4],[5]

2.1 Carbon Chemistry

In diffuse clouds, the initiating reaction is thought to be the radiative association

$$C^+ + H_2 \rightarrow CH_2^+ + h\nu \tag{1}$$

which is an intrinsically slow process, but remains important because it involves the most abundant positive ion and the most abundant neutral molecule. In the opaque regions, there is an additional source of hydrocarbon ions through

$$H_3^+ + C \rightarrow CH^+ + H_2 \tag{2}$$

which is followed rapidly by

$$CH^+ + H_2 \rightarrow CH_2^+ + H\ . \tag{3}$$

The products of (1) and (3) are usually followed by

$$CH_2^+ + H_2 \rightarrow CH_3^+ + H \tag{4}$$

which terminates the sequence, because the next such hydrogen–atom abstraction reaction is endothermic. In the diffuse clouds, the fractional ionization, $x_e \approx 10^{-4}$, is high enough that dissociative recombination

$$CH_3^+ + e \rightarrow \text{products} \tag{5}$$

is a more important sink for the CH_3^+ than the relatively rapid radiative association process[6],[7]

$$CH_3^+ + H_2 \rightarrow CH_5^+ + h\nu \tag{6}$$

or photodissociation. The branching ratios $\eta(CH)$ and $1 - \eta(CH)$ for formation of the products CH and CH_2 in reaction (5) appear to have observable consequences on the chemistry in diffuse and translucent clouds. The reactions

$$C^+ + CH \rightarrow C_2^+ + H \tag{7}$$

and

$$C^+ + CH_2 \rightarrow C_2H^+ + H \tag{8}$$

both lead eventually to formation of C_2. The abundance of CH_2 is directly proportional to $(1 - \eta(CH))$; on the other hand, the sensitivity of the CH abundance to $\eta(CH)$ is much smaller because it forms either directly through reaction (5) or by photodissociation of CH_2. The production of C_2 is enhanced through reaction (8) when the abundance of CH_2 is large. Both CH and C_2 can be observed by absorption line techniques. A detailed comparison of observed CH, C_2, and CN abundances with chemical models indicates that the observed C_2 abundance can be reproduced only if dissociative recombination of CH_3^+ favors production of CH_2 over CH.[5] Specifically, the astronomical evidence suggests a small value $\eta(CH) \approx 0.1$. There are enough other uncertainties in the models that it is difficult to be dogmatic about this, but enterprising theorists and experimenters could take the small CH branching ratio as a challenge. Within a few years, we will probably be able to determine the value of $\eta(CH)$ reliably from astronomical observations once it becomes possible to measure the abundance of CH_2 directly. The observable abundance ratio of C_2H and C_2 must also be sensitive to the product branching ratio in dissociative recombination of $C_2H_2^+$. Another very interesting feature of the interstellar carbon chemistry concerns the recently discovered molecules with cyclic (ring) structures: C_3H_2 and C_3H. It is likely that the product branching ratio in dissociative recombination of $C_3H_3^+$ governs the relative abundances of these species. Moreover, the observation of both cyclic and linear forms of interstellar C_3H requires even more detailed knowledge of the dissociative recombination process for its interpretation.[3]

The high observed abundance of CH^+ in diffuse interstellar clouds has provided a major, unsolved mystery for decades, because the known chemical sources appear to be inadequate to match the known destruction mechanisms. Dissociative recombination of CH^+ is one of the removal processes, and there remain interesting questions about whether it is indeed rapid under interstellar conditions (specifically, for ground–state ions). Although this is a very intriguing problem for molecular physics, dissociative recombination is unlikely to affect the CH^+ abundance appreciably in neutral regions regardless of the size of its rate coefficient, because reactions of CH^+ with the most abundant species H and H_2 are so rapid.

2.2 Oxygen Chemistry

In the case of oxygen chemistry, cosmic ray ionizations of hydrogen lead to H^+ and H_3^+, which drive a sequence of reactions

$$H^+ + O \rightarrow O^+ + H \tag{9}$$

$$O^+ + H_2 \rightarrow OH^+ + H \tag{10}$$

$$H_3^+ + O \rightarrow OH^+ + H_2 \tag{11}$$

$$OH^+ + H_2 \rightarrow H_2O^+ + H \tag{12}$$

$$H_2O^+ + H_2 \rightarrow H_3O^+ + H \tag{13}$$

rapidly toward the saturated molecular ion H_3O^+. The product branching ratios in dissociative recombination

$$H_3O^+ + e \rightarrow OH + H_2 \qquad \eta(OH) \tag{14}$$

$$H_3O^+ + e \rightarrow H_2O + H \qquad 1 - \eta(OH) \tag{15}$$

influence the abundances of OH and H_2O, although the effects are partly mediated by the main destruction mechanism of H_2O in diffuse clouds,

$$H_2O + h\nu \rightarrow OH + H \quad , \tag{16}$$

which is also a source of OH. The existing measurements of OH and the best upper limits on H_2O in diffuse clouds are in harmony with almost any value of η(OH).[1],8] Again, the eventual detection of H_2O through ultraviolet absorption spectroscopy with the *Hubble Space Telescope* may provide better empirical constraints on the value of η(OH).

2.3 The Hydrogen Molecular Ions

Hydrogen molecular ions deserve particular attention, in part because of the controversy surrounding the rate of dissociative recombination of H_3^+. As already indicated, cosmic ray ionizations of H_2 produce H_2^+ (primarily) at a rate $\zeta_0 \approx 7\times10^{-17}$ s^{-1}.[8] As long as the number densities of hydrogen satisfy $n(H_2) > n(H) \gg n(e)$, this is followed almost invariably by rapid H–atom abstraction

$$H_2^+ + H_2 \rightarrow H_3^+ + H \tag{17}$$

to form H_3^+. This ion is removed by dissociative recombination

$$H_3^+ + e \rightarrow \text{products} \tag{18}$$

and perhaps more slowly by radiative recombination

$$H_3^+ + e \rightarrow H_3^* \rightarrow \text{fragments} + h\nu \tag{19}$$

that may produce the spectrum of the transient H_3 identified by Herzberg[9–11] and sought in astronomical sources by Pritchet and Grillmair[12]. The H_3^+ ion is also destroyed by reactions with various abundant carbon– and oxygen–bearing species, as in reactions (2) and (11), and

$$H_3^+ + CO \rightarrow HCO^+ + H_2 \ \ , \tag{20}$$

by reactions with D and HD

$$H_3^+ + D \rightarrow H_2D^+ + H \tag{21}$$

$$H_3^+ + HD \rightarrow H_2D^+ + H_2 \ \ , \tag{22}$$

and perhaps by dissociative charge transfer with various neutral atoms of low ionization potential, but *not* by photodissociation in the typical interstellar radiation field.

The extent to which dissociative recombination (18) limits the abundance of H_3^+ depends upon the conditions in interstellar clouds and upon the value of the dissociative recombination rate coefficient, k_{DR}, that is appropriate for those conditions. Unfortunately, the relevant rate coefficient for H_3^+ in its vibrational ground state is still uncertain. Theoretical calculations of Michels and Hobbs[13] and measurements of vibrationally relaxed ions in a flowing afterglow of Smith and Adams[14] both suggest that the rate is very low for ground–state ions, the experimental upper limit being $k_{DR} \leq 2 \times 10^{-8}$ cm^3 s^{-1} at $T = 95$ and 300 K. Another recent measurement that may not completely isolate vibrationally relaxed H_3^+ indicates $k_{DR} > 10^{-7}$ cm^3 s^{-1} at electron temperatures below 500 K [15]. The experiments of Mitchell's group[16] produce H_3^+ in $v = 0$ and $v = 1$ only and suggest a rate coefficient at $T = 100$ K of $k_{DR} = 2 \times 10^{-8}$ cm^3 s^{-1}. Even more recently, Amano[17] has reported a recombination rate coefficient of 1.8×10^{-7} cm^3 s^{-1} based upon direct spectroscopic detection of $H_3^+(v = 0)$ in a decaying plasma. However, Smith (elsewhere in this volume) has suggested that Amano's plasma was dense enough that the measured rate was compromised by three–body effects. A definitive measurement of k_{DR} for ground–state H_3^+ is urgently needed.

In molecular regions in steady state, the concentration of H_3^+ can be written

$$n(\mathrm{H}_3^+) \approx \frac{0.9\varsigma_0 n(\mathrm{H}_2)}{k_{DR}n(e) + 8 \times 10^{-10}n(\mathrm{O}) + 1.7 \times 10^{-9}n(\mathrm{CO}) + \cdots} \quad \mathrm{cm}^{-3} \tag{23}$$

where ς_0 is the cosmic ray ionization rate (s^{-1}) and k_{DR} is the rate coefficient of dissociative recombination, reaction (18), in cm^3 s^{-1}. More specifically, in diffuse clouds, the concentration of H_3^+ is nearly independent of the total density

$$n(\mathrm{H}_3^+) \approx 10^{-5}\left(\frac{\varsigma_0}{10^{-17}}\right)\left[0.04\left(\frac{k_{DR}}{10^{-10}}\right)\left(\frac{x_e}{1.5 \times 10^{-4}}\right) + 1\right]^{-1} \mathrm{cm}^{-3}. \tag{24}$$

Equation (24) is appropriate for the centers of diffuse clouds where $n(\mathrm{H}_2) \approx n_H/2$, where $n(e) \approx (1-2) \times 10^{-4} n_H$, and where the atomic oxygen, $n(\mathrm{O}) \approx 5 \times 10^{-4} n_H$, is rather more abundant than C, CO, D, or HD. In some well observed diffuse clouds like those in the directions of the stars ς Per and ς Oph,[8] the cosmic ray ionization rate implied by the observed abundances of OH and HD is $\varsigma_0 \approx 7 \times 10^{-17}$ s^{-1} and the line–of–sight dimension is $L \approx N_H/n_H \approx 3 \times 10^{18}$ cm. Thus the column density of H_3^+ in such regions will be $N(\mathrm{H}_3^+) \approx n(\mathrm{H}_3^+)L \approx 2 \times 10^{14}$ cm^{-2} if dissociative recombination of cold H_3^+ is slow, $k_{DR} < 10^{-9}$ cm^3 s^{-1}, but it will be significantly smaller, $N(\mathrm{H}_3^+) \approx 3 \times 10^{12}(2 \times 10^{-7}/k_{DR})$ cm^{-2}, if the recombination rate for ground–state H_3^+ is large. In dense, dark molecular clouds where the fractional ionization is low, $x_e \leq 10^{-6}$, reactions with CO and O control

TABLE 2. Potential Interstellar Absorption Lines of H_3^+

Transition $J'G'U' \leftarrow J''K''$	$\tilde{\nu}$ (cm^{-1})	f
0 1 +1 ← 1 1	2457.290	9.85(-6)
1 0 −1 ← 1 0	2529.724	9.07(-5)
1 1 +1 ← 1 1	2545.418	4.61(-5)
2 1 −1 ← 1 1	2691.444	8.90(-5)
2 0 +1 ← 1 0	2725.898	1.66(-4)
2 1 +1 ← 1 1	2726.219	1.02(-4)

the removal of H_3^+ and the term in brackets in equation (24) differs from unity by 50% or less for any value of $k_{DR} < 2 \times 10^{-7}$ cm^3 s^{-1}.[18] Column densities of H_3^+ of the order of 10^{14} cm^{-2} or higher are predicted in these clouds. The H_3^+ ion is thought to play a pivotal role in interstellar chemistry in general[19] and in isotope fractionation in many molecules in particular.[20] We see that its abundance is rather insensitive to the dissociative recombination process deep inside dense clouds, but that the abundance is tied directly to the value of k_{DR} in diffuse and translucent clouds.

As pointed out by Oka[21], interstellar H_3^+ is observable by infrared absorption line techniques in the directions of bright stars that happen to lie in or behind molecular clouds. In fact, H_3^+ column densities smaller than 10^{14} cm^{-2} are detectable with existing telescopes and spectrometers. In this way, information on the cosmic ray ionization rate could be inferred for thick dense clouds, provided that the absorbing column could be characterized further by corresponding measurements of species such as CO.[22] In the diffuse and translucent clouds, continuing uncertainty about the value of k_{DR} limits the usefulness of H_3^+ measurements as diagnostics of ζ_0. On the other hand, in those cases where ζ_0 is already estimated from analysis of OH and HD abundances, the observation of H_3^+ could be used to place observational constraints on the value of k_{DR}.

Two other considerations that enter the discussion of the observability of interstellar H_3^+ are the absorption line oscillator strengths and the degree of vibrational and rotational excitation of H_3^+. Fortunately, Miller and Tennyson[23] have recently determined accurate transition probabilities for lines of the ν_2 fundamental vibrational band and for forbidden, distortion–induced pure rotational transitions from *ab initio* calculations. The potential interstellar absorption lines arising in the 1_1 and 1_0 levels of the vibrational ground–state of H_3^+ (the respective ground levels of the para and ortho modifications) are listed in Table 2. The transition frequencies $\tilde{\nu}$ and assignments are from the work of Majewski *et al.*[24], which extends the earlier work of Oka[21],[25]. The absorption oscillator strengths f have been calculated from the *ab initio* transition probabilities.[23] A full treatment of the excitation of interstellar H_3^+ has not been done, but some general comments can be made. As is the case with interstellar H_2 [26] and as suggested by Oka[21], it is likely that proton–interchange reactions

$$H_3^+(J'K') + H_2(J') \rightleftarrows H_2(J'') + H_3^+(J''K'') \qquad (25)$$

bring the populations of the lowest rotational levels into equilibrium at the kinetic temperature of the gas, since the rates of (25) probably exceed those of the chemical destruction processes and of infrared pumping in the background starlight. Some of the excited rotational levels (*viz.*, 2_2 and 2_1) will probably be subthermally excited because their radiative lifetimes are short ($\sim 3 \times 10^5$ s) compared with the lifetimes against reaction (25) in low density regions. The radiative lifetimes of excited vibrational levels are short enough that they will not be significantly populated under normal interstellar conditions, even if the formation process of H_3^+, reaction (17), favors vibrationally excited products. Thus in normal clouds, we expect that almost all H_3^+ will exist in the 1_0 and 1_1 levels with relative populations

$$\frac{n(1_0)}{n(1_1)} = 2\exp(-32.8/T) \quad . \qquad (26)$$

Although a submillimeter emission line of H_2D^+ in a single molecular cloud has been reported[27], the interstellar lines of H_3^+ have not yet been detected despite the efforts of at least two groups. For example, preliminary results for one line of H_3^+ in the molecular cloud toward the infrared source AFGL 2591 imply a column density in the 1_1 level $< 8 \times 10^{13}$ cm^{-2} and an upper limit on the total column density $N(H_3^+) < 1.3 \times 10^{14}$ cm^{-2} for an adopted temperature $T \leq 30$ K.[28]

The uncertainty in the H_3^+ dissociative recombination rate also affects our understanding of other astronomical phenomena. For example, it has been suggested that H_3^+ rather than H^+ will be the dominant ion in the ionosphere of Jupiter if k_{DR} is small.[29] Models of the Jovian ionosphere with low values of k_{DR} suggest vertical column densities of H_3^+ that approach 10^{14} cm^{-2}. The ions lie entirely above the altitude at which the infrared continuum arises; therefore, H_3^+ may be observable by infrared absorption line spectroscopy just as discussed above for interstellar clouds.

3. NEBULAE AND PHOTODISSOCIATION REGIONS

Very hot stars are able to maintain ionized nebulae in the interstellar gas surrounding them. Such a nebula typically has a very sharp boundary within which virtually all of the hydrogen–ionizing photons are contained. Ultraviolet photons of energy insufficient to ionize ground–state H, however, can penetrate more deeply into surrounding neutral gas and produce a "photodissociation region" of very active chemistry.[30]

There is an interesting point about H_2^+ in photodissociation regions and ionized nebulae. The rate of fluorescent excitation of H_2 can be high enough to produce appreciable amounts of vibrationally excited molecules, enough so that

$$H_2(v \geq 4) + h\nu \rightarrow H_2^+(v') + e \qquad (27)$$

becomes a major source of H_2^+ through the ultraviolet starlight ($\lambda > 912$ Å) that penetrates through the photodissociation region. Although reactions of H_2^+ with H and H_2 will tend to control its destruction in predominantly neutral regions, dissociative recombination will be a major sink in the partly ionized transition zones at the boundaries of the ionized nebulae. Moreover, since all of the formation and destruction rates of H_2^+ appear to depend strongly on its vibrational state, it is of interest to have accurate information on state–dependent rate coefficients, especially for dissociative recombination.[31]–[35] These effects are particularly noticeable for an ion like H_2^+, for which the excited vibrational states are long–lived. Differences in dissociative recombination rates for $v = 0$ and 1, for example, will affect the population distribution over these levels.

Although the conditions in the ionized nebulae themselves are inimical to molecules, some hardy species may persist in detectable amounts.[36] In particular, it has been suggested that HeH^+ may be present in nebulae.[37] An unsuccessful search for infrared emission lines of HeH^+ in the spectrum of the planetary nebula NGC 7027 hes recently been reported.[38] The observed upper limit is lower than the predicted abundance of HeH^+ and one interpretation of this is that a major destruction process for the ion has been overlooked or its rate underestimated. Dissociative recombination of HeH^+ is another case in which there appear to be no favorable potential curve crossings for the process to occur at all through a direct channel, and thus its rate has been assumed to be small. A detailed theoretical treatment and laboratory measurements would be useful.

4. SUMMARY.

We have seen that many dissociative recombination reactions are important in interstellar chemistry and that some of these processes have directly observable consequences. In particular, the product branching ratios in the recombination of CH_3^+ and H_3O^+ affect the relative abundances of CH, CH_2, C_2, OH, and H_2O. The behavior of H_3^+ recombination for ground–state ions affects the abundance of this ion directly and much of the rest of the ion–molecule chemistry indirectly in diffuse and translucent clouds. This process urgently needs to be understood in detail.

This work has been supported in part by NASA through astrophysical theory grant NAGW–763 to the University of Arizona and by NSF grant RII 86–20342 to Princeton University.

REFERENCES

1] van Dishoeck, E. F., and Black, J. H. To appear in *Rate Coefficients in Astrochemistry*, T. J. Millar and D. A. Williams, editors, (Reidel, 1988).

2] Winnewisser, G., and Herbst, E. *Topics Curr. Chem.*, **139**, 119 (1987).

3] van Dishoeck, E. F. To appear in *Millimetre and Submillimetre Astronomy*, R. D. Wolstencroft and W. B. Burton, editors, (Reidel, 1988).

4] Crutcher, R. M. *Astrophys. J.*, **288**, 604 (1985).

5] van Dishoeck, E. F., and Black, J. H. *Astrophys. J.*, submitted (1988).

6] Barlow, S. E., Dunn, G. H., and Schauer, M. *Phys. Rev. Letters*, **52**, 902 (1984).

7] Bates, D. R. *Astrophys. J.*, **312**, 363 (1987).

8] van Dishoeck, E. F., and Black, J. H. *Astrophys. J. Suppl.*, **62**, 109 (1986).

9] Herzberg, G. *J. Chem. Phys.*, **70**, 4806 (1979).

10] Dabrowski, I., and Herzberg, G. *Canad. J. Phys.*, **58**, 1238 (1980).

11] Herzberg, G., and Watson, J. K. G. *Canad. J. Phys.*, **58**, 1250 (1980).

12] Pritchet, C. J., and Grillmair, C. J. *Publ. Astron. Soc. Pacific*, **96**, 349 (1984).

13] Michels, H. H., and Hobbs, R. H. *Astrophys. J. Letters*, **286**, L27 (1984).

14] Smith, D., and Adams, N. G. *Astrophys. J. Letters*, **284**, L13 (1984).

15] Macdonald, J. A., Biondi, M. A., and Johnsen, R. *Planet. Space Sci.*, **32**, 651 (1984).

16] Hus, H., Youssif, F., Sen, A., and Mitchell, J. B. A. *Phys. Rev. A*, in press (1988).

17] Amano, T. *Astrophys. J. Letters*, **329**, L121 (1988).

18] Lepp, S., Dalgarno, A., and Sternberg, A. *Astrophys. J.*, **321**, 383 (1987).

19] Herbst, E., and Klemperer, W. *Astrophys. J.*, **185**, 505 (1973).

20] Dalgarno, A., and Lepp, S. *Astrophys. J. Letters*, **287**, L47 (1984).

21] Oka, T. *Philos. Trans. R. Soc. London*, **A303**, 543 (1981).

22] Black, J. H., and Willner, S. P. *Astrophys. J.*, **279**, 673 (1984).

23] Miller, S., and Tennyson, J. *Astrophys. J.*, **335**, in press (1988).

24] Majewski, W. A., Marshall, M. D., McKellar, A. R. W., Johns, J. W. C., and Watson, J. K. G. *J. Molec. Spectrosc.*, **122**, 341 (1987).

25] Oka, T. *Phys. Rev. Letters*, **45**, 531 (1980).

26] Dalgarno, A., Black, J. H., and Weisheit, J. C. *Astrophys. Letters*, **14**, 77 (1973).

27] Phillips, T. G., Blake, G. A., Keene, J., Woods, R. C., and Churchwell, E. *Astrophys. J. Letters*, **294**, L45 (1985).

28] Black, J. H., Willner, S. P., van Dishoeck, E. F., Woods, R. C., and Churchwell, E. B. in preparation (1988).

29] McConnell, J. C., and Majeed, T. *J. Geophys. Res.*, **92**, 8570 (1987).

30] Tielens, A. G. G. M., and Hollenbach, D. *Astrophys. J.*, **291**, 722 and 747 (1985).

31] Auerbach, D., Cacak, R., Caudano, R., Gaily, T. D., Keyser, C. J., McGowan, J. W., Mitchell, J. B. A., and Wilk, S. F. J. *J. Phys. B*, **10**, 3797 (1977).

32] Giusti–Suzor, A., Bardsley, J. N., and Derkits, C. *Phys. Rev. A*, **28**, 682 (1983).

33] Hickman, A. P. *J. Phys. B*, **20**, 2091 (1987).

34] Nakashima, K., Takagi, H., and Nakamura, H. *J. Chem. Phys.*, **86**, 726 (1987).

35] Hus, H., Yousif, F., Noren, C., Sen, A., and Mitchell, J. B. A. *Phys. Rev. Letters*, **60**, 1006 (1988).

36] Black, J. H. In *Planetary Nebulae*, D. R. Flower, editor, p. 91, (Reidel, 1983).

37] Roberge, W., and Dalgarno, A. *Astrophys. J.*, **255**, 489 (1982), and references therein.

38] Moorhead, J. M., Lowe, R. P., Maillard, J.–P., Wehlau, W. H., and Bernath, P. F. *Astrophys. J.*, **326**, 899 (1988).

DISSOCIATIVE ELECTRONIC RECOMBINATION IN ASTROPHYSICS AND ASTROCHEMISTRY

B. E. Turner
National Radio Astronomy Observatory

ABSTRACT

Branching ratios and total rates for dissociative electronic recombination (DER) reactions are generally needed to predict abundances of all interstellar molecules which are formed via ion-molecule reactions. Certain key rates, however, affect a broader range of questions. We discuss three of these. Deuterium fractionation in interstellar molecules has important implications for cosmology, and for the ionization fraction in dense interstellar clouds, hence for mechanisms of star formation in general. Recombination rates for H_3^+ and CH_3^+ are vital here. In astrochemistry, the carbon-chain cyanopolyynes trace the chemistry of the hydrocarbons themselves, and hence probably of large-molecule formation in general. A key test of current cyanopolyyne formation schemes involves the recombination rate for HC_3NH^+, as yet unmeasured. Finally, we discuss astrophysical tests of the Bates theory of DER, in particular the prediction that only one H atom will detach in the recombination with a polyatomic ion which contains several H atoms.

I. DEUTERIUM FRACTIONATION IN INTERSTELLAR MOLECULES: IONIZATION IN DENSE CLOUDS

Although the fractional ionization is well known in diffuse interstellar clouds, which are penetrated by the interstellar UV field, it is rather poorly known in dense clouds where it depends on the uncertain rates of cosmic ray ionization and of transfer of charge via uncertain reaction rates ultimately to the metals, whose gas phase abundances are equally poorly known. The fractional ionization is important because it determines whether magnetic fields may exist in appreciable strength in such clouds, and whether significant heating of the cloud may occur via ambipolar diffusion. Thus the thermodynamic balance of the clouds is at issue, as well as the rate of star formation which depends upon the rate of gravitational contraction of the clouds, a rate which may depend strongly upon the existence of magnetic forces.

Several interstellar molecules (HCO^+, N_2H^+, NH_3, HCN, HNC, C_2H...) are known to contain remarkable amounts of deuterium fractionation. While the "cosmic" abundance ratio D/H is known optically in the solar vicinity to be ~ 3(-5), the highly fractionated species are characterized by ~1(-3) ≤ AD/AH ≤ 1(-1). Besides having served as an important test of ion-molecule chemistry, the high fractionation determines the (very small) ionization fraction in dense clouds, as follows. Since the gas is molecular, most deuterium reposes in the form HD, since H_2 is by far the most abundant species. Deuterium may transfer from HD to some other species XH^+ via the reactions

$$HD + X^+ \dashrightarrow XD^+$$

$$H_2 + X^+ \dashrightarrow XH^+$$

$$HD + XH^+ \quad XD^+ + H_2 + \Delta E \quad \text{(rate k)}$$

$$XH^+,\ XD^+ + e \dashrightarrow \text{products}$$

In steady state, these processes give

$$\frac{XD^+}{XH^+} = \frac{1}{2}\ \frac{HD}{H_2}\ \frac{k}{k\ \exp(-\Delta E/T) + \alpha(XH^+)\ [e] + \Sigma k_i [M_i]} \qquad (1)$$

where ΔE is the difference in zero-point energies of XH^+ and XD^+, inhibiting the back-reaction, $\alpha(XH^+)$ is the DER coefficient for XH^+ (and XD^+), and M_i are species which destroy XH^+ (and XD^+) at rates k_i. [ ] denotes fractional abundances (relative to H_2). In principle XD^+/XH^+ may be >> HD/H_2 if all terms in the denominator of Equation (1) are sufficiently small.

The most fundamental interstellar species to which this scheme applies is H_3^+, since it forms directly from the cosmic ray ionization of H_2 (formed on grains). Thus:

$$\frac{H_2D^+}{H_3^+} = \frac{1}{2}\ \frac{HD}{H_2}\ \frac{k}{k\ \exp(-150/T) + \alpha(H_3^+)\ [e] + \Sigma k_i [M_i]} \qquad (2)$$

Unfortunately neither H_2D^+ nor H_3^+ are easily observable (H_2D^+ because it lacks transitions at low interstellar energies, H_3^+ because it has no permanent dipole moment). We turn therefore to the simplest observable species, HCO^+, for which

$$H_3^+ + CO \dashrightarrow HCO^+ + H_2$$
$$H_2D^+ + CO \quad DCO^+ + H_2 + \Delta E$$

from which

$$\frac{DCO^+}{HCO^+} \approx \frac{1}{3} \; \frac{H_2D^+}{H_3^+} \qquad (3)$$

DCO^+ and HCO^+ are readily observed in a wide variety of cold dense clouds ($T \sim 10K$) where the exp(-150/T) term in Equation (2) is negligible. If the term $\Sigma k_i[M_i]$ were $<< \alpha(H_3^+)[e]$ (or were precisely known), the observed ratio DCO^+/HCO^+ (~ 1(-3) to 3(-2)) would allow [e] to be determined, provided $\alpha(H_3^+)$ were known.

Unfortunately, the consensus of this Conference is that $\alpha(H_3^+) <$ 2(-8) and may be considerably smaller. Such a low value permits only the limit [e] > 1(-5), at best; other astrophysical arguments already establish that [e] is lower than this. Another species is needed, which avoids use of $\alpha(H_3^+)$, yet whose chemistry is simple. The ions involved must react with H_2 (and HD) but not rapidly, otherwise their abundances will be too low to observe. Besides H_3^+, only CH_3^+ and $C_2H_2^+$ satisfy these needs.

Consider CH_3^+. It is deuterated via

$$CH_3^+ + HD \quad CH_2D^+ + H_2 + \Delta E \quad (\text{rate } k_1)$$

and also forms HCN via

$$CH_3^+ + N \dashrightarrow HCNH^+ + H$$
$$HCNH^+ + e \dashrightarrow HCN,\ HNC + H$$

so that

$$\frac{CH_2D^+}{CH_3^+} = \frac{1}{2}\,\frac{HD}{H_2}\,\frac{k_i}{k_i\ \exp(-370/T) + \alpha(CH_3^+)\ [e] + \Sigma k_i[M_i]}$$

and, taking various branching ratios into account,

$$\frac{DCN}{HCN} = \frac{2}{15}\ \frac{CH_2D^+}{CH_3^+}$$

These reactions dominate the formation of DCN at low temperatures (T<30K). DCN and HCN are readily observed interstellar species in a variety of cold and warm clouds. $\alpha(CH_3^+)$ has been measured to be ~2(-6) $cm^3\ s^{-1}$ at T = 10 K [1]. The large value is fortunate for present purposes since $\Sigma k_i[M_i]$ is also large for CH_3^+, being dominated by reaction with H_2 and CI. Equations (4) and (5) allow us to place significant limits on [e], < 1(-8) in cold clouds. Some minor complexities have been ignored, however. DCN can also form via H_2D^+ + HCN --> $DCNH^+$, which brings in the H_2D^+/H_3^+ ratio with its possible dependence on $\alpha(H_3^+)$, and with a different T-dependence. Other reactions, e.g., $C^+ + NH_3$ --> might also affect the DCN/HCN ratio, although their effect is expected to be small. Detailed modelling [2] is needed to refine these estimates, but sensitive determinations of [e] nevertheless rest predominantly on accurate values for the DR coefficients of H_3^+ and CH_3^+. Species which derive from the $C_2H_2^+$ ion (e.g., C_2H) have more complex chemistry, so that their deuterated fractions do not define [e] independently of the abundances of many other species, which have limited precision.

II. DEUTERIUM FRACTIONATION IN INTERSTELLAR MOLECULES: COSMOLOGICAL IMPLICATIONS

The species H,D,He^3,He^4 were all formed in the first minute or so after the big bang origin of the universe. Big-bang nucleosynthesis models may be extrapolated forward in time to current densities in the universe, conserving baryon number, to yield predictions for the mass fractions of D, He^3, He^4 as a function of current mass density. The currently accepted value of D/H ~ 1.5(-5) determined locally corresponds to a mass density of 4(-31) gm/cm^3, insufficient by a factor of ~75 to close the universe. Astronomers have argued for three decades whether various forms of "hidden" mass exist in sufficient quantity to close the universe (required density 3(-29) gm/cm^3).

The problem is that the locally determined D/H ratio is not an unambiguous measure of current mass densities, because the deuterium is easily destroyed by processes of astration and may also be formed in strong explosions such as occur in supernovae. Thus the observed D may not reflect the primordial component. To settle this question, one needs to observe D/H as a function of distance from the galactic center. The number density of both stars and supernovae increases toward the G.C.

Thus if D is actually created in supernova explosions, D/H should increase toward the G.C. If the observed D is primordial, D/H should decrease toward the G.C. because of increasing astration.

Interstellar molecules radiating at radio wavelengths provide the only means of detecting deuterium across the Galaxy. The ratios DCO^+/HCO^+, DCN/HCN, DNC/HNC, are all observed to decrease toward the G.C. Does this reflect a decreasing D/H, or is it a gradient produced by chemical effects in the fractionation?

Consider DCO^+/HCO^+. A more exact expression than Equation (3) is needed:

$$\frac{DCO^+}{HCO^+} = \frac{1}{3}\,\frac{H_2D^+}{H_3^+}\,\frac{\Sigma k_i[M_i]}{k_2\exp(-100/T) + \Sigma k_i[M_i]}$$

Here, k_2 is the rate for $HD + HCO^+ \quad DCO^+ + H_2 + \Delta E$, $\Delta E = 100$ K, and k_i, M_i refer to destruction processes for HCO^+ (and DCO^+). There is evidence that T for molecular clouds increases toward the G.C. Coupled with the small value of ΔE, this may produce an apparent decrease in DCO^+/HCO^+ toward the G.C., so that the observed gradient in DCO^+/HCO^+ can not be attributed to the actual D/H ratio.

Again, we turn to alternative species. At the higher temperatures which occur in clouds near the G.C., DCN is formed primarily by $N + CHD \rightarrow DCN + H$, but since CHD, CH_2 result from dissociative recombination of CH_2D^+, CH_3^+, we have

$$\frac{DCN}{HCN} = \frac{1}{6}\,\frac{CH_2D^+}{CH_3^+} = \frac{1}{12}\,\frac{HD}{H_2}\,\frac{k_i}{k_i\exp(-370/T) + \alpha(CH_3^+)\,[e] + \Sigma k_i[M_i]}$$

Now the large energy defect (370 K) makes the T-dependence negligible at all interstellar cloud temperatures. The large value of $\alpha(CH_3^+)$ creates another possible problem, however, namely the possibility that [e] increases toward the G.C. In principle careful measurements of DCO^+/HCO^+ and DCN/HCN, as well as $HDCO/H_2CO$ which derives from yet another set of slightly more complex reactions [2] should allow one to disentangle the effects of gradients in T and [e] from actual gradients in D/H. The best evidence at present is that there is indeed a decrease in D/H toward the G.C., implying that the observed D is primordial, and that there is insufficient hidden mass to close the universe. The crucial role of a few

DER rates in these considerations, which range far beyond astrochemistry, should be clear.

III. THE FORMATION OF CYANOPOLYYNES

Cyanopolyynes (HC_nN, currently observed up to n=11) appear to trace the chemistry of interstellar hydrocarbons, which in turn may be fundamental to complex astrochemistry in general. The importance of cyanopolyynes to astrochemistry has spawned several ideas about their formation, which we have combined in Figure 1.

They are:

(a) addition of HCN to hydrocarbon ions (long diagonals);

(b) addition of $C_2H_2^+$ or $C_2H_3^+$ to preceding (lower order) cyano polyynes (short diagonals);

(c) addition of C_2H_2 to protonated cyanopolyynes (a "secondary" scheme, since protonated species must form via (a), (b), or (d)); and

(d) addition of N to hydrocarbon ions.

In Figure 1 we have indicated reaction rates in units 1(-9) $cm^3\ s^{-1}$. With the exception of the DER rate on HC_3NH^+, all fast rates (> 10) are either measured or calculated at 10 K. Otherwise, rates refer to 300 K. The rates for reaction of C_2H_2 with $HCNH^+$ and HC_3NH^+ have been measured by Amano to be < 0.01 at 300 K [3].

There are several difficulties in building up higher-order cyanopolyynes by these gas phase reactions. Rates for (a), (b), and (c) are clearly too slow. Rates for (d) MAY be slow, and may become negligible for order C5 and higher [4]. The observed abundance of HC_3N (and higher orders) is considerably higher than calculated via these schemes.

A decided problem is that one observes $HCNH^+/HCN \sim 0.1$ and $HC_3NH^+/HC_3N < 0.007$ in interstellar clouds. According to Figure 1, for any ratio HC_nNH^+/HC_nN the determining processes are the fast protonation and electron recombination reactions, and those destruction processes for HC_nNH^+ and HC_nN which do NOT produce HC_nN and HC_nNH^+ respectively (the processes (d) do not affect the ratio to first order). In steady state,

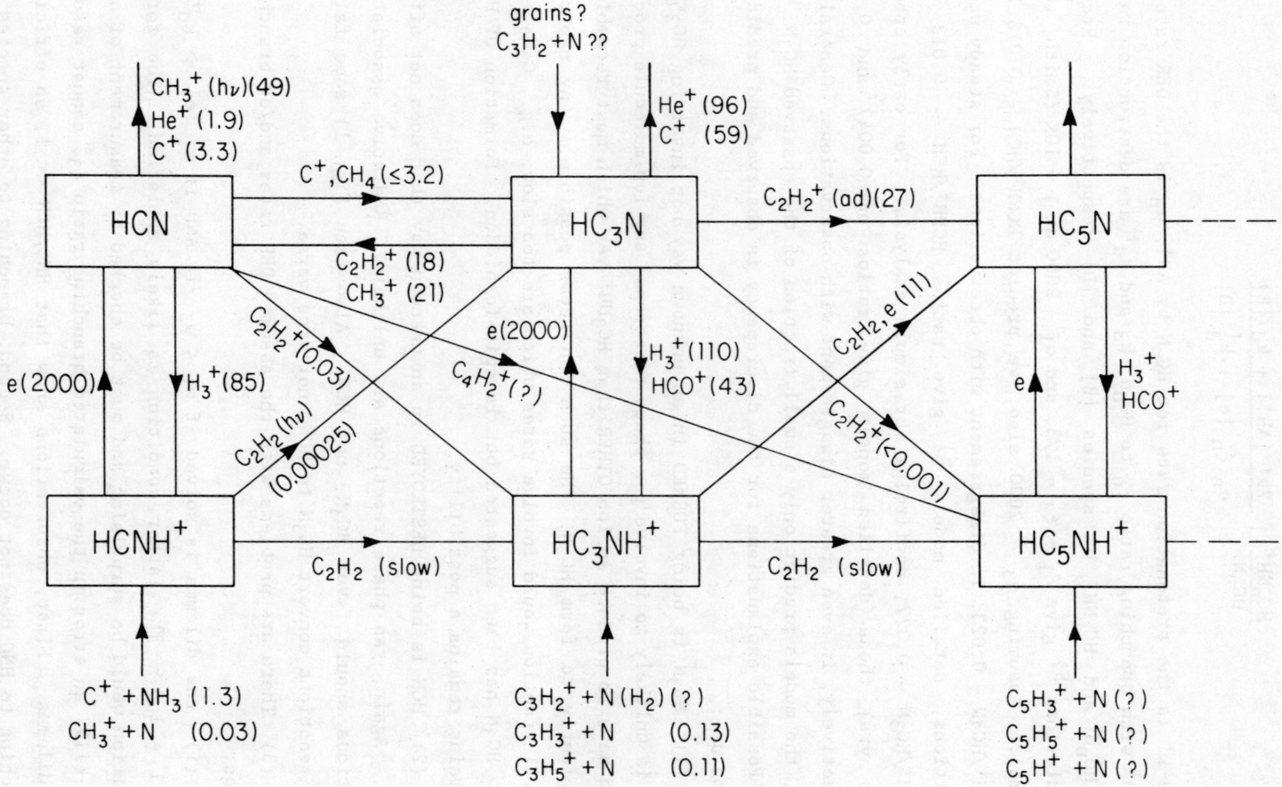

Figure 1. Reactions That Govern the Chemistry of Interstellar Cyanopolyynes.

$$\frac{HC_nNH^+}{HC_nN} = \frac{2k'\ [AH^+] + k_d\ [B]}{\alpha_n\ \xi_n\ [e] + k_d^+\ [D]}$$

where k′ is the protonation rate for HC_nN by AH^+, α_n is the DER rate of HC_nNH^+ with branching ratio ξ_n to HC_nN, k_d and $k_d{}^+$ are destruction rates for HC_nN and HC_nNH^+ by species [B] and [D] respectively. Recent calculations [5] give $\xi_1 = \xi_3 = 1/3$, and $\alpha_1 \sim 2000$ at T = 10K (Smith, this volume). Assuming $\alpha_3 = 2000$ also, we predict $HCNH^+/HCN \sim 0.018$ and $HC_3NH^+/HC_3N \sim 0.023$, in disagreement with observations. For steady state conditions, detailed models [6] give ratios $HCNH^+/HCN \sim 0.013$ and $HC_3NH^+/HC_3N \sim 0.027$, confirming our simple analysis. For early epochs (1(5) yrs) these detailed models give ratios of 0.0047 and 0.019 respectively, in even greater disagreement with observations. Generally, also, the models predict only a small fraction of the observed HC_3N.

Possible explanations for the discrepancy in observed and predicted ratios are:

(1) HC_3N is being FORMED in an unknown way not involving HC_3NH^+. This is unlikely to involve gas phase reactions: any ion-molecule process would have to produce an ion OTHER than HC_3NH^+ yet which must recombine efficiently to form HC_3N. No such ion seems possible. Any neutral-neutral reaction would involve rates which are too slow; e.g., C_3H_2 + N --> HC_3N has been suggested but is spin-forbidden. Formation of HC_3N on grains remains a possibility.

(2) HCN is being DESTROYED in an unknown way that does not affect HC_3N. Again, gas phase reactions are unlikely. Radiative associative reactions should favor HC_3N, not HCN. Also (see Fig. 1) some fairly fast reactions convert HC_3N to HCN, not vice versa.

(3) There are problems with the assumed DER rates and/or branching ratios:

(i) the dilemma is solved if $\alpha_1 < \alpha_3 /14$ AND if $\alpha_i\ [e] \gg k_d{}^+[D]$ for i = 1,3. The latter condition is likely satisfied. The former condition would be surprising and must be checked by measurement of α_3.

(ii) an error in the calculated branching ratio ξ_3, cannot resolve the dilemma. First, observations show that HNC/HCN << 1, so efficient branching to HNC does not occur. Second, branching to other species

($\xi_1 << 1/3$) simply adds to the $k_d^+[D]$ term and keeps the predicted ratio $HCNH^+/HCN$ small.

(iii) an error in the calculated branching ratio ξ_3 also cannot solve the problem. If $\xi_3 << 1/3$, the discrepancy between observed and predicted $[HC_3N]$ becomes even larger.

It is of central importance for astrochemistry that the dilemma of the HC_nNH^+/HC_nN ratios be resolved, because the cyanopolyynes act as tracers of the hydrocarbon ion chemistry, which in turn is believed to be the basis of all complex molecule formation in interstellar space. The problem most likely lies with the DER α_3 and not with the branching ratios. An experimental determination of $\alpha_3 \equiv \alpha(HC_3NH^+)$ is needed.

IV. THE INTERSTELLAR KETYL RADICAL: A TEST OF BATES' THEORY OF D.E.R.

Bates [7,8] has advocated a new method of estimating the likely products of D.R. reactions. The polyatomic ion is pictured as formed by localized valence bonds, and the bonds involving the charged atom are converted into anti-bonds by the incoming electron, and rupture. Only one bond ruptures at a time. Strong bonds between heavy atoms may break, but H-bonds are more susceptible because they are weaker and because the fragments separate more rapidly.

We describe an astrophysical test of the Bates prediction that when a parent ion containing several H atoms is dissociated, only one H atom is detached. The requirements of the test are to observe a family of species AH_n, AH_{n-1}, AH_{n-2},... all of which form ONLY via DER of the parent ion AH^+_{n+1}. The prediction is that AH_n should be observed but AH_{n-1}, AH_{n-2},... should not be.

Of the ~ 80 known interstellar molecules, only one such sequence seems suitable for the test:

$$H_3C_2O^+ + e \longrightarrow H_2CCO + H \quad \text{(ketene)}$$
$$\longrightarrow HCCO + H_2 \quad \text{(ketyl)}$$
$$\longrightarrow CCO + H_2 + H$$

By observation we find $HCCO/H_2CCO < 0.02$ in several types of interstellar clouds (SgrB2, TMC-1, etc.) [9]. CCO is also not detected, but upper limits are modest ($CCO/H_2CCO < 0.2$).

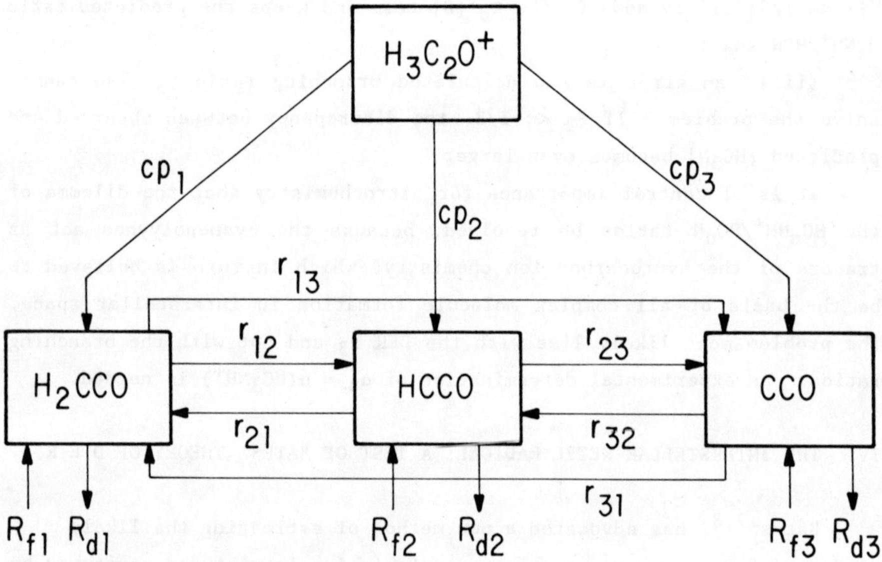

Figure 2. Reaction Pathways Affecting Inter stellar CCO, HCCO, H_2CCO.

Figure 2 shows schematically the chemical network governing H_2CCO, HCCO, and CCO. The p_i are probabilities that recombination of $H_3C_2O^+$ produces the indicated neutral species; c is the overall recombination rate. If all rates r_{ij}, as well as alternative formation rates R_{fi} were negligible, and if all destruction rates R_{di} were similar, then $H_2CCO/HCCO/CCO = p_1/p_2/p_3$, hence by observation $p_2 \ll p_1$, consistent with Bates' conjecture. Under steady state conditions, the solution to the network shown in Figure 2 is

$$\frac{HCCO}{H_2CCO} = \frac{cp_2 + R_{f2}}{cp_1 + R_{f1}} \frac{d_{12} + d_1}{2d_{21} + d_2} + \frac{d_{12}}{2d_{21} + d_2} \tag{6}$$

where we have assumed $d_{21}[HCCO] \ll cp_1 + R_{f1}$. Here $d_{ij} = k_{ij}[A]$ where k_{ij} is the rate constant and [A] is the fractional abundance of species A which reacts with species i (H_2CCO or HCCO). In analyzing only the $HCCO/H_2CCO$ ratio because of the poor limits on CCO, we consider r_{23} as part of R_{d2}, r_{31} as part of R_{f1}, etc. A search of the literature for all possible interstellar reactions which actually or potentially contribute to the network yields only the following:

d_{21}	none
d_{12}	$AH^+ + H_2CCO \rightarrow H_3C_2O^+ \rightarrow HCCO$ (probability p_2)
	$X^+ + H_2CCO \rightarrow HCCO + XH^+$?
d_1	$C^+ + H_2CCO \rightarrow$ products
	$H + H_2CCO \rightarrow CH_3 + CO$?
	$AH^+ + H_2CCO \rightarrow H_3C_2O^+ \rightarrow$ products
d_2	$C^+ + HCCO \rightarrow$ products
	$H + HCCO \rightarrow CCO + H_2$, etc ?
R_{f2}	$O + C_2H_2 \rightarrow HCCO + H$?
R_{f1}	$O + C_2H_4 \rightarrow H_2CCO + H_2$?

Reactions marked "?" are not measured and are unlikely either because of activation barriers or because they do not form the indicated products. Evaluation of the terms in Equation (6), using detailed models for the abundances (e.g., Herbst and Leung 1988) yields $d_{21} \ll d_2$, $d_{12} \ll d_1, d_2$, and $d_1 > d_2$. We also find $R_{f2} \gg R_{f1}$, c which implies $HCCO/H_2CCO \gg 1$, contrary to observation. Therefore the reaction $O + C_2H_2 \rightarrow HCCO + H$ evidently does not proceed at low interstellar temperatures. Thus we obtain $cp_2 + R_{f2} < (cp_1 + R_{f1})r$ where r is the observed upper limit of the $HCCO/H_2CCO$ ratio. Hence $cp_2 < (cp_1 + R_{f1})r$. A recent determination of the rate for $O + C_2H_4 \rightarrow H_2CCO$ [10] gives $R_{f1} \ll cp_1$, hence $p_2/p_1 < r = 0.02$. If the reactions in d_1 and d_2 involving H also do not proceed, then we find $d_1 \gg d_2$ and $d_{12} < d_1$, in which case evaluation of Equation (6) yields an even better limit: $p_2/p_1 < 5(-4)$. Our analysis of interstellar HCCO and H_2CCO is therefore consistent with Bates' conjecture that, among DER reactions that detach H atoms, the detachment of more than one H atom is highly unlikely.

At present there are no counter-examples to this rule. For example, cyclic C_3H has recently been identified in interstellar space, joining cyclic C_3H_2 as the only two ring compounds. If both form via recombination of $C_3H_3^+$, Bates' conjecture would be violated. However there are several alternative ways to form cyclic C_3H. [11]. Similarly the recently identified cyanomethane radical (CH_2CN) probably does not form via recombination of CH_3CNH^+ (the progenitor of interstellar CH_3CN). Rather, CH_2CN appears to form via $C_2H_4^+ + N \rightarrow CH_2CNH^+$.

A possible second test of Bates' conjecture may be the vinoxy radical, CH_2CHO, which has not yet been searched in interstellar space, although its microwave spectrum is known. Both vinoxy and CH_3CHO (acetaldehyde, a well-known interstellar species) would appear to have a common progenitor, CH_3CHOH^+, which may form in several ways.[12)]

REFERENCES

1. Mul, P. M., Mitchell, J.B.A., D'Angelo, V. S., Defrance, P., McGowan, P. W., and Froelich, H. R. 1981, J. Phys. B, 14, 1353.
2. Millar, T. J., Bennett, A., and Herbst, E. 1988, Ap. J., submitted.
3. Turner, B. E., Bally, J., Amano, T., Lee, S., and Feldman, P. A. 1988, in MOLECULAR CLOUDS IN THE MILKY WAY AND EXTERNAL GALAXIES, Ed. J. S. Young and R. L. Dickman, Springer-Verlag.
4. Federer et al. 1986, Chem. Phys. Lett. 123, 12.
5. Millar, T. J., DeFrees, D. J., McLean, A. D., and Herbst, E. 1988 Astron. & Astrophys. in press.
6. Herbst, E., and Leung, C. M. 1988, Ap. J. (Suppl.), submitted.
7. Bates, D. R. 1986, Ap. J. (Letters) 306, L45.
8. Bates, D. R. 1987, "Interstellar Cloud Chemistry Revisited" in MODERN APPLICATIONS OF ATOMIC AND MOLECULAR PROCESSES, Ed. A.E. Kingston, London, Plenum.
9. Turner, B. E., and Sears, T. J. 1988, Ap. J., submitted.
10. Cventanovic, R. J. 1987, J. Phys. Chem. Ref. Data 16, 264.
11. Herbst, E. 1988 "Dense Interstellar Cloud Chemistry" in REACTIVE RATE COEFFICIENTS IN ASTROPHYSICS, Ed. T. J. Millar, D. Reidel
12. Huntress, W. J., and Mitchell, G. F. 1977, Ap. J. 231, 456.

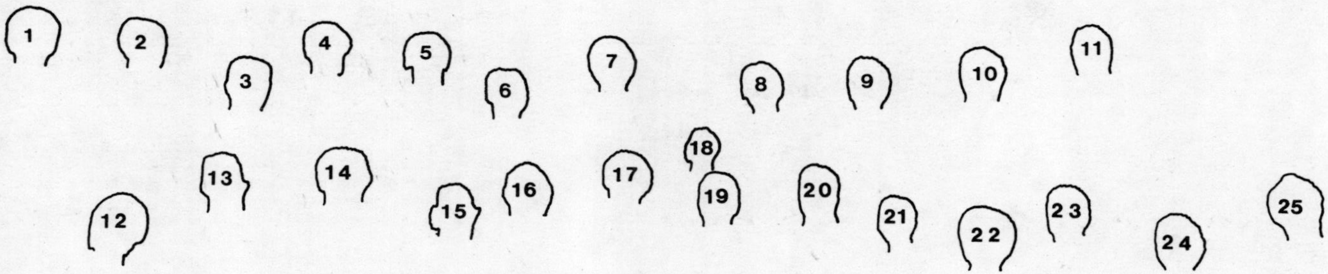

1. Peet Hickman
2. Charles Herd
3. Alan Garscadden
4. David Smith
5. Brendan McLaughlin
6. Bertran Rowe
7. John Bahns
8. Tom Morgan
9. Sam Yee
10. Wolfgang Kraemer
11. Barry Turner
12. Annick Giusti
13. Bill Graham
14. Steve Guberman
15. Jim Peterson
16. John Hiskes
17. Steve Lundeen
18. Polly Mitchell
19. Brian Mitchell
20. Chris Bottcher
21. Tom O'Malley
22. Jane Fox
23. John Black
24. Winifred Huo
25. Rainer Johnsen